Co-design Approaches
for Dependable Networked Control Systems

Co-design Approaches for Dependable Networked Control Systems

Edited by
Christophe Aubrun
Daniel Simon
Ye-Qiong Song

First published 2010 in Great Britain and the United States by ISTE Ltd and John Wiley & Sons, Inc.

ISTE Ltd
27-37 St George's Road
London SW19 4EU
UK

www.iste.co.uk

John Wiley & Sons, Inc.
111 River Street
Hoboken, NJ 07030
USA

www.wiley.com

Library of Congress Cataloging-in-Publication Data

Co-design approaches for dependable networked control systems / edited by Christophe Aubrun, Daniel Simon, Ye-Qiong Song.
    p. cm.
  Includes bibliographical references and index.
  ISBN 978-1-84821-176-6
  1. Feedback control systems--Reliability. 2. Feedback control systems--Design and construction. 3. Sensor networks--Reliability. 4. Sensor networks--Design and construction. I. Aubrun, Christophe. II. Simon, Daniel, 1954- III. Song, Ye-Qiong.
  TJ216.C62 2010
  629.8'3--dc22
                                                                                            2009041851

British Library Cataloguing-in-Publication Data
A CIP record for this book is available from the British Library
ISBN 978-1-84821-176-6

Edited and formatted by Aptara Corporation, New Delhi, India
Printed and bound in Great Britain by CPI Antony Rowe, Chippenham and Eastbourne

Contents

Chapter 2. Computing-aware Control   63
Mongi BEN GAID, David ROBERT, Olivier SENAME, Alexandre SEURET and
Daniel SIMON

Chapter 3. QoC-aware Dynamic Network QoS Adaptation 105
Christophe AUBRUN, Belynda BRAHIMI, Jean-Philippe GEORGES, Guy JUANOLE,
Gérard MOUNEY, Xuan Hung NGUYEN and Eric RONDEAU

Chapter 4. Plant-state-based Feedback Scheduling 149
Mongi BEN GAID, David ROBERT, Olivier SENAME and Daniel SIMON

Chapter 5. Overload Management Through Selective Data Dropping . . . 185
Flavia FELICIONI, Ning JIA, Françoise SIMONOT-LION and Ye-Qiong SONG

Chapter 7. Implementation: Control and Diagnosis for an Unmanned Aerial Vehicle 267

Cédric BERBRA, Sylviane GENTIL, Suzanne LESECQ and Daniel SIMON

Foreword

Modeling, analysis and control of networked control systems (NCS) have recently emerged as topics of significant interest to the control community. The defining feature of any NCS is that information (reference input, plant output, control input) is exchanged using a digital band-limited serial communication channel among control system components (sensors, controller, actuators) and usually shared by other feedback control loops. The insertion of a communication network in the feedback control loop makes the analysis and design of an NCS more challenging. Conventional control theory with many ideal assumptions, such as synchronized control and non-delayed sensing and actuation, must be revisited so that the limitations on communication capabilities within the control design framework can be integrated.

Furthermore, the new trend is to implement the realization of fault diagnosis (FD) and fault tolerant control (FTC) systems that employ supervision functionalities (performance evaluation, fault diagnosis) and reconfiguration mechanisms by using cooperative functions that are also distributed on a networked architecture. A critical issue, therefore, which must always be considered in the design of any networked process control system, is its robustness with respect to failure situations, including system component failures as well as network failures. By network failure, we mean a total breakdown in the communication between the control system components as a result of, for example, some physical malfunction in the networking devices or severe overloading of the network resources that cause a network shut down.

In this framework, dependability of NCS represents the emergence of an important research field. Dependability groups together with three properties that the NCS must satisfy in order to be designed: safety, reliability, and availability. Therefore, the design of a dependable NCSs implies a multidisciplinary approach; more precisely, dealing with a deep knowledge of both fault tolerant control and computer science (mainly real time scheduling and communication protocols).

The content of this book gives an overview of the main results obtained after three years of research work within the safe-NECS project funded by the French "Agence Nationale de la Recherche—ANR". During these three years, five research groups have cooperated intensively to propose a framework for the design of dependable networked control systems. In this context, the research of safe-NECS took into consideration process control functions, FDI/FTC and their implementation over a network as an integrated system. In particular, the project aim was to develop, in a coordinated way, a "co-design" approach that integrates several kinds of parameters: the characteristics modeling the Quality of Control (QoC), the dependability properties required for a system and the parameters of real-time scheduling (tasks and messages). Issues such as network-induced delays, data losses and signal quantization as well as sensor and actuator faults represent some of the more common problems that have motivated the extensive research work developed within the safe-NECS project.

The research work in the safe-NECS project aimed at enhancing the integration of control, real-time scheduling and networking. The safe-NECS project was clearly a multi-disciplinary project in that it brought together partners from the control and computer science communities. A major difficulty for this project came from the fact that both communities manipulate objects and formalisms of very different natures. We consider that one of the major contributions of this project was to promote a synergistic approach, which is a necessary condition for achieving the objectives of the co-design which is demonstrated in this project.

Safe-NECS project focuses on the following points, which are reported in this book:

– specification of a dependable system and performance evaluation;

– modeling the effect of a real-time distributed implementation (i.e. scheduling parameters and networking protocols) on Quality of Control (QoC) and dependability;

– an integrated control and scheduling co-design method realized by developing feedback scheduling algorithms, taking into account both QoC parameters and dependability constraints;

– fault tolerance and the on-line re-configuration of NECS with distributed diagnostic and decision-making mechanisms;

– testing and validation using a UAV-type quadrotor.

Dominique SAUTER

Introduction and Problem Statement

Networked control systems (NCS) are feedback control systems wherein the control loops are closed over a shared network. Control and feedback signals are transmitted among the system's components as information flows through a network, as depicted in figure I.1. A fully featured NCS is made up of four kinds of components to close the control loops: sensors to collect information on the controlled plant's state, controllers to provide decisions and commands, actuators to apply the control signals and communication networks to enable communications between the NCS components.

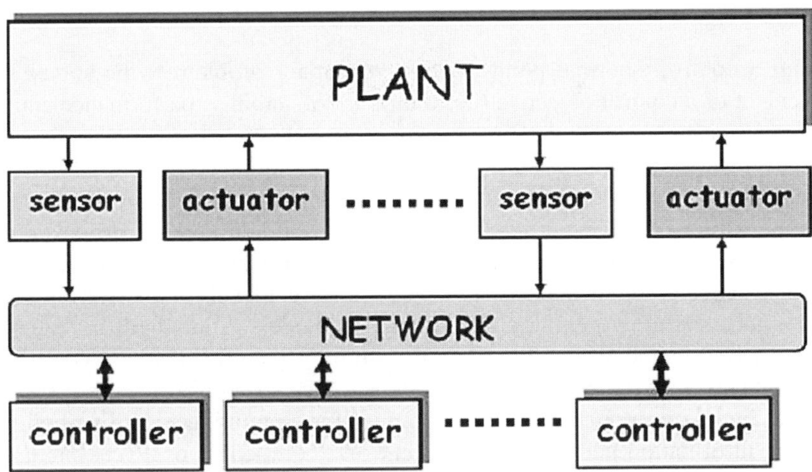

Figure I.1. *Typical NCS architecture*

Introduction written by Christophe AUBRUN, Daniel SIMON and Ye-Qiong SONG.

Compared with conventional point-to-point control systems, the advantages of NCS are lighter wiring, lower installation costs, and greater abilities in diagnosis, reconfigurability and maintenance. Furthermore, the technologies used in computer and industrial networks, both wired and wireless, have progressed rapidly providing high bandwidth, quality of service (QoS) guarantees and low communication costs. Because of these distinctive benefits, typical application of these systems nowadays ranges over various fields of industry and services. X-by-wire automotive systems, coordinated control of swarms of mobile robotics and advanced aircraft on-board control and housekeeping are examples of NCS usage in the field of embedded systems. Large-scale utility systems are deployed in order to control and monitor water, gas, energy networks, and transportation services on roads and railways. The capabilities of wireless sensor networks, which are widespread and cost very little to cooperatively monitor physical or environmental conditions, can be enhanced by adding some actuation capacities to make them control systems, such as in smart automated buildings.

I.1. Networked control systems and control design challenges

The design of NCS combines the domains of control systems, computer networks, and real-time computing. Historically, tools for the design and analysis of systems related to these disciplines have been designed and used with limited interaction. The increasing complexity of modern computer systems and the rapidly evolving technology of computer networks require more integrated methodologies, specifically suited to NCS.

From a control-theoretic point of view, the main problem to be solved is the achievement of a control objective (i.e. a mixture of stability, performance and reliability requirements), despite the disturbances induced by the distribution of the control system over a network. For instance, the sharing of common computing resources and communication bandwidths by competing control loops along with other general purpose applications introduces random delays and even data losses. Moreover, as the computations are supported by heterogenous computers and the communication between distributed components may spread over different levels of area networks, NCS become more and more complex and difficult to model. A key built-in feature of a multi-layer NCS is the availability of redundant information pathways, as well as the distributed nature of the overall task that the NCS has to perform. Hence the distribution of devices on nodes provides the ability for computing load distribution and re-routing information pathways in the event of a component or a subsystem of the network malfunctioning. This leads to the ability to reconfigure a new system from a nominally configured NCS.

Control is playing an increasing role in the design and run-time management of large interconnected systems to enhance high performance in nominal modes and safe reconfiguration processes in the occurrence of faults and failures. In fact, it appears that the achievement of system-level requirements by far exceed the achievable reliability of individual components [MUR 03]. The deep intrication of components and sub-systems, which come from different technologies, and which are subject to various constraints, calls for a joint design to solve potentially conflicting constraints early on. The fact that control has the ability to cope with uncertainty and disturbances, due to closed loops based on sensing the current system's state, makes it a basic methodology to be used in such complex systems design.

Besides reaching the specified performance in normal situations, reliability and safety-related problems are of a constant concern for system designers. A general and integrated concept is *dependability*, which is the system property that includes various attributes such as availability, reliability, safety, confidentiality, integrity, and maintainability [LAP 92]. Being confronted with faults, errors and failures, a system's dependability can be achieved in different ways, i.e. fault prevention, fault tolerance, fault removal and fault forecasting [AVI 00]. While these concepts have been formalized for systems in a broad sense, it appears that the existing control toolbox already provides concepts, e.g. robust and fault tolerant control, which are likely to achieve control system dependability. A better integration of advanced control, computing power and redundancy based on resource distribution is expected to further enhance this capability.

Except in the case of failures due to hardware or software components, most processes usually run with nominal behavior: however, even in the nominal modes, neither the process nor the execution resource parameters are ever perfectly known or modeled. A very conservative viewpoint consists of allocating system resources to satisfy the worst case, but this results in the execution resources being over-provisioned and thus wasted. From the control viewpoint, specific deficiencies to be considered include poorly predictable timing deviations, delays, and data loss.

Control usually deals with modeling uncertainty, dynamic adaptation, and disturbance attenuation. More precisely, as shown with recent results obtained on NCS [BAI 07], control loops are often robust and can tolerate computing and networking performance induced disturbances, up to a certain extent. Therefore, timing deviations such as jitter or data loss, as long as they remain inside the bounds which are compliant with the control specification, may be considered as features of the nominal system, not exceptions. Relying on control robustness allows for provisioning the execution resources according to average needs rather than for worst cases, and to consider system reconfiguration only when the failures exceed the capabilities of the running controller tolerance.

An NCS is made up of a heterogenous collection of physical devices, falling within the realm of continuous time, and information sub-systems basically working with discrete timescales. During an NCS design process, many conflicting constraints must be simultaneously solved before reaching a satisfactory and implementable so-lution. For example, trade-offs must be negotiated between processing and network-ing speed, control tracking performance, robustness, redundancy and reconfigurabil-ity, energy consumption, and overall cost effectiveness. These conventional design process problems from the different domains in succession prevent any coherent and effective integration of methodologies, technologies or associated constraints.

Traditionally control usually deals with a single process and a single computer, and it is often assumed that the limitations of communication links and computing resources do not significantly affect performance, or they are taken into account in a limited way. Existing tools dealing with modeling and identification, robust control, fault diagnosis and isolation, fault tolerant control and flexible real-time scheduling need to be enhanced, adapted and extended to cope with the networked characteristics of the control system.

Finally, the concept of a *co-design system approach* has emerged to allow progress in the integration of control, control, and communications in the NCS design [MUR 03] and to develop *implementation-aware* control system co-design approaches [BAI 07].

According to the systems scientist and philosopher, C.W. Churchman, "the system approach begins when first you view the world through the eyes of another" [CHU 79]. The basic aspect of co-design applied to NCS is that the design of controllers and the design of the execution resources, i.e. the real-time computing and communication sub-systems, are integrated right from the early design steps to jointly solve the con-straints arising from all sides.

I.2. Control design: from continuous time to networked implementation

In the early age of control, analog computers were used to work out the control signals, firstly from mechanical devices, and then from electronic amplifiers and in-tegrators: both the plant and the controller remained in the realm of continuous time, while frequency analysis and Laplace transform were the main tools at hand. The main drawbacks of analog computing come from limited accuracy and bandwidth, drift and noise, and from limited capabilities to handle nonlinearities. Pure and known delays could, however, be handled at control synthesis time by using the well-known Smith predictor.

Then, due to the increasing power and availability of cheap numerical processors, digital controllers gradually took over the analogue technology. However, controlling a continuous plant with a discrete digital system inevitably introduces timing distor-tions. In particular, it becomes necessary to sample and convert the sensors measure-ments to binary data, and conversely to convert them back to physically related values and hold the control signals to actuators. The sampling theory and the z transform became the standard tools for digital control systems analysis and design. A smart property of the z transform is that it keeps the linearity of the system through the sampling process. As the underlying assumption behind the z transform is equidistant sampling, periodic sampling became the standard for the design and implementation of digital controllers.

Note that, at the infancy of digital control, where computing power was weak and memory was expensive, it was important to minimize the controllers' complexity and needed operating power. It is not obvious that the periodic sampling assumption is always the best choice: for example, [DOR 62] show that adaptive sampling, where the sampling frequency is changed according to the value of the derivative of the error signal, can be more effective than equidistant sampling in terms of the number of computed samples (but possibly not in terms of disturbance rejection [SMI 71]). [HSI 72] and [HSI 74] provide a summary of these efforts. However, due to the constantly increasing power and decreasing costs of computing, interest in sampling adaptability and the related computing power savings has progressively vanished, while the linearity preservation property of equidistant sampling has helped it to remain the indisputable standard for years.

From the computing side, real-time scheduling modeling and analysis were introduced in the illustrious seminal paper [LIU 73]. This first schedulability analysis was based on restrictive assumptions, one of them being the periodicity of all the real-time tasks in the system. Even if more general assumptions have been progressively introduced to cope with more realistic problems and tasks sets [AUD 95; SHA 04], the periodicity assumption remains very popular, e.g. see today's success of rate monotonic analysis (RMA) based tools in industry, e.g. [SHA 90; DOY 94; HEC 94]. The combination of these popular modeling and analysis methods and tools has likely reinforced the understanding that control systems are basically periodic and hard real-time systems.

More recently, again due to the progress in electronic devices technology, it has become possible to distribute control loops over networks. Networking allows for the dissemination of sensors, actuators, and controllers on different physical nodes. Moreover, the topology of the network can be time varying, thus allowing the control devices to be mobile: hence, the whole control system can be highly adaptive in a dynamic environment. In particular, wireless communications allow for a cheap deployment of sensor networks and remotely controlled devices. However, networking also induces disturbances in control loops, such as variable and potentially long delays, data corruption, message desequencing and occasional data loss. These timing uncertainties and disturbances are in addition to those coming from the digital implementation of the controllers in the network nodes.

I.3. Timing parameter assignment

Digital control systems can be implemented as a set of tasks running on top of a commercial off-the-shelf real-time operating system (RTOS) using fixed-priority and pre-emption. The performance of a control loop, e.g. measured by the tracking error, and even more importantly its stability, strongly relies on the values of the sampling rates and sensor-to-actuator latencies (the latency considered for control purposes is the delay between the instant when a measure q_n is taken at a sensor and the instant when the control signal $U(q_n)$ is received by the actuators [ÅST 97]. Therefore, it is essential that the implementation of the controller respects an adequate timing behavior to meet the expected performance. However, implementation constraints such as multi-rate sampling, pre-emption, synchronization, and various sources of delays make the run-time behavior of the controller very difficult to accurately predict. Dealing with closed-loop controllers may take advantage of the *robustness* and *adaptivity* of such systems to design and implement flexible and adaptive real-time control architectures.

Closed-loop digital control systems use a computer to sample sensors, calculate a control law and send control signals to the actuators of a physical process. The control algorithm can be either designed in continuous time and then discretized or directly synthesized in discrete time, taking into account a model of the plant sampled by a zero-order holder. A control theory for linear systems sampled at fixed rates was established a long time ago [ÅST 97].

Assigning an adequate value for the sampling rate is a decisive duty, as this value has a direct impact on the control performance and stability. While an absolute lower limit for the sampling rate is given by Shannon's theorem, in practice, rules of thumb are used to give a useful range of control frequencies according to the process dynamics and the desired closed loop bandwidth. Among others, such a rule of thumb is given in [ÅST 97] as $\omega_c h \approx 0.15 \dots 0.5$, where ω_c is the desired closed-loop pulsation and h is the sampling period. Note that such rules only give preliminary information about the sampling rate to be actually implemented, and sampling rate selection needs to be further refined by simulations and experiments. In particular, it appears that the actual sampling rate to be used with some nonlinear systems, as those described in sections 1.4.2.3, 2.4.4, and in Chapter 7, must be far faster to achieve closed-loop stability. However, most often, it can be stated that the lower the control period and latencies are, the better the control performance is, e.g. measured by the tracking error or disturbance rejection. This assumption can be reinforced by providing a suitable control structure and parameter tuning, as shown in section 2.3 with the discussion on weakly hard real-time constraints and accelerable control tasks.

While timing uncertainties have an impact on the control performance, the actual scheduling parameters are difficult to model accurately or constrain within precisely known bounds. Thus, it is worth examining the sensitivity of control systems w.r.t.

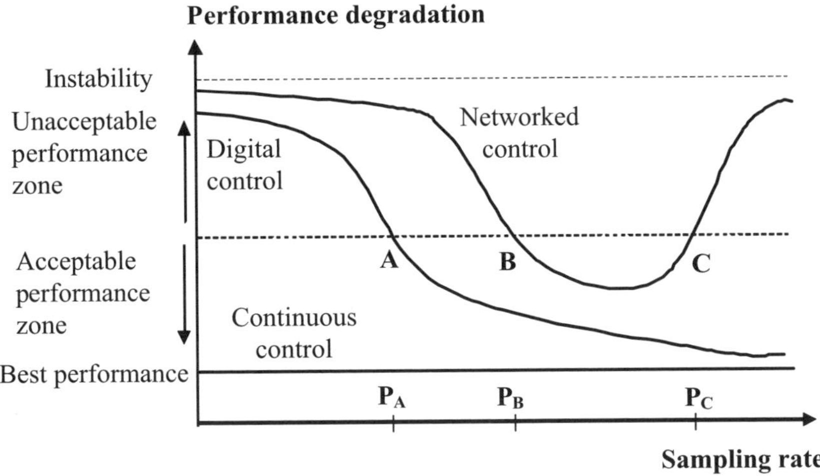

Figure I.2. *Performance comparison of continuous control, digital control, and networked control vs. sampling rate*

timing fluctuations. The accepted wisdom is that the lower the control period is, the better the control performance is. However, the underlying implementation system is a limited resource and cannot accommodate arbitrarily high sampling rates; thus, there must be a trade-off between the control performance and the execution resource utilization. This is particularly true for distributed embedded system design with limited resources due to weight, cost and energy consumption constraints. In [MOY 07] and [LIA 02], the illustrative chart in Figure I.2 is given to show the importance of choosing a good sampling period which gives a trade-off between control performance and the related network load.

From Figure I.2, it can be seen that the control performance is acceptable for a sampling rate range from P_B to P_C. Increasing the sampling rate beyond P_C will increase the network load and lead to longer network-induced delays. This results in a control performance degradation. Note that for a given control application, the affordable network bandwidth can also vary accordingly. This can provide a network QoS designer with a larger solution space for designing the QoS mechanisms with more flexibility.

The hard real-time assumption must be softened to better cope with the reality of closed-loop control, for instance, changing the hard timing constraints for "weakly-hard" constraints [BER 01]. For example, hard deadlines may be replaced by statistical models, e.g. to specify the jitter characteristics compliant with the requested control performance. They may also be changed for deadline miss or data-loss patterns, e.g. to specify the number of deadline misses allowed over a specified time window according to the so-called (m, k)-firm model [HAM 95]. More precisely, a task meets the (m, k)-firm constraint if at least m among any k consecutive task instances meet their deadline. Note that to be fully exploited, weakly hard constraints should be associated with a decisional process: tasks missing their deadline can, for example, be delayed, aborted or skipped according to their impact on the control law behavior, e.g. as analyzed in [CER 05].

Finding the values of such weakly hard constraints for a given control law is currently out of the scope of current control theory, in general. However, the intrinsic robustness of closed-loop controllers allows for relying on softened timing constraint specification and flexible scheduling design, leading to an adaptive system with graceful performance degradation during system overloads. Chapter 5 gives an example of finding the values of m and k when the (m, k)-firm constraint is applied to the control-loop task execution requirement.

I.4. Control and task/message scheduling

From the implementation point of view, real-time systems are often modeled as a set of periodic tasks assigned to one or several processors. In distributed real-time systems, messages are exchanged among related tasks through a network. To ensure the execution of tasks on a processor, the worst-case response time analysis technique is often used to analyze fixed-priority real-time systems. Well-known scheduling policies, such as rate monotonic for fixed priorities and EDF for dynamic priorities, assign priorities according to timing parameters, respectively, sampling periods and deadlines. They are said to be "optimal" as they maximize the number of task sets which can be scheduled with respect to deadlines, under some restrictive assumptions. Unfortunately, they are not optimized for control purposes. They hardly take into account precedence and synchronization constraints, which naturally appear in a control algorithm. The relative urgency or criticality of the control tasks can be unrelated with the timing parameters. Thus, the timing requirements of control systems w.r.t. the performance specification do not fit well into scheduling policies based purely on schedulability achievement.

It has been shown through experiments, e.g. [CER 03], that a blind use of such traditional scheduling policy can lead to an inefficient controller implementation; on the other hand, a scheduling policy based on an application's requirements, associated with a smart partition of the control algorithm into real-time modules may give better results. It may be that improving some computing related feature is in contradiction with another one targeted to improve the control behavior. For example, the case studies examined in [BUT 07] show that an effective method to minimize the output control jitter consists of systematically delaying the output delivery at the end of the control period: however, this method also introduces a systematic one period input/output latency, and therefore most often provides the worst possible control performance among the set of considered strategies.

Another example of unsuitability between computing and control requirements arises when using priority inheritance or priority ceiling protocols to bypass priority inversion due to mutual exclusion, e.g. to ensure the integrity of shared data. While they are designed to avoid deadlocks and minimize priority inversion lengths, such protocols jeopardize the initial schedule at run time, although it was carefully designed with latencies and control requirements in mind. As a consequence, latencies along some control paths can be largely increased, leading to a poor control performance or even instability.

In a distributed system design, not only should the task set schedulability be ensured, but also message transmission through the network, since sensor to actuator latencies heavily depend on transmission delay. Also, the schedulability tests must consider tasks and messages as a whole, since a message is produced by a task execution and it may be needed to trigger another task execution. In [TIN 94], a holistic approach is proposed. It consists of applying the worst-case response time analysis technique to evaluating the end-to-end response time of a set of tasks distributed over a controller area network (CAN). Periodic messages over CAN are scheduled under a non pre-emptive, fixed priority policy. Their release jitters are caused by local task scheduling. The worst case is considered where all messages are assumed to be released at the same time. This approach can effectively be used to validate distributed control applications, but suffers from the resource over-provisioning problem because of the obligation to consider the worst-case. The inherent robustness of the closed-loop control application is not exploited.

Finally, off-line schedulability analysis relies on correctly estimating the tasks' worst-case execution time (WCET). Even in embedded systems the processors use caches and pipelines to improve the average computing speed, but this decreases the timing predictability. When a network is included, additional timing uncertainty is emphasized. Depending on the network protocols, the network-induced delay and data loss can be very different. For instance, let us just take as an example the two widely used industrial networks CAN and switched Ethernet.

A CAN uses a global priority-based medium access control protocol. High priority messages have lower transmission latencies while low priority ones can suffer from longer transmission delays [LIA 01]. An Ethernet switch can also deal with messages using quite different scheduling policies (e.g. FIFO, WRR, fixed priority), resulting in very different transmission delays. For real-time computing and networking designers, the new challenge is how to design the quality of service (QoS) mechanisms in scheduling both tasks in multitasking computing and messages managed by network protocols, e.g. message scheduling at the MAC level, routing, etc. to meet specified control-loop requirements. In fact, traditional QoS approaches assume that there is a deadline for each application, and a static resource allocation principle based on the worst-case situation is often used to provide delay/deadline guarantees. This approach also leads to a resource over-provisioning problem since a worst-case scenario is considered.

Another source of uncertainty may come from some parts of the estimation and control algorithms themselves. For example, the duration of a vision process highly depends on incoming data from a dynamic scene. Also, some algorithms are iterative, with a poorly predictable convergence rate, so the time before reaching a predefined threshold is unknown (and must be bounded by a timeout associated with a recovery process). In a dynamic environment, some of the less important control activities can be suspended or resumed in the case of transient overload, or alternative control algorithms with different costs can be scheduled according to various control modes, leading to large variations in the computing load.

Thus, real-time control design based on worst-case execution time, maximum expected delay, and strict deadlines inevitably lead to a low average usage of the computing resources and to a poor adaptability w.r.t. a complex execution environment. All these drawbacks call for a better integration of control objectives with computing and communication capabilities through a co-design approach taking into account both actual application requirements and the implementation system characteristics.

I.5. Diagnosis and fault tolerance in NCS

Due to an increasing complexity of dynamic systems, as well as the need for reliability, safety and efficient operation, model-based fault diagnosis has became an important subject in modern control theory and practice, e.g. [WIL 76; FRA 90; GER 98]. Different techniques of model-based methods include observer based-, parity relation- and parameter estimation approaches [CHE 99; MAN 00; ZHA 03]. When sampling and control data are transmitted over the network, many network-induced effects such as time delays and packet losses will naturally arise. Owing to the network-induced effects, the theories for traditional point-to-point systems should be revisited when dealing with NCSs. Different studies have shown the importance of taking into account characteristics of networks in the design of a fault diagnosis system [DIN 06; LLA 06]. The main idea of these approaches is to minimize the false alarms caused by transmission delays. In this case, a network-induced delay is considered when designing the FDI filter. On the other hand, FDI algorithms require specific information on the process, thus the implementation of such algorithms increases the network load and consequently, affects its QoS. The number of signals to be transmitted may be reduced by allowing only a part of sensors and actuators to have access to the network. In this case, less information is available for FDI at each sampling time.

In [ZHA 05], observability conditions are established for the reduced communication pattern.

Based on the fault diagnosis algorithm for NCSs, fault-tolerant control of NCSs can be obtained. The existing methods of fault-tolerant control techniques against actuator faults can be categorized into two groups: passive [SEO 96; CHE 04] and active approaches [ZHA 02; ZHA 06]. [ZHE 03] proposed a passive controller for NCSs considering random time delays. Although the passive controllers are easy to implement, their performances are relatively conservative. The reason is that this class of controllers, based on the alleged set of component failures and with a fixed structure and parameters, is used to deal with all the different failure scenarios possible. If a failure occurs out of those considered in the design, the stability and performance of the closed-loop system is unanticipated. Such potential limitations of passive approaches are behind the motivation for the research on active FTC (AFTC). AFTC procedures require an on-line and real-time fault diagnosis process and a controller reconfiguration mechanism. Because AFTC approaches propose a kind of flexibility to select different controllers according to different component failures, better performance of the closed loop system is expected. However, the above case holds only if the fault diagnosis process provides a correct, or synchronous, decision.

Some preliminary results have been obtained on AFTC which tend to make the reconfiguration mechanism immune from imperfect fault diagnosis decisions, as in [MAH 03] and [WU 97]. [MAK 04] further discussed the above issue by using the guaranteed cost control approach and on-line controller switching in such a way that the closed-loop system was stable at all times. However, [MAK 04] did not consider the plant controlled over the network.

I.6. Co-design approaches

NCS encompasses the control-loop application and the implementation system (CPU and the networks managed by operating systems and protocols). Research work on NCS mainly focuses on the robust control-loop design which takes into account the implementation-induced delays and data loss. Network delays are assumed to be either constant (can be realized by input data buffering) or randomly distributed, following a well-known probability distribution. Data losses are assumed to follow a Bernoulli process [ANT 07a; ZAM 08]. From these works, it appears that the assumptions on network delay or data loss patterns seldom consider the actual network characteristics nor the possible QoS mechanisms which are specific for each type of network. In fact, using a prioritized bus like CAN, a switched Ethernet or a wireless sensor network will result in fundamentally different QoS characteristics. This point is of primary importance especially when the network is shared by several control loops and other applications whose exact characteristics are often unknown at the control loop design step. In this case, the traffic scheduling has a great impact on both delay variations and packet loss, which in turn impacts the control quality (stability and performance) of the control loops. One solution to this problem relies on a tight coupling between the control specification and the implementation system at design time.

Generally speaking, two ways to achieve an efficient NCS design can be distinguished. One way that is currently being explored by the control theorists can be called "*implementation-aware* control law design." The idea is to make on-line adaptations to the control loops parameters by adjusting, for example, the control loop sampling period using a LQ approach in [EKE 00; CER 03], sample-time and/or delay dependent gain scheduling in [MAR 04] and [SAL 05], a robust H_∞ design in [SIM 05], a hybrid rate adaptation in [ANT 07b], or both the actuation intervals and control gains using model predictive control and hybrid modeling as in [Ben 06], or LQ control associated with a (m, k)-firm policy as in [JIA 07] and [FEL 08].

Another is the so-called "*control-aware* QoS adaptation," which is being explored by the network QoS designers. The idea is to re-allocate the implementation system resources on-line to maintain or increase the QoS level required by the control application. In [JUA 07], a hybrid CAN message priority allocation scheme is proposed. When there is an urgent transmission need, a dynamic priority field can be used to give it even higher priority. This work exhibits a link between the hybrid priority scheme and the control loop performance. In [DIO 07], the dynamic allocation of bandwidth sharing in Ethernet switches with weighted round-robin (WRR) scheduler based on both observed delay and the variation in the control quality (i.e. the difference between the reference and the process state) is presented. For this purpose, bandwidth sharing (i.e. the weight assigned to each data flow or Ethernet switch port) is defined as a function of the sensor to actuator delay and the current quality of control (QoC) level.

Of course, a combination of both approaches will contribute to a more efficient NCS design. Another very important aspect is the diagnosis and fault tolerance in NCS. This aspect is to be integrated in the co-design approach for achieving efficient and dependable NCS design.

I.7. Outline of the book

This book intends to provide an introduction to the problems that arise in NCS design and to present the different co-design approaches.

The first chapter provides preliminary concepts, together with state of the art techniques and existing solutions to implement distributed process control and diagnosis. In particular, control-aware static computing and network resource allocation schemes are first reviewed. Then, the real-time adaptive resource allocation approach through feedback scheduling is described, and its feasibility is assessed through a robot control application. Techniques for diagnosis and fault tolerance for networked control systems are also reviewed.

A first step toward dependable control systems consists of using robust controllers, e.g. controllers which are weakly sensitive to both process model and execution resource incertitude. Chapter 2 deals with implementation-aware or more precisely computing-aware robust control. Computation durations, pre-emption between tasks and communication over networks provide various sources of delays in the control loops. Although the control systems with constant delays have been studied earlier, taking into account more realistic variables or badly known delays recently brought new and powerful results which are surveyed in section 2.2. Traditionally, real-time control systems have been considered as "hard real-time", i.e. systems where timing deviations such as deadline misses are forbidden. In fact, closed-loop systems have some intrinsic robustness against timing uncertainties, which can be enhanced, as shown in section 2.3, by being devoted to weakly hard control tasks. Robustness can be considered as a passive approach w.r.t. timing uncertainties which are not measured and which are only assumed to have known bounds. In the case where the control intervals can be controlled, an adaptation of the controller gains w.r.t. the varying control interval can be used in combination with robust control design, as developed in section 2.4.

In addition to the controllers design, different approaches to enhance the QoC consist of techniques enabling to adjust the QoS offered by a network. Chapter 3 deals with control-aware dynamic network QoS adaptation. The key point here relies on the determination of the relation between QoC and QoS. QoC might be formulated in terms of overshoot or damping for instance, whereas QoS is often expressed in terms of delays. As the QoS adaptation mechanisms differ with the kinds of networks being considered, two approaches are illustrated in this chapter. One approach is based on CAN bus, the second is based on switched Ethernet architectures. These two protocols are widely used in industrial networks. In the case of the CAN protocol, a dynamic hybrid message-priority allocation, taking into account the control application needs, is proposed. In the case of switched Ethernet networks, bandwidth allocation control strategies are defined. For both approaches, the network feedback control is based on both the current QoS and the dynamic application-related parameters such as process state output deviation or certain control loop cost functions. Finally, since networks might have to support several applications, the QoS offered to each one is adjusted according to its specific needs.

Feedback scheduling, as presented in the previous chapters, provides an effective but limited adaptability of execution resource allocation w.r.t. varying operation conditions because the control performance is not directly taken into account in the scheduling parameter tuning. Chapter 4 describes other elaborate control and scheduling co-design schemes, coupling control performance, and scheduling parameters more tightly, through the means of a few case studies. In section 4.2 the varying sampling control laws are used as building blocks for such co-design schemes, with no guarantees for the stability of the global loops. Section 4.3, summarizes a suboptimal solution for the case of control/scheduling co-design using a slotted timescale based on the model predictive control approach. A convex optimization approach in the framework of linear systems and linear quadratic control is provided in section 4.4. Finally, section 4.5 describes a LPV-based joint control and scheduling approach applied to the control of a robot arm.

During system and network overload, excessive delays, or even data loss, may occur. To maintain the QoC of an NCS, the implementation system overload must be dealt with. As shown in Chapters 1 and 4, a common approach to deal with this overload problem is to dynamically change the sampling period of the control loops. In Chapter 5, an alternative to the explicit sampling period adjustment is proposed as an indirect sampling period adjustment. It is based on selective sampling data drops according to the (m, k)-firm model [HAM 95]. The interest of this alternative is its ease of implementation, despite reduced adjustment quality since only the multiples of the basic sampling period are used.

Chapter 6 deals with the processing of faults that may occur during the NCS operation. These faults may affect the system's physical components, or its sensors and actuators, or even the network itself. From a safety point of view, it is important to detect the occurrence of such faults, to determine the faulty physical components by fault detection and isolation (FDI). FDI allows some automatic reaction to the supervision system (for instance automatic shut down in case of danger) or reaction to the human operators in charge of the installation (for instance, manual control in open loop). Sometimes, when these faults exceed the capabilities of robust control loops, they can be accommodated by advanced control algorithms, compensating for the system's deficiencies by means of FTC. Section 6.2 recalls the basic results in FDI/FTC for the centralized case. Then the drawbacks of networking for fault detection and isolation are highlighted and some pragmatic solutions are given in section 6.4. More advanced techniques for FDI/FTC over networks subject to delays and data loss, some of which are still in the research field, are finally given in section 6.5.

Finally, Chapter 7 is devoted to the experimental validation of the approaches exposed along the previous chapters, to assess their feasibility and effectiveness. A quad-rotor miniature drone has been used throughout the SafeNecs project[1] as a common setup to integrate contributions from the team's partners, including variable sampling control, control under (m, k)-firm scheduling constraints, diagnosis and FDI

---

1. SafeNecs is a research project funded by the French "Agence Nationale de la Recherche" under grant ANR-05-SSIA-0015-03

over a network, dynamic priorities on a CAN bus and fault-tolerant control of the set of actuators. The development process starts with simulation under Matlab/Simulink; real-time constraints are then added and evaluated using the TrueTime toolbox. The next step consists of setting up a "hardware_in_the_loop" real-time simulation before performing tests on the real process.

I.8. Bibliography

[ANT 07a] ANTSAKLIS P., AND BAILLIEUL J., Guest editorial, special issue on Technology of networked control systems, *Proceedings of the IEEE*, vol. 95, p. 5–8, 2007.

[ANT 07b] ANTUNES A., PEDREIRAS P., ALMEIDA L., AND MOTA A., Dynamic rate and control adaptation in networked control systems, *5th International Conference on Industrial Informatics*, Vienna, Austria, p. 841–846, June 2007.

[ÅST 97] ÅSTRÖM K. J., AND WITTENMARK B., *Computer-Controlled Systems*, Information and System Sciences Series, Prentice Hall, Englewood Cliffs, NJ, 3rd edition, 1997.

[AUD 95] AUDSLEY N., BURNS A., DAVIS R., TINDELL K., AND WELLINGS A., Fixed priority preemptive scheduling: an historical perspective, *Real-Time Systems*, vol. 8, p. 173–198, 1995.

[AVI 00] AVIŽIENIS A., LAPRIE J.-C., AND RANDELL B., Fundamental concepts of dependability, *Report 01145*, LAAS, Toulouse, France, 2000.

[BAI 07] BAILLIEUL J., AND ANTSAKLIS P.-J., Control and communication challenges in networked real-time systems, *Proceedings of the IEEE*, vol. 95, p. 9–28, 2007.

[Ben 06] BEN GAID M., Optimal scheduling and control for distributed real-time systems, PhD thesis, University of Evry Val d'Essonne, France, 2006.

[BER 01] BERNAT G., BURNS A., AND LLAMOSÍ A., Weakly hard real-time systems, *IEEE Transactions on Computers*, vol. 50, p. 308–321, 2001.

[BUT 07] BUTTAZZO G., AND CERVIN A., Comparative assessment and evaluation of jitter control methods, *15th International Conference on Real-Time and Network Systems*, Nancy, France, March 2007.

[CER 03] CERVIN A., Integrated control and real-time scheduling, PhD thesis, Department of Automatic Control, Lund Institute of Technology, Sweden, April 2003.

[CER 05] CERVIN A., Analysis of overrun strategies in periodic control tasks, *16th IFAC World Congress*, Prague, Czech Republic, July 2005.

[CHE 99] CHEN J., AND PATTON R., *Robust Model Based Fault Diagnosis for Dynamic Systems*, Kluwer Academic Publishers, Dordrecht, 1999.

[CHE 04] CHENG C., AND ZHAO Q., Reliable control of uncertain delayed systems with integral quadratic constraints, *IEE Proceedings Control Theory Applications*, vol. 151, p. 790–796, 2004.

[CHU 79] CHURCHMAN C., *The Systems Approach and its Enemies*, Basic Books, New York, 1979.

[DIN 06] DING S., AND ZHANG P., Observer based monitoring for distributed networked control systems, *6th IFAC Symposium SAFEPROCESS'06*, Beijing, China, August 2006.

[DIO 07] DIOURI I., GEORGES J., AND RONDEAU E., Accommodation of delays for NCS using classification of service, *International conference on networking, sensing and control*, London, UK, April 2007.

[DOR 62] DORF R., FARREN M., AND PHILLIPS C., Adaptive sampling frequency for sampled-data control systems, *IEEE Transactions on Automatic Control*, vol. 7, p. 38–47, 1962.

[DOY 94] DOYLE L., AND ELZEY J., Successful use of rate monotonic theory on a formidable real time system, *Proceedings of the 11th IEEE Workshop on Real-Time Operating Systems and Software*, Seattle, USA, p. 74–78, 1994.

[EKE 00] EKER J., HAGANDER P., AND ARZEN K.-E., A feedback scheduler for real-time controller tasks, *Control Engineering Practice*, vol. 8, p. 1369–1378, 2000.

[FEL 08] FELICIONI F., JIA N., SIMONOT-LION F., AND SONG Y.-Q., Optimal on-line (m,k)-firm constraint assignment for real-time control tasks based on plant state information, *13th IEEE ETFA*, Hamburg, Germany, September 2008.

[FRA 90] FRANK P. M., Fault diagnosis in dynamic systems using analytical and knowledge-based redundancy: a survey and some new results, *Automatica*, vol. 26, p. 459–474, 1990.

[GER 98] GERTLER J., *Fault Detection and Diagnosis in Engineering Systems*, Marcel Dekker Inc., New York, 1998.

[HAM 95] HAMDAOUI M., AND RAMANATHAN P., A dynamic priority assignment technique for streams with (m, k)-firm deadlines, *IEEE Transactions on Computers*, vol. 44, p. 1443–1451, December 1995.

[HEC 94] HECHT M., HAMMER J., LOCKE C., DEHN J., AND BOHLMANN R., Rate monotonic analysis of a large, distributed system, *IEEE Workshop on Real-Time Applications*, Washington, DC, USA, p. 4–7, July 1994.

[HSI 72] HSIA T. C., Comparisons of adaptive sampling control laws, *IEEE Transactions on Automatic Control*, vol. 17, p. 830–831, 1972.

[HSI 74] HSIA T. C., Analytic design of adaptive sampling control law in sampled data systems, *IEEE Transactions on Automatic Control*, vol. 19, p. 39–42, 1974.

[JIA 07] JIA N., SONG Y.-Q., AND SIMONOT-LION F., Graceful degradation of the quality of control through data drop policy, *Proceedings of the European Control Conference*, Kos, Greece, July 2007.

[JUA 07] JUANOLE G., AND MOUNEY G., Networked control systems: definition and analysis of a hybrid priority scheme for the message scheduling, *IEEE RTCSA*, Daegu, Korea, August 2007.

[LAP 92] LAPRIE J.-C. (ed.), *Dependability: Basic Concepts and Terminology*, Springer-Verlag, New York, 1992.

[LIA 01] LIAN F.-L., MOYNE J., AND TILBURY D., Performance evaluation of control networks: Ethernet, ControlNet, and DeviceNet, *IEEE Control Systems Magazine*, vol. 21, p. 66–83, February 2001.

[LIA 02] LIAN F.-L., MOYNE J., AND TILBURY D., Network design consideration for distributed control systems, *IEEE Transactions on Control Systems Technology*, vol. 10, p. 297–307, 2002.

[LIU 73] LIU C., AND LAYLAND J., Scheduling algorithms for multiprogramming in hard real-time environment, *Journal of the ACM*, vol. 20, p. 40–61, February 1973.

[LLA 06] LLANOS D., STAROSWIECKI M., COLOMER J., AND MELENDEZ J., H_∞ detection filter design for state delayed linear systems, *6th IFAC Symposium SAFEPROCESS'06*, Beijing, China, August 2006.

[MAH 03] MAHMOUD M., JIANG J., AND ZHANG Y., Active fault tolerant control systems: stochastic analysis and synthesis, vol. 287 of *Lecture Notes in Control and Information Sciences*, Springer, Berlin, 2003.

[MAK 04] MAKI M., JIANG J., AND HAGINO K., A stability guaranteed active fault-tolerant control against actuator failures, *International Journal of Robust and Nonlinear Control*, vol. 14, p. 1061–1077, 2004.

[MAN 00] MANGOUBI R., AND EDELMAYER A., Model based fault detection: the optimal past, the robust present and a few thoughts on the future, *Proceedings of the fourth IFAC symposium on fault detection supervision and safety for technical processes, SAFEPRO-CESS'00*, Budapest, Hungary, p. 64–75, June 2000.

[MAR 04] MARTI P., YEPEZ J., VELASCO M., VILLA R., AND FUERTES J., Managing quality-of-control in network-based control systems by controller and message scheduling co-design, *IEEE Transactions on Industrial Electronics*, vol. 51, p. 1159–1167, December 2004.

[MOY 07] MOYNE J., AND TILBURY D., The emergence of industrial control networks for manufacturing control, diagnostics, and safety data, *Proceedings of the IEEE*, vol. 95, p. 29–47, 2007.

[MUR 03] MURRAY R. M., ÅSTRÖM K. J., BOYD S. P., BROCKETT R. W., AND STEIN G., Future directions in control in an information-rich world, *IEEE Control Systems Magazine*, vol. 23, April 2003.

[SAL 05] SALA A., Computer control under time-varying sampling period: an LMI gridding approach, *Automatica*, vol. 41, p. 2077–2082, 2005.

[SEO 96] SEO C., AND KIM B., Robust and reliable H_∞ control for linear systems with parameter uncertainty and actuator failure, *Automatica*, vol. 32, 1996.

[SHA 90] SHA L., AND GOODENOUGH J. B., Real-time scheduling theory and Ada, *IEEE Computer*, vol. 23, p. 53–62, 1990.

[SHA 04] SHA L., ABDELZAHER T., ÅRZÉN K.-E., CERVIN A., BAKER T., BURNS A., BUTTAZZO G., CACCAMO M., LEHOCZKY J., AND MOK A. K., Real time scheduling theory: a historical perspective, *Real Time Systems*, vol. 28, p. 101–156, 2004.

[SIM 05] SIMON D., ROBERT D., AND SENAME O., Robust control / scheduling co-design: application to robot control, *Proceedings of the 11th IEEE Real-Time and Embedded Technology and Applications Symposium*, San Francisco, USA, March 2005.

[SMI 71] SMITH M., An evaluation of adaptive sampling, *IEEE Transactions on Automatic Control*, vol. 16, p. 282–284, 1971.

[TIN 94] TINDELL K., AND CLARK J., Holistic schedulability analysis for distributed hard real-time systems, *Microprocessors and Microprogramming*, vol. 40, p. 117–134, 1994.

[WIL 76] WILLSKY A., A survey of design methods for failure detection in dynamic systems, *Automatica*, vol. 12, p. 601–611, 1976.

[WU 97] WU N., Robust feedback design with optimized diagnostic performance, *IEEE Transactions on Automatic Control*, vol. 42, p. 1264–1268, 1997.

[ZAM 08] ZAMPIERI S., Trends in networked control systems, *17th IFAC World congress*, Seoul, Korea, p. 2886–2894, July 2008.

[ZHA 02] ZHANG Y.-M., AND JIANG J., An active fault-tolerant control system against partial actuator failures, *IEE Proceedings, Control Theory and Applications*, vol. 149, p. 95–104, 2002.

[ZHA 03] ZHANG Y., AND JIANG J., Bibliographical review on reconfigurable fault-tolerant control systems, *5th IFAC Symposium SAFEPROCESS'03*, Washington, DC, USA, June 2003.

[ZHA 05] ZHANG L., AND HRISTU-VARSAKELIS D., Stabilization of networked control systems: designing effective communication sequence, *16th IFAC world congress*, Prague, Czech Republic, July 2005.

[ZHA 06] ZHANG J., AND J.JIANG, Modeling of vertical gyroscopes with consideration of faults, *Proceedings of Safeprocess'2006*, Beijing, China, 2006.

[ZHE 03] ZHENG Y., Fault diagnosis and fault tolerant control of networked control systems, PhD thesis, Huazhong University of Science and Technology, 2003.

Chapter 1

Preliminary Notions and State of the Art

1.1. Overview

Basically, co-design needs to share, compare, and gather the knowledge and perspectives brought by the stakeholders involved in the design process. Indeed, the design of safe networked control systems involves many basic methodologies and technologies. The essential methodologies involved here are feedback control, real-time scheduling, fault detection and isolation (FDI), filtering and identification, networking protocols, and QoS metrics: each of them relies on theoretic concepts and specific domains of applied mathematics such as optimization and information theory. On the other hand, these concepts are implemented via various technologies and devices, for example, involving mechanical or chemical engineering, continuous and digital electronics and software engineering.

These basic domains are explained in existing literature; hence, this chapter is not meant to give an exhaustive overview of all the methodologies and technologies used further in the book. This book is also not meant to provide an exhaustive state of the art nor to be a definitive treatise on the open topic of safe NCS; it is aimed at recording and disseminating the experience gathered by the authors during the joint SAFENECS academic research project. The team brought together people from different horizons, with basic backgrounds in control or computer science, and expertise in various domains and technologies such as digital control design, modeling of dynamic systems, real-time scheduling, identification, and diagnosis. Fault-tolerant control (FTC), networking protocols, quality of service (QoS) analysis in networks, and model-based software development, among others.

Chapter written by Christophe AUBRUN, Daniel SIMON and Ye-Qiong SONG.

Besides the knowledge provided by basic education in control and computer science, it appears that some topics that are useful in the joint design of control systems over networks are too specific, or too new and not disseminated enough, to be currently a part of basic education in control or industrial computing. So the next sections provide additional knowledge about such topics and will be useful in what follows.

Section 1.2 gives preliminary notions about real-time scheduling as well as some popular real-time scheduling policies. A particular focus is given on the so-called (m, k)-firm scheduling policy, which is, in particular, the groundwork for the control/networking co-design methodology that is developed in Chapter 5. Then, section 1.3 provides basic considerations and describes the current solutions for control-aware computing, i.e. providing computing architecture designs able to improve the quality of control of the system. One very appealing solution for the control of computing and networking resources subject to variable and/or badly known operating conditions uses a feedback-scheduling loop, whose basic design and implementation are described in section 1.4. Finally, section 1.5 provides a brief state of the art about fault diagnosis in control systems subject to network-induced effects.

1.2. Preliminary notions on real-time scheduling

When taking into account the implementation aspect of the control applications, one of the fundamental problems is to ensure timely execution of the tasks and transmission of messages related to control loops, e.g. transmission of a sampling data from a sensor to a controller, execution of the control task on a multitasking operating system (OS), sending the command from the controller to the actuator.

Control applications are typical real-time applications. The execution of a task or transmission of data is under time constraint (often under deadline constraint) in order to ensure the reactivity of the system and thus guarantee the stability and desired control performance. Real-time scheduling theory has been developed for studying how to effectively schedule the access to a shared resource of the concurrent tasks (through scheduling algorithm development) and to guarantee that the designed system can meet time constraints (through schedulability analysis).

This section is not intended to give a comprehensive review of the real-time scheduling theory, but rather provides the necessary basic background to facilitate the understanding of the remaining chapters of this book. Readers interested in more detail may refer to [LIU 00], and also to [LEU 04] for a broader view on scheduling.

The notion of priority is commonly used to order access to the shared resources such as a processor in multitask systems and a communication channel in networks. In the following, except in case of necessity, we will always use the term *task* which may represent either a task execution on a processor or a packet/message transmitted on a network channel.

A classic periodic task model is proposed by Liu and Layland [LIU 73]. Each periodic task of priority i, denoted by τ_i, is characterized by its worst-case execution time (WCET) C_i, its period T_i with which its execution is requested, and its relative deadline D_i. The problem is how to schedule a set of n independent periodic tasks $\Gamma = \{\tau_1, \tau_2, ..., \tau_n\}$ on one processor to ensure that the deadline of each *instance* is met (i.e. executed before the deadline). This is called hard real-time guarantee. In priority-based scheduling, it is usual to use the value $i = 1$ for the highest priority and larger integer of i for lower priority. During the execution of a task instance of priority i, if a higher priority one arrives, two scheduling policies exist: *pre-emptive* and *non-pre-emptive*. In the pre-emptive case, on-going lower priority task execution is interrupted by the higher priority one and its execution is resumed after the end of the execution of the higher priority one if there is no other released higher priority ones. Pre-emption is often not allowed when dealing with packet transmission in a communication channel, or when the pre-emption overhead is too high.

This classic task model can be used for representing the control task execution on a processor where the deadline is deduced from the sampling period of the control loop. When several control tasks (or one control task and several other tasks) share the same processor, scheduling policies must be studied for ensuring the deadlines are met and consequently the Quality of Control (QoC). However, as mentioned earlier in the Introduction and Problem Statement, guaranteeing deadlines are met for all the task instances (i.e. hard real-time guarantee) generally requires huge resource reservation leading to the over-provisioning problem. While we know that feedback control loops have certain robustness with respect to timing uncertainty. Occasional deadline miss or instance non-execution can often be tolerated if they do not occur in a long-term consecutive way. In this case, the (m, k)-firm model introduced by [HAM 95] seems more suitable. In fact, a task meets the (m, k)-firm constraint if there are at least m among any k consecutive task instances meet their deadline. This can thus be used to specify how the deadline miss or instance discarding is tolerated.

In what follows, we will give some basic results on priority-based classic task model scheduling and (m, k)-firm one.

1.2.1. *Some basic results on classic task model scheduling*

In this part two scheduling algorithms are presented: rate monotonic (RM) and earliest deadline first (EDF). RM is a fixed-priority scheduling algorithm where the priorities are assigned to the tasks according to their periods (or appearing rates). The task with the smallest period has the highest priority. Note that the same principle can be used to get some variants such as deadline monotonic (DM) or generally speaking fixed priority according to whatever importance criteria. EDF is a typical example of dynamic priority scheduling algorithms. Priorities assigned to the tasks are inversely proportional to the absolute deadlines of the active tasks. That is, the earlier the deadline, the higher the priority. The priority assigned to a task is of course dynamic and recalculated every time there is a new active task or an execution completion.

Let us consider a set of n independent periodic tasks $\Gamma = \{\tau_1, \tau_2, ..., \tau_n\}$ on one processor. For this system, the total normalized workload or processor utilization is $U = \sum_{i=1}^{n} \frac{C_i}{T_i}$, and the system is feasible when $U \leq 1$.

1.2.1.1. *Fixed priority scheduling*

Let the priority of the tasks τ_i be classified in decreasing orders: $i < j \Rightarrow$ the priority of τ_i is higher than that of τ_j; in the case of RM or DM, the priority of τ_i is $\frac{1}{\min(D_i, T_i)}$.

In [LIU 73], under pre-emptive RM, the following sufficient condition is established on the feasibility of the task set for $\forall i$, $D_i = T_i$:

$$\sum_{i=1}^{n} \frac{C_i}{\min(D_i, T_i)} \leq n \cdot \left(2^{\frac{1}{n}} - 1\right). \tag{1.1}$$

With n tending to infinity, $n(2^{1/n} - 1)$ approaches $ln2 \approx 69.31\%$.

In [LIU 73], it has also been shown that the worst-case response time is obtained when the first instances of all the tasks are synchronized.

A sufficient and necessary condition for the non-concrete task set has been given in [JOS 86] based on the technique called worst response time analysis (RTA).

Formally, for $\forall i$, $R_i \leq D_i$, R_i is iteratively calculated by taking into account the interference caused by the higher priorities. For $U \leq 1$ and $\forall i$, $D_i \leq T_i$, R_i is obtained with the following fixed point calculation:

$$R_i^0 = C_i$$

$$\forall k \geq 1, \ R_i^k = C_i + \sum_{j<i} C_j \cdot \left\lceil \frac{R_i^{k-1}}{T_j} \right\rceil.$$

The computing stops when the iteration can no longer progress ($R_i^k = R_i^{k-1} = R_i$) or when $R_i > D_i$.

In a general case with unrelated D_i and T_i, and especially for the case of $D_i > T_i$, this RTA technique has been extended [TIN 94].

Note that this technique is also applicable to the non pre-emptive case by including the blocking factor due to the on-execution low-priority task [TIN 94].

1.2.1.2. *EDF scheduling*

For a set of independent periodic tasks $\Gamma = \{\tau_1, \tau_2, ..., \tau_n\}$ with $\forall i, D_i = T_i$ and under pre-emptive EDF, the necessary and sufficient condition of the schedulability is [LIU 73]:

$$U = \sum_{i=1}^{n} \frac{C_i}{T_i} \leq 1. \tag{1.2}$$

It is proved [BAR 90] that this condition is still true for the case $\forall i, D_i \geq T_i$. For a task set with $\forall i, D_i \leq T_i$, the previous condition is no longer sufficient. In [BAR 90] and [SPU 96], two necessary and sufficient conditions are given for task set with arbitrary D_i and T_i.

Under non pre-emptive EDF, in the case of non-concrete tasks, a sufficient and necessary schedulability test with pseudo-polynomial complexity is given in [JEF 91]. The test is based on the processor demand calculation. When the tasks are synchronous, the same condition becomes only sufficient. It has been shown in [JEF 91] and [GEO 95] that determining the schedulability is an NP-hard problem.

1.2.1.3. *Discussion*

Fixed priority scheduling is now supported by most of commercial off-the-shelf (COTS) OS. It can also be found in some networks. For instance, CAN network uses a priority-based MAC protocol so that CAN messages schedulability can be analyzed using the RTA technique [TIN 94]. This scheduling algorithm is also present in some Ethernet switches. Chapter 3 will study the control and network QoS co-design of the NCS distributed around a CAN and switched Ethernet network, respectively.

EDF is known as an optimal scheduling algorithm. However its implementation can induce unacceptable high overhead due to the frequent context changes related to the dynamic priority assignment. Today, few COTS OS support EDF.

Ensuring hard real-time constraint by the schedulability analysis may lead to resource over-provisioning problem since in practice it is difficult to get a tight upper bound on the WCET, especially when dealing with packet transmission in a network. The worst-case workload scenario may never happen. To cope with these uncertainties, a system designer may try to either prevent overloads by making safe assumptions about workload or tolerate overloads (but still providing reduced-but-acceptable level of service). The latter is particularly interesting for feedback control applications thanks to the robustness of the control loops. For the overload management, fixed-priority scheduling has an advantage over EDF. In fact, during an overload situation, only low-priority tasks are affected in fixed-priority scheduling. EDF has very bad behavior during overload since it tries to always give the highest priority to the task that will miss its deadline, resulting in a general deadline miss of all tasks.

Many other approaches such as imprecise computation model, skippable model, (m, k)-firm model have been developed to deal with overloads. We introduce in the following some basic notions on (m, k)-firm. This model is then used in Chapters 2 and 5. Chapter 5 also gives further details before applying it to the overload management when several control loops share the same processor.

1.2.2. *(m,k)-firm model*

A system meeting (m, k)-firm constraint requires a minimum QoS of m out of any k consecutive deadlines to meet in the worst case, where m and k are two positive integers (the case where $m = k$ is equivalent to the ideal case, which is noted by (k, k)-firm and corresponds to the hard real-time constraint). In general cases, more than m deadlines are met as the system does not always run at the worst-case condition. This is to say that if the (m, k)-firm constraint is respected, during whatever window of k consecutive instance occurrences, there exist at least m instances that meet their deadline. Note that in general the k consecutive instance occurrences are not necessarily periodic, so the window of k consecutive instances does not necessarily have a constant time duration [LI 09]. However, in the networked control applications, most of the cases are periodic ones due to the sampling principle. Therefore in this book, a task τ_i is characterized by $\{C_i, T_i, D_i, m_i, k_i\}$, with $i = 1, 2, ..., n$ representing the index of tasks (but not necessarily their priority). A task could be a stream of messages to transmit or a periodic task to execute.

A task under (m, k)-firm constraint can be found in one of the two following states: normal and dynamic failure [HAM 95]. Figure 1.1 shows an example of the state-transition diagram for $(2, 3)$-firm: 1 denotes that a deadline is met and 0 denotes a deadline missed, respectively. These states are evaluated according to the past situation of the system; every state that is either normal or dynamic failure depends on the last three deadlines services. The next deadline's meet or miss will cause the system

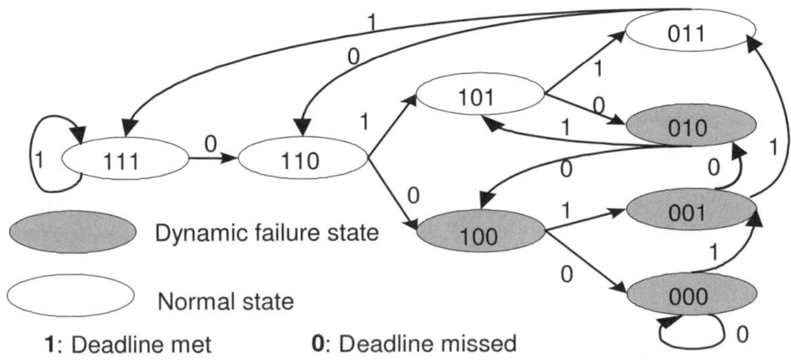

Figure 1.1. *State-transition diagram with $(2, 3)$-firm*

to transit to another state. If there is more than 1 missed deadline, the system is in a dynamic failure state. Otherwise, the system is in a normal state.

For the efficient overload management, we adopt dropping strategy: any instance that cannot be executed before its deadline is dropped. So whenever talking about (m, k)-firm in this book, a missed deadline is equivalent to an instance drop.

If a control system can accept control performance degradation until $k - m$ deadlines misses (or equivalent packet losses) among any k consecutive ones, the system can then be designed according to the (m, k)-firm approach to offer the variable levels of control performance between (k, k)-firm (ideal case) and (m, k)-firm (worst case) with as many intermediate levels as the possible values there are between k and m. This results in a control system with graceful degradation of the control performance.

The problem of scheduling tasks under (m, k)-firm constraint has drawn particular attention in real-time community. Some important results have been obtained.

The first category concerns the development of the specific scheduling algorithms. The well-known one is distance-based priority (DBP) proposed in [HAM 95]. Considering n tasks sharing a common server (may be a processor or a communication channel) and each has its own (m_i, k_i)-firm constraint, the principle is to dynamically assign priorities to the different tasks according to the distance to the dynamic failure state. The closer the task to a failure state, the higher its priority. A failure state occurs when the task's (m_i, k_i)-firm requirement is violated, i.e. there are more than $k_i - m_i$ deadlines missed within the last k-length window. So to know the current state of a task we should examine the execution history of the last k instances. If we associate 1 with an instance with deadline met and 0 with an instance with deadline missed, this history is then entirely described by a word of k bits called the *k-sequence*. The k-sequence is a word of k ordered bits in which each bit keeps memory of whether the deadline is missed (bit= 0) or met (bit=1). In the k-sequence, the bit is ordered the most recent to the oldest task instance where the leftmost bit represents the oldest one. Each newly arrived instance causes a shift of all the bits toward the left, the leftmost exits the word and is no longer considered, while the rightmost will be a 1 if the instance has met its deadline (i.e. it has been served within) or a 0 otherwise. Figure 1.2 gives an example with (3,5)-firm constraint.

Thus for each task τ_i under (m_i, k_i)-firm constraint, the priority is assigned based on the number of consecutive deadline misses that leads the task to violate its (m_i, k_i)-firm requirement. This number of missed deadlines is referred to as *the distance to failure state* from the current state. DBP assigns priority to a given instance by the distance from the current k-sequence to a failure state. Considering the above example with $(3, 5)$-firm constraint, the current instance is assigned the priority of 2 if the current *5-sequence* is (11011), and is set the priority of 3 if the current *5-sequence* is (10111). Note that in case of equal priority, EDF is used to break the tie. For

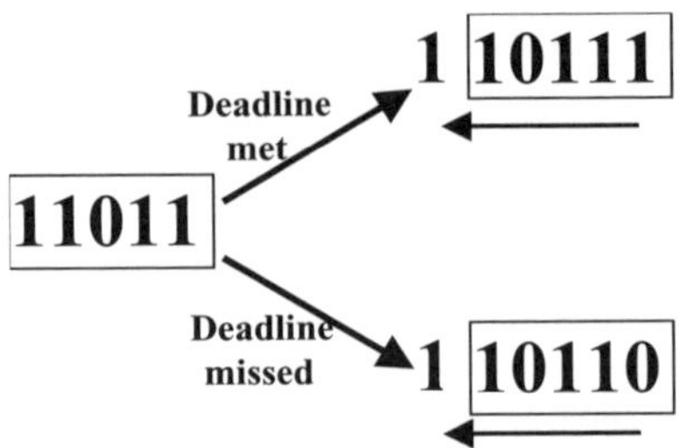

Figure 1.2. *Evolution of the k-sequence*

non-pre-emptive DBP, a first necessary schedulability condition has been given in [POG 03]. A sufficient schedulability condition is presented in [LI 04a]. Dynamic window constrained-scheduling (DWCS) is another similar algorithm proposed in [WES 99] with its schedulability analysis in [WES 04]. Other scheduling algorithms have also been developed specifically for control applications such as Markov chain-driven algorithm (MDA), dropout-rate-driven algorithm (DDA), and feedback-driven algorithm (FDA) [LIU 06].

The second category concerns the adaptation of the existing scheduling algorithms to the (m, k)-firm constraint. In [RAM 99], the RM algorithm is adapted to the (m, k)-firm model by defining the notion of (m, k)-pattern which is in fact a fixed k-sequence. This allows the instances to be classified in two priorities: mandatory and optional. The guarantee of the execution of the mandatory instances ensures meeting of the (m, k)-firm constraint. The sufficient schedulability condition is also presented. Enhanced fixed-priority (EFP) algorithm [QUA 00] proposes an improvement by introducing a heuristic rotation algorithm to reduce the effect of the worst-case interference point due to the superposition of the mandatory instances of tasks. This part will be further detailed in Chapter 5.

1.3. Control aware computing

Control and real-time computing have been associated for a long time, with the control of industrial plants and in embedded or mobile systems, e.g. automotive and robotics. However, both parts, control and computing, are often designed with poor interaction and mutual understanding. From the control design point of view, a constant and unique period is usually assumed. Delays are supposed negligible or constants, and jitter is ignored. The implementation design then follows, trying to meet these assumptions.

Real-time scheduling has mainly focused on how to dimension resources to meet deadlines, or equivalently, on the schedulability analysis for a given resource. Indeed, the real-time community has usually considered that control tasks have fixed periods, hard deadlines, and worst-case execution times. This assumption has served the separation of control and scheduling designs, but has led to under-utilization of CPU resources and inflexible design.

The hard and costly way consists in building a highly deterministic system, from the hardware, operating system and communication protocols sides, so that the actual implementation parameters meet the ideal ones. This extreme solution is used if, for instance, determinism is requested for formal verification and/or certification purpose, e.g. as in the synchronous programming approach [BEN 91] or in the time-triggered paradigm [KOP 03]. However, trying to nullify (even virtually) latencies and jitter generally leads to worst-case-based resources provisioning and tends to needlessly overconstraint the system's design and implementation.

In fact, the hard real-time constraints can be often relaxed in a controlled way, e.g. considering the intrinsic robustness provided by the closed-loop paradigm. In a real control implementation, latencies and sampling jitter inevitably exist, in particular when actuators, sensors, and controllers are distributed over a network. A smart organization and use of network and processor resources, with control features in mind, may lead to serious improvements in the control performance, resources usage and overall cost.

1.3.1. *Off-line approaches*

A first set of methods consists of computing off-line the set of scheduling parameters which (ideally) maximize the control performance under schedulability constraints. The first step consists in getting a model of the control performance function of the execution parameters. The problem of optimal sampling period selection, subject to schedulability constraints, was first introduced in [SET 96]. Considering a bubble control system benchmark, the relationship between the control cost (corresponding to a step response) and the sampling periods was approximated using convex exponential functions. Using the Karush–Kuhn–Tucker (KKT) first-order optimality conditions, the analytic expressions of the optimal off-line sampling periods were established. The problem of the joint optimization of control and off-line scheduling has been studied in [REH 04; LIN 02a; BEN 06].

In a multitasking system, several control tasks share a common computing resource: the resulting pre-emption induces latencies due to the computations themselves, but also due to the interleaving between their executions. Models of the control behavior based on linear systems theory are used in [RYU 97] and [SAK 98] to derive cost functions which depict the control performance, e.g. the rise time, as a function of two execution parameters, the control period and loop delay. Then an optimization iterative algorithm (simplex) is used to tune the execution parameters in order to maximize the overall control performance with respect to the implementation feasibility. However, due to the complexity of the optimization process this method can be used only off-line.

Often the lazy way to implement a controller consists of programming a single real-time task when all the components of the controller are executed in sequence in a single loop. However, it appears that all the components of a control algorithm do not require the same timing parameters, and do not have the same weight in the final performance and stability. Some parts of the controller are more critical w.r.t. latencies, or require more frequent updating than others. Therefore, the controller can be split into modules according to these timing requirements, so that latencies can be minimized along some critical data paths, or to enforce the execution of safety critical functions even in the case of transient overload.

For example, it is possible to split the controller of a linear system in several parts according to their relative urgency, as shown in the following piece of pseudocode [ÅST 97]:

```
loop{
    Wait_Clock();    //waiting periodic request
    Get_Sensors();    //read y(k)
        Calculate_Output();    //u(k) = f(y(k), x̂(k − 1), ...)
    Send_Control();    //send u(k)
        Update_State();    //x̂(k) = g(y(k), x̂(k − 1), ...) }
```

Here the input/output latency is minimized, as the control signals are computed and sent to the actuators immediately after updating the measures, while updating the model and internal state of the controller can be delayed until the end of the control period.

This method is, for example, used in [EKE 99] where this control task split is applied to the control of a set of concurrent inverted pendulums: compared with the naive implementation where all computations are made before sending the control signals, it provides an impressive increase in the control performance with no additional computing cost. A more complex and nonlinear system can also benefit from such separation of the control algorithm between fast and critical control paths (e.g. low-level stabilization loop) and slower components, e.g. vision-based navigation. Obviously the operating system and associated run-time framework must allow for such multi-task/multi-rate implementation [SIM 05a]. This modular timing analysis seems to be an essential starting point for flexible and efficient real-time control implementation, as in the example depicted in section 1.4.2.3.

1.3.2. *Quality of Service and flexible scheduling*

Other approaches define a QoS criterion to depict, e.g. the relations between the performance and the controller's period. This performance model can be used to configure an admission controller managing the overall system load [ABD 97], or to perform an on-line negotiation involving periods and priorities as in [SAN 00].

Besides control considerations, flexible and control aware solutions have also been provided by the computer science side. For example, let us cite the "Elastic Tasks" paradigm [BUT 00], where the sensitivity of the QoS relative to the execution period for every task is modeled by a "stiffness" and takes into account bounds in the allowed execution period. To make the task set schedulable, the task stack is "compressed" until the accumulated execution load fit with the allocated CPU capacity. Although this implementation is in open loop w.r.t. the actual QoS, it allows for an improved adaptation against transient overloads.

As sharing the computing resource between controllers is a central issue, some variants of the Control Bandwidth Server (CBS) approach [ABE 98] have been used to enforce protection between competing control activities. For example in [CAC 00] the nominal control periods of the competing controllers are computed thanks to the optimization process of [SET 96] aiming at maximizing the control performance under scheduling constraints. The on-line execution time variations of the controllers are locally processed inside the computing budget allocated by the CBS server.

Let us cite also the control server ([CER 03b]) where a fraction of the total CPU power is statically reserved to each control thread. Then the system behaves as if each controller was isolated using its own computation resource, in particular an overloaded controller does not disturb its neighbors. Inside each computing segment the individual controllers are organized to minimize their *I/O* latency and jitter. In the case of transient overload, the missing computing budget for one controller is postponed to its next reserved slice, with no impact on the others.

This mainly concerns the integration of control performance knowledge in the scheduling parameters assignment. Indeed, once a control algorithm has been designed, a first job consists of assigning timing parameters, i.e. period of tasks and deadlines, so that the controller's implementation meets the control objective. This may be done off-line or on-line.

In off-line control/scheduling co-design, the task of setting adequate values for the timing parameters rapidly fall into case studies based on simulation and experiments. For instance in [RYU 97] off-line iterative optimization is used to compute an adequate setting of periods, latencies, and gains resulting in a requested control performance according to the available computing resource and implementation constraints. Also in [SAN 02] the temporal requirements of the control system are described using complex temporal attributes (e.g. nominal period and allowed variations, precedence constraints, etc.): this model is then used by an off-line iterative heuristic procedure to assign the scheduling parameters (e.g. priorities and offsets) to meet the constraints.

Concerning co-design for on-line implementation, recent results deal with varying sampling rates in control loops in the framework of linear systems: for example

[SCH 02] show that, while switching between two stable controllers, too frequent control period switches may lead to instability. Unfortunately, most real-life systems are nonlinear and the extrapolation of timing assignment through linearization often gives rough estimations of allowable periods and latencies or they can even be meaningless. In fact, as shown later in the examples, knowledge of the plant's behavior is necessary to get an efficient control/scheduling co-design.

1.4. Feedback-scheduling basics

Besides traditional assignment of fixed scheduling parameters, more flexible scheduling policies have been investigated. The main idea is that fast sampling and computing are costly, so running the controllers only when useful or necessary is expected to save computing power, network bandwidth and energy. Networked control systems can be made of autonomous and/or mobile devices connected by wireless communications. As these devices may have a limited on-board energy storage, optimizing the cost of computations and communications induced by the control activities is again gaining interest.

As already mentioned in section 0.2, the on-line adaptation of the sampling interval as a function of the system's behavior and state has been studied from the beginning of computer-controlled systems [DOR 62; HSI 74]. Besides the time-triggered approach and variations around the sampling period adaptation, an even more radical approach is the so-called "Event-based control" concept. Indeed, this approach is natural in some application domains, as in engine control where the basic events scale is linked to the crank-shaft position turning at a variable speed rather than to clocks. However, it has been proposed as an alternative to time-triggered sampling when the execution resources used by the controller are constrained.

Within the event-triggered control approach the decision to compute and apply a new control action is based on the level crossing of some signal of interest, e.g. the error signal between the desired set-point and actual measurement as in [ÅRZ 99] and [DUR 09]. Note that even if the controller is sleeping most of the time waiting for awaking events, the controlled plant must be continuously observed for event detection at a rate fast enough to allow for fast reactions and effective disturbance rejection.

Anyway, effective real-time management of the computing and networking resources needs to close the loop between the execution resource utilization and the actual scheduling parameters. The feedback-scheduling approach has been initiated both from the real-time computing side [LU 00; LU 02] and from the control side [CER 00; EKE 00; CER 02]. The idea consists of adding to the process controller an outer sampled feedback loop ("scheduling regulator") to control the scheduling parameters as a function of a QoC (Quality of Control) measure. It is expected that an on-line adaptation of the scheduling parameters of the controller may increase its

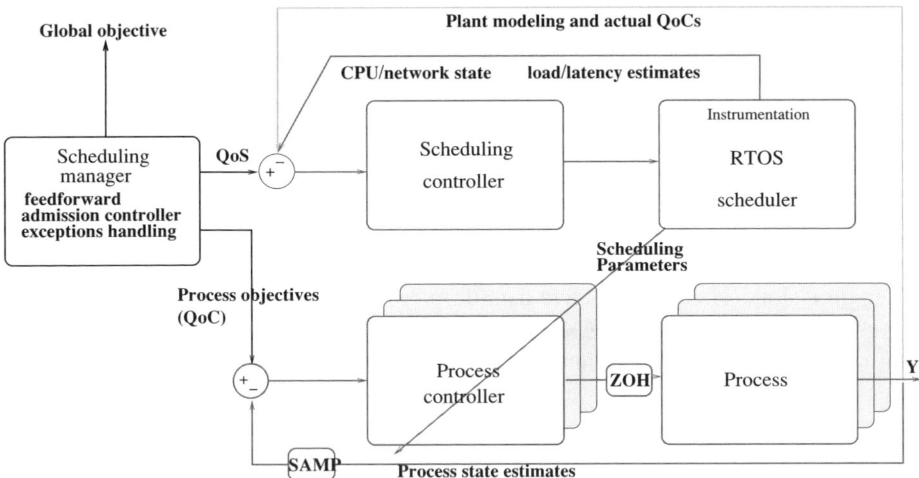

Figure 1.3. *Hierarchical control structure*

overall efficiency w.r.t. timing uncertainties coming from the unknown controlled environment. Also we know from control theory that closing the loop may increase performance and robustness against disturbances when properly designed and tuned (otherwise it may lead to instability).

Figure 1.3 gives an overview of a feedback scheduler architecture where an outer loop (the *scheduling controller*) adapts in real time the scheduling parameters from measurements taken on the computer's activity, e.g. the computing load. Ideally it would be also fed by measures related to the quality of control, thus really providing integrated control and scheduling, which is the topic of Chapter 4. Besides this controller working periodically (at a rate larger than the sampling periods of the plant control tasks), the system's structure may evolve along a discrete time scale upon occurrence of events, e.g. for new task admission or exception handling. These decisional processes may be handled by another real-time task, the *scheduling manager*, which is not further detailed in this paper. Notice that such a manager may give a reference to the controller resource utilization.

The design problem can be stated as control performance optimization under constraint of available computing resources. Early results come from [EKE 00] where a problem of optimal control under computation load constraints is theoretically solved by a feedback scheduler, but leads to a solution too complex to be implemented in real time. Then [CER 03a] shows that this optimal control problem can often be simply implemented by computing the new task periods by the re-scaling:

$$h_i^{k+1} = h_i^k \frac{U}{U_{sp}},$$

where U_{sp} is the utilization set-point and U the estimated CPU load. The feedback scheduler then controls the processor utilization by assigning task periods that optimize the overall control performance. This approach is well suited for a "quasi-continuous" variation of the sampling periods of real-time tasks under control of a pre-emptive real-time operating system (RTOS).

Another approach has been used in the framework of the so-called (m, k)-*firm* schedulability policy, where the scheduling strategy ensures the successful execution of at least m instances of a given task (or message sending) for each time window of length k slots. Hence a selective data drop policy (as in [JIA 07]) or a computing power allocation to selected tasks (as in [BEN 06]) can be used to perform optimal control of a plant under constraint of computing or communication limitations. This latter approach is well suited for non pre-emptive scheduling of control tasks and for networked control systems subject to message loss: the tasks or messages are scheduled to jointly perform congestion avoidance and optimal control.

Indeed, in all cases the adaptive behavior of a feedback scheduler, associated with the relative tolerance of the control system w.r.t. the implementation induced timing uncertainties, allows for the design and implementation of real-time control systems based on their average execution behavior rather than on pessimistic worst-case estimates.

1.4.1. *Control of the computing resource*

Feedback scheduling is a dynamic approach allowing a better use of the computing resources, in particular when the workload changes e.g. due to the activation of an admitted new task. Indeed, the CPU activity will be controlled according to the resource availability by adjusting scheduling parameters (i.e. period) of the plant control tasks.

In the approach proposed here, a way to take into account the resource sharing over a multitasking process is developed. In what follows, the control design issue is described including the control structure, the specification of control inputs and measured outputs, as well as the modeling step.

1.4.1.1. *Control structure*

In Figure 1.4 scheduling is viewed as a dynamic system between control task frequencies and processor utilization. As far as the adaptation of the control tasks is concerned, the load of the other tasks is seen as an output disturbance.

1.4.1.2. *Sensors and actuators*

As stated in section 1.3, priorities must be assigned to control tasks according to their relative urgency; this ordering remains the same in the case of a dynamic scheduler. Dynamic priorities, e.g. as used in EDF, only alter the interleaving of running tasks and will fail in adjusting the computing load w.r.t. the control requirements.

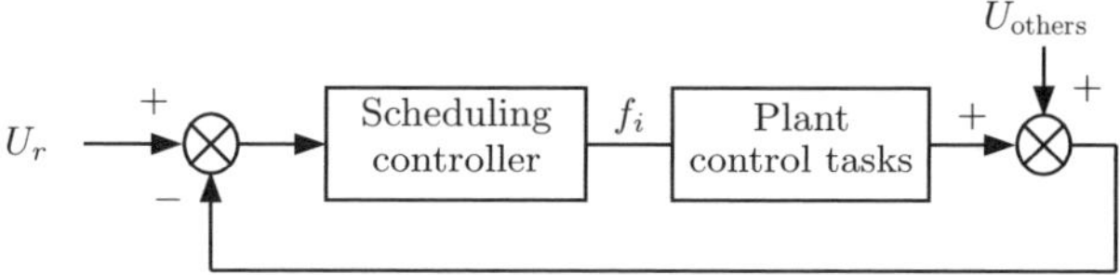

Figure 1.4. *Feedback-scheduling block diagram*

Consequently, we have elected the task periods to be the primary actuators of the system running on top of a fixed-priority scheduler. Note that if the control timing setting, based only on the scheduling adaptation, becomes out of reach (e.g. because the requested intervals would be out of bounds), possible secondary actuators are variants of the control algorithms, with different computing costs and QoS contributions to the whole system. Such variants must be handled by the scheduling manager working on a discrete events time scale.

As the aim is to adjust on-line the sampling periods of the controllers in order to meet the computing resource requirements, the control inputs are thus the periods of the control tasks. The measured output is the CPU utilization. Let us first recall that the scheduling is here limited to periodic tasks. In this case the processor load induced by a task is defined by $U = \frac{c}{h}$ where c and h are the execution time and period of the task. Hence, processor load induced by a task is estimated, in a similar way [CER 02], for each period h_s of the scheduling controller, as:

$$\hat{U}_{kh_s} = \lambda\, \hat{U}_{(k-1)h_s} + (1 - \lambda)\, \frac{\overline{c}_{kh_s}}{h_{(k-1)h_s}} \tag{1.3}$$

where h is the sampling frequency currently assigned to the plant control task (i.e. at each sampling instant kh_s) and $\overline{c}$ is the mean of its measured job execution-time. λ is a forgetting factor used to smooth the measure.

1.4.1.3. *Control design and implementation*

The proposed control design method for feedback scheduling is here developed. First one should note that, as shown in [SIM 03], if the execution times are constant, then the relation, $U = \sum_{i=1}^{n} C_i f_i$ (where $f_i = 1/h_i$ is the frequency of the task) is a linear function (while it would not be the case if expressed as a function of the task periods). Therefore, using (1.3), the estimated CPU load is given as:

$$\hat{U}(kh_S) = \frac{(1 - \lambda)}{z - \lambda} \sum_{i=1}^{n} \overline{c}_i(kh_S) f_i(kh_S) \tag{1.4}$$

An illustration, for the case of a single control task system, is given in Figure 1.5 where the estimated execution-times are used on-line to adapt the gain of the controller for

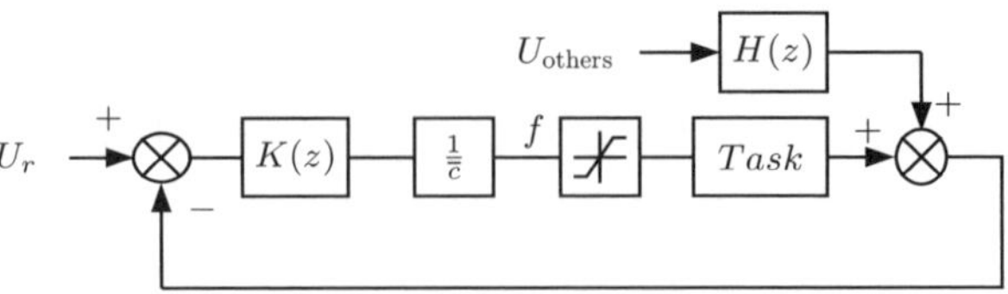

Figure 1.5. *Control scheme for CPU resources*

the original CPU system (1.4) (this allows us to compensate the variations of the job execution time).

As $\bar{c}$ depends on the run-time environment (e.g. processor speed) a "normalized" linear model of the task i (i.e. independent of the execution time), G_i, is used for the scheduling controller synthesis where $\bar{c}$ is omitted and will be compensated by on-line gain scheduling $(1/\bar{c})$ as shown below:

$$G_i(z) = \frac{\hat{U}(z)}{f_i(z)} = \frac{1 - \lambda}{z - \lambda}, \quad i = 1, \ldots, n. \tag{1.5}$$

According to this control scheme, the design of the controller K can be made using any control methodology at hand. In fact all the control toolbox resources may be adapted for feedback-scheduling purpose, e.g. as reviewed in [XIA 08].

One of the most frequently used is the well known P.I.D. control: it has been for example used for the on-line regulation of purely computing systems as web and mail servers, as shown in section 1.4.2.1. Another popular approach is the linear quadratic (LQ) control method, whose application to scheduling control has also been investigated as shown in section 1.4.2.2.

Model predictive control is known to cope well with control of complex systems under control and/or state constraints. As scheduling control deals with control under computing and/or communication limitations, this control design has also been investigated, as shown by the example in section 4.3.

Finally, as a digital control system combines uncertainties and modeling errors from both the plant and the control implementation, robustness seems to be a crucial issue: the well known H_∞ control theory, which can lead to a robust controller w.r.t modeling errors (see [ZHO 96] for details on H_∞ control), is also a good candidate to perform. Moreover, it provides good properties in presence of external disturbance, as emphasized in the robot control example below (1.4.2.3).

1.4.2. *Examples*

1.4.2.1. *Feedback scheduling a web server*

One of the most popular and widely used controller for SISO systems is the so-called proportional integral derivative (PID). The basic formulation for a continuous time PID controller is [ÅST 97]

$$U = K_p.e + K_v.\frac{de}{dt} + K_i.\int_0^t e(\tau).d\tau,$$

where U is the control signal to be applied to the process input and e is the error signal between the desired set-point y_d and the measured output y. It is largely used in industry as it can be applied to many SISO systems with an easy tuning and a minimal modeling effort.

This simple design has been used to control and tune the behavior of computation devices submitted to QoS constraints, for example web (Figure 1.6) or mail servers ([LU 01], [PAR 02]). Design basics, control oriented models for computing devices, and case studies for feedback control of computing systems can be found in [LU 00; LU 02], and [HEL 04] among other references.

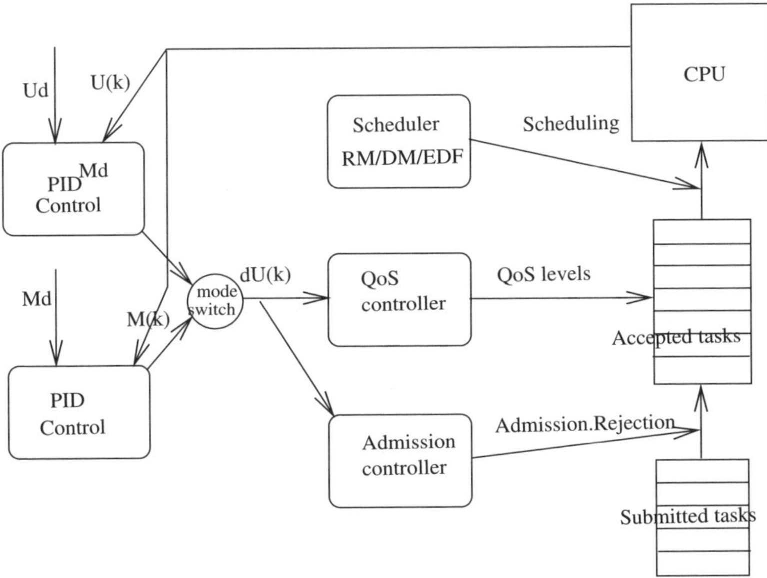

Figure 1.6. *Web server closed-loop regulation*

For each period of the scheduling controller, the measures are the total CPU load $U(k)$ and the miss ratio $M(k)$. The corresponding gains G_A and G_M are images of the modeling uncertainties.

The execution of requests T_i is modeled by at least two levels of quality, i.e. couples (QoS contribution, execution cost). The actuation provided is the choice of the execution mode corresponding to a given cost, at every sampling period. The accumulated regulated cost is finally the global CPU load.

If the sampling period h is large enough, the transfer function of the CPU load submitted to computing requests Δu can be modeled by an integrator, where G_A and G_M are the weakly known gains of the open loop process:

$$\begin{cases} U(k) = U(k-1) + G_A.\Delta u(k-1) \text{ if CPU under-loaded,} \\ M(k) = M(k-1) + G_M.\Delta u(k-1) \ \text{ if CPU over-loaded.} \end{cases}$$

Thus, a simple proportional regulator is able to control the server load

$$\begin{cases} \Delta u(k) = K_{pu}.E_U(k) \text{ où } E_U(k) = U_s - U(k) \ \text{ if } U \leq 1, \\ \Delta u(k) = K_{pm}.E_M(k) \text{ où } E_M(k) = M_s - M(k) \ \text{ if } M > 0. \end{cases}$$

Figure 1.7 (borrowed from [LU 02] with permission of the author) shows the steady state behavior (CPU load $U(k)$ and deadlines missed $M(k)$) as a function of the desired load U_d. The role of the underlying scheduling policy can be observed: using EDF (on the right picture) allows for nullifying the deadlines missed up to $U_d = 1$, but exhibits degradation in the case of permanent overload faster than a static DM priority policy. However, in all cases, the server shows some ability for automatic adaptation to the arrival of sporadic requests and for recovery against sporadic overloads, thus leading to a kind of self-administration at a very low-computing cost.

1.4.2.2. *Optimal control-based feedback scheduling*

The aforementioned PID regulation approach is very simple to design and tune. The downside of the very limited number of tuning parameters is the limited capabilities of, e.g. shaping robustness templates or decoupling several transfer modes.

Let us come back to the initial problem, which may consist formally in the optimization of a control performance under constraints of limited computing resources. This problem has been analytically solved by [EKE 00] and [CER 03a] for the following case study. A given computing resource is shared by n real-time control tasks, each one is used to control a linear stochastic process; each controller has an h_i period and a C_i execution time. A sampling frequency dependent quality criterion $J_i(h_i)$ is

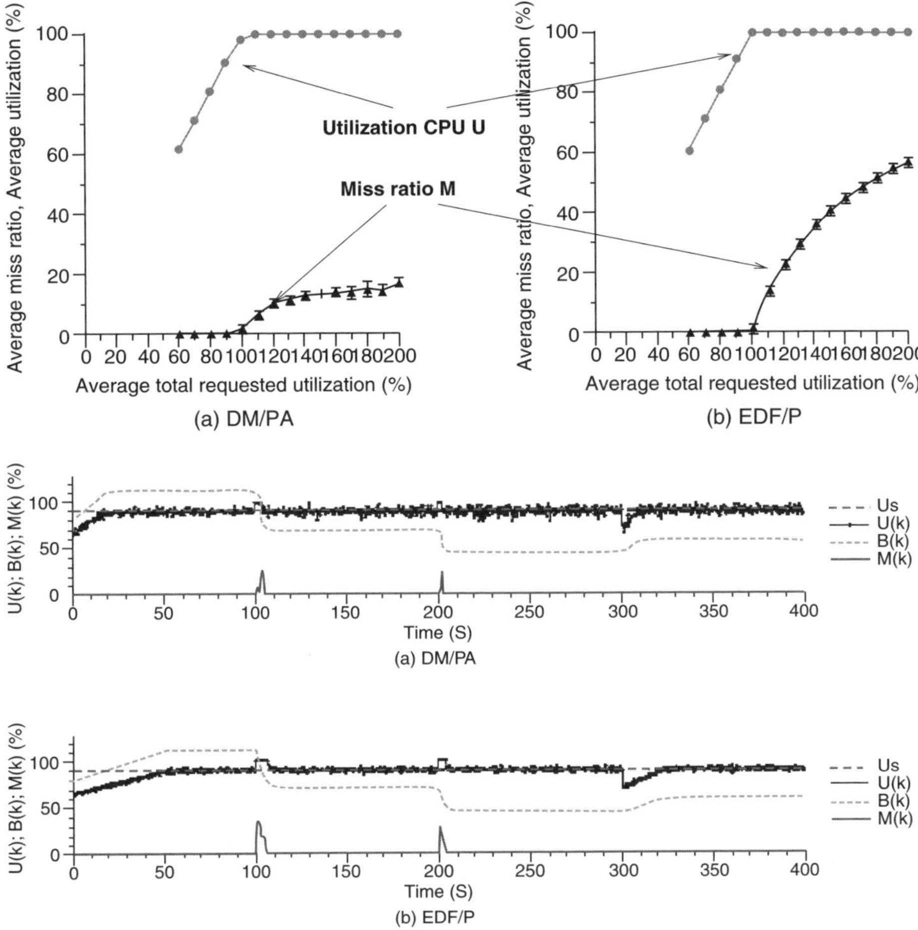

Figure 1.7. *Load response of the server (steady state and dynamic response)*

attached to each controller. The control goal is the maximization of a global cost function over the set of controller, with respect of a desired computing load U_d:

$$\min_n J = \sum_{i=1}^{n} J_i(h_i) \text{ under constraint } \sum_{i=1}^{n} C_i/h_i \leq U_d.$$

The control variables are the control periods h_i. The problem is solved using the cost function

$$J(h) = \frac{1}{h} \int_0^h \begin{bmatrix} x^T(t) & u^T(t) \end{bmatrix} Q \begin{bmatrix} x(t) \\ u(t) \end{bmatrix} dt.$$

This particular cost function allows for a theoretical state feedback controller performing the optimization. However, the execution of such a controller would require to solve Lyapunov and Ricatti equations at each sample which is clearly two expensive to be executed in real time with reasonable computing resources.

Fortunately, approximations can often be found. The cost functions can be often approached by linear $J_i(h) = \alpha_i + \gamma_i h$ or quadratic $J_i(h) = \alpha_i + \beta_i h^2$ functions. Computing the optimal values for the h_i periods becomes particularly easy if *all* the cost functions are either linear or quadratic [CER 02].

In this case, the periods are given by the following algorithm:

– the initial control frequencies $f_i = 1/h_i$ are chosen proportionally to $(\beta_i/C_i)^{1/3}$ (quadratic costs) or to $(\gamma_i/C_i)^{1/2}$ (linear costs);

– these values provide a nominal computing load $\hat{U}_0 = \sum_{i=1}^{n} \frac{\hat{C}_i}{h_{0i}}$;

– estimation of the execution times and filtering with the λ forgetting factor $\hat{C}_i(k) = \lambda\hat{C}_i(k-1) + (1-\lambda)c_i$;

– for a different CPU desired load the new periods are given by a simple re-scaling $h_i = h_{0i}\frac{U_{sp}}{\hat{U}_0}$;

– the new control gains can be either computed on-line, or extracted from a pre-calculated table;

– the values desired for CPU loads U_{sp} are elaborated by a supervision process called *feed-forward*, whose role is similar to the admission controller of the previous section.

Figure 1.8 shows some simulation results using TrueTime, a toolbox for Matlab/Simulink dedicated to models of real-time systems and networks ([OHL 07]).

In this experiment, four control tasks share a common computing resource. Each task $T_i, i = 1,\ldots,4$ controls an inverted pendulum, under a fixed-priority ordering $T_1 \prec T_2 \prec T_3 \prec T_4$. The performance criterion is the classic quadratic cost $J_i = \int_0^{T_{sim}} (y_i^2(t) + u_i^2(t))dt$. T_1 and T_2 are executed when the system starts, then T_3 is admitted at $t = 2$ s and T_4 at $t = 4$ s.

Without adaptation (Figure 1.8(a)) all controllers remain executed at their nominal frequency, the computer becomes overloaded, and finally the lowest priority tasks T_1 and T_2 are so disturbed by pre-emption that they can no longer stabilize their pendulum (and their criterion becomes ∞).

The controlled scheduler (Figure 1.8(b)) adapts on-line the control periods; thus, it avoids the processor overload and keeps stability for all the process. (Note that the

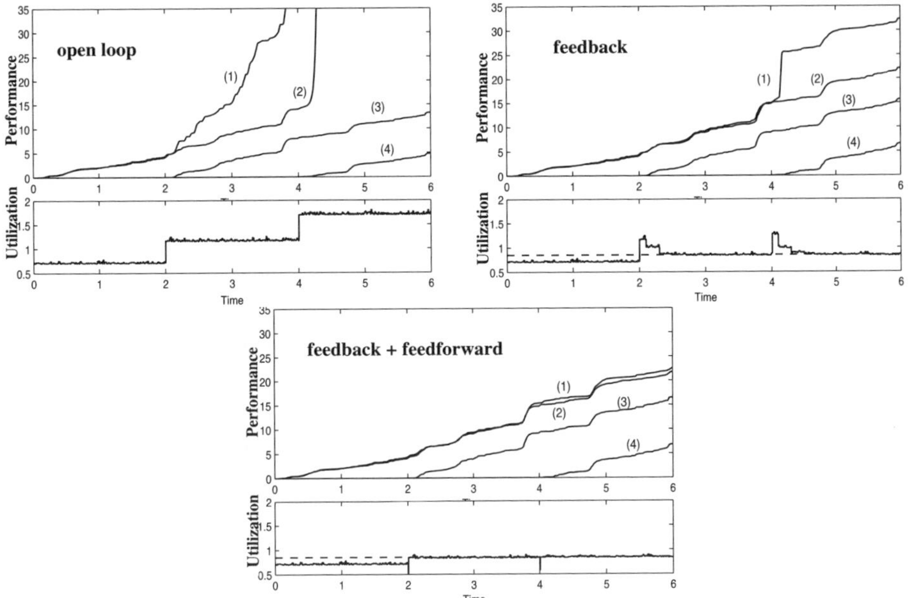

Figure 1.8. *Simulations under TrueTime, from [CER 03a]*

control quality decreases for lower priority process). Adding a *feed-forward* admission controller (Figure 1.8(c)) allows for future tasks cost anticipation and for enhanced transient behavior.

Note that here the optimality of the process control performance relies on open-loops pre-computed cost functions and that robustness issues are not taken into account. Also nothing is done here to analyze the effect of on-the-fly switching periods on the system's stability, as studied in section 2.4.

1.4.2.3. *Feasibility: feedback-scheduler implementation for robot control*

We consider here a seven degrees of freedom Mitsubishi PA10 robot arm that has been previously modeled and calibrated [SIM 05b].

1.4.2.3.1. Plant modeling and control structure

The problem under consideration is to track a desired trajectory for the position of the end-effector. Using the Lagrange formalism, the following model can be obtained:

$$\Gamma = M(q)\ddot{q} + \text{Gra}(q) + C(q, \dot{q}), \tag{1.6}$$

where q stands for the positions of the joints, M is the inertia matrix, Gra is the gravity forces vector, and C gathers Coriolis, centrifugal, and friction forces.

The structure of the (ideal) linearizing controller includes a compensation of the gravity, Coriolis/centrifugal effect and inertia variations as well as a proportional-derivative (PD) controller for the tracking and stabilization problem, of the form

$$\Gamma = \mathrm{Gra}(q) + C(q, \dot{q}) + K_p(q_d - q) + K_d(\dot{q}_d - \dot{q}), \tag{1.7}$$

leading to the linear closed-loop system $M(q)\ddot{q} = K_p(q_d - q) + K_d(\dot{q}_d - \dot{q})$.

This controller is divided into four tasks, i.e. a specific task is considered for the PD control, for the gravity, inertia and Coriolis compensations, in order to use a multi-rate controller. In this first cautious feedback-scheduling scheme, only the periods of the compensation tasks will be adapted, as they are time-consuming compared with the PD task while being less critical for the stability.

1.4.2.3.2. Scheduling controller design

The block diagram of Figure 1.9 is considered for the H_∞ design where $G'(z)$ is the model of the scheduler, the output of which is the vector of all task loads. To get the sum of all task loads, we use $C' = [1\ 1\ 1]$. The $H(z)$ transfer function represents the sensor dynamic behavior which measures the load of the other tasks. It may be a first-order filter. The template W_e specifies the performances on the load-tracking error as follows:

$$W_e(s) = \frac{s/M_s + \omega_b}{s + \omega_s \epsilon}, \tag{1.8}$$

with $M_s = 2$, $\omega_s = 10$ rad s^{-1}, $\epsilon = 0.01$ to obtain a closed-loop settling time of 300 ms, a static error less than 1 % and a good robustness margin. Matrix M is defined as $M = [1\ -1\ -1]$.

The contribution of each of the compensation tasks to the controller performance w.r.t to its execution period has been evaluated via numerous simulations. However,

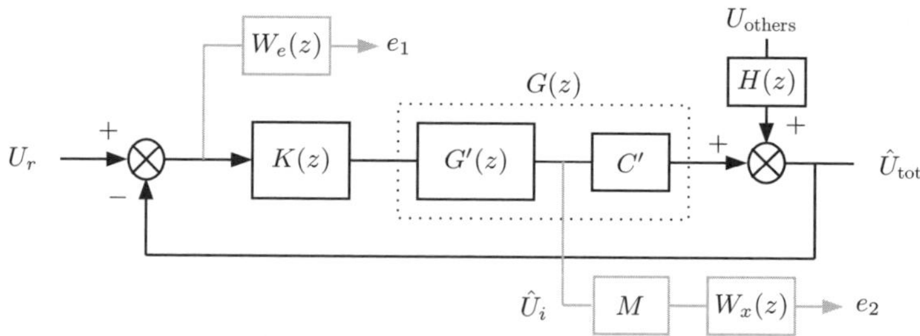

Figure 1.9. H_∞ *design block diagram*

due to the nonlinear nature of the robot arm, only a very rough cost function could be identified, leading to static relative costs.

The template W_x allows the load allocation between the control tasks to be specified. With a large gain in W_x, it leads to

$$U_{\text{gravity}} \approx U_{\text{Coriolis}} + U_{\text{inertia}},$$

i.e. we allocate more resources for the gravity compensation.

All templates are discretized with a sampling period of 30 ms. Finally, a discrete-time H_∞ synthesis produces a discrete-time-scheduling controller of order 4.

1.4.2.3.3. Implementation of the feedback scheduler

After preliminary simulations using TrueTime [CER 03a], we have developed a feedback-scheduler prototype running in real time inside a "hardware-in-the-loop" simulator: a well-calibrated model of the robot arm is numerically integrated in parallel with the execution of the controller, on top of a real-time, pre-emptive, and fixed priorities operating system.

The process controller uses the so-called *computing torque controller* which is split into several computing modules to implement a multi-rate controller as in [SIM 98] (Figure 1.10). The system is implemented using only the basic features of an

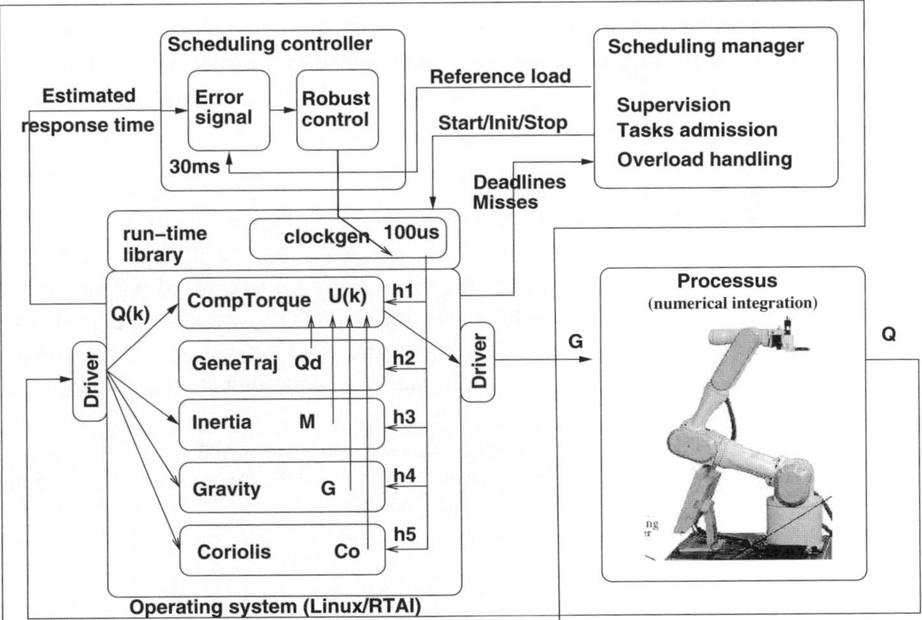

Figure 1.10. *Feedback-scheduling experiment*

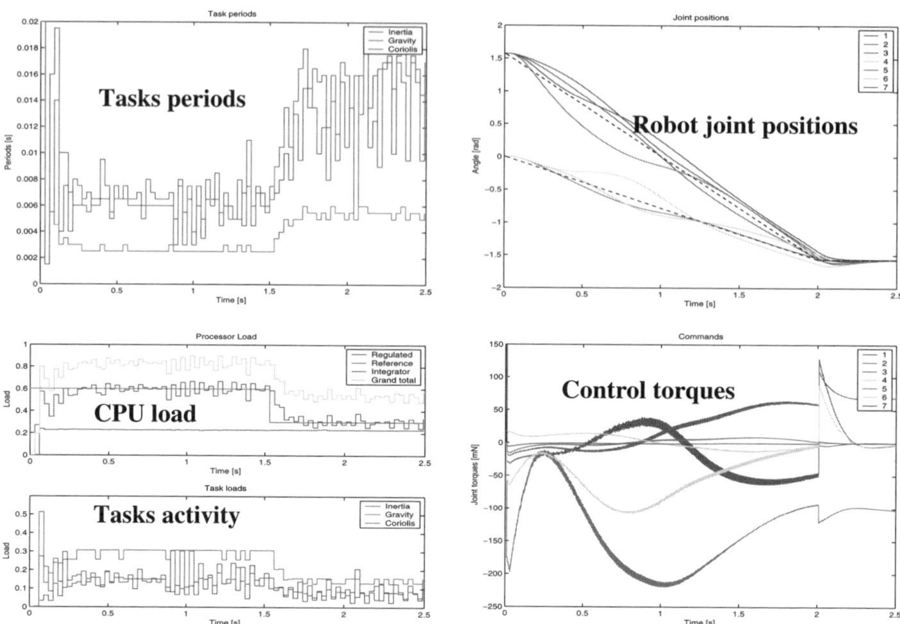

Figure 1.11. *Hardware in the loop simulation: periods and load*

off-the-shelf RTOS, which anyway must be instrumented with a task-execution-time operator[1]. In this application, the period of the feedback scheduler has been fixed to 30 ms to be larger than the robot control tasks (whose limits have been set here from 1 ms to 30 ms).

In this experiment, due to the poor quality of the cost functions which were identified, the feedback scheduler directly controls the CPU usage rather than taking into account the state of the physical system as in an ideal case. In the experiment depicted in Figure 1.11, the desired CPU usage is initially set to 60% of the maximum usage and then lowered to 40% after 1.5 s. The upper plots show the tasks periods and CPU usage. Note that the processor also executes the robot arm numerical integration which induces a high and varying load, inducing some unpredictable overloads.

These first experimental results are encouraging: they show that such a feedback-scheduling architecture can be quite easily designed and implemented on top of an off-the-shelf RTOS with fixed priority and pre-emption.

---

1. As in the several real-time variants of Linux we have used, i.e. RTAI (www.rtai.org) and Xenomai (www.xenomai.org)

In this particular case, the scheduling controller is a low-order state feedback, which, moreover, is executed at a slow rate: hence, its computing cost is very low (about 75 µs every 30 ms on a 400 Mhz Pentium 2), i.e. less than 1% of the total control cost.

Indeed, compared with a fixed rate controller, the gain in control performance measured by the integrated tracking error is not impressive: this is due to the very rough modeling of the performance/control rate relationships of this nonlinear system. The real improvement lies in the robustness of the system against transient overloads, and in the automatic setting of the tasks periods: the designer only needs to set reasonable initial values based on easily measured average execution times. As stated in [CER 05], it must be noticed that the recovery strategies used in the case of CPU overload can be selected in a set of predefined behaviors to improve the overall control performance. Here we used the *Skip* overrun processing, where the overrunning task finishes its current job but prevents its next expected schedule to be executed. Note that, thanks to the robustness of the closed-loop system w.r.t. jitter and occasional data loss, overruns must not be considered as fatal events as they would be in a system specified as "hard real time".

From the sampled control point of view, it may be observed that abrupt and/or frequent period switches may lead to control instability, even if each periodic controller is stable for each constant sampling period [SCH 02]. Adding a low-pass filtering template in the H_∞ scheduling controller here provides period variations smoother than the one provided by a simple period re-scaling; however, this does not guarantee stability and a wiser solution is looked for in section 2.4.

1.5. Fault diagnosis of NCS with network-induced effects

The introduction of communication networks in the control loops makes the analysis of NCS complex. There are several network-induced effects that arise when dealing with the NCS, such as time-delays, packet losses and limited communication. Because of the inherent complexity of such systems, the control issues of NCS have attracted most attention of many researchers, taking into account network-induced effects. For instance, the stability and stabilization problems of NCS were investigated in [HAL 88; NIL 98; BRA 00; ZHA 01b; LI 05] for network-induced delays, [LIN 02b; SEI 05] for packet losses, [HU 03; YUE 05; LI 06] for network-induced delays and packet losses, [NAI 97; HRI 99; ISH 02] for limited communication. We refer the readers to the survey in [TIP 03; HOK 04] and an up-to-date supplement [YAN 06] for more information of NCS on modeling, design and analysis from the viewpoint of estimation and control.

On the other hand, due to an increasing complexity of dynamic systems, as well as the need for reliability, safety and efficient operation, the model-based fault diagnosis and fault-tolerant control has been becoming an important subject in modern control theory and practice, see [MAN 00; ZHA 03]. Disturbances decoupling FDI methods include the works done by [FRA 94; PAT 00]. Extended reviews on FDI have been given in [GER 98; QIN 01]. Owing to the network-induced effects, the theories for traditional point-to-point systems should be reevaluated.

1.5.1. *Fault diagnosis of NCS with network-induced time delays*

Time-delays in the NCS system consist of: a) communication delay between sensors and controllers τ^{sc}, b) communication delay between controllers and actuators τ^{ca}, c) computational time in controllers τ^{c}. Generally speaking, computational time of controllers can be included in communication delay between controllers and actuators. Under the assumption that there is no packet dropout during signal transmission, the sensor-to-controller delay and controller-to-actuator delay can be lumped together as [ZHA 01b], $\tau = tau^{sc} + tau^{ca}$ where τ is supposed to be smaller than the sampling period h. An extended survey on FDI of NCS can be found in [FAN 07].

1.5.1.1. *Low-pass post-filtering*

Consider the random and unknown network-induced delay shorter than one sampling period, then the NCS with unknown inputs d and faults f can be modeled as [ZHA 01a]:

$$
\begin{aligned}
x(k+1) &= \bar{A}x(k) + \bar{\Gamma}_0 u(k) + \bar{\Gamma}_1 u(k-1) + \bar{B}_d d(k) + \bar{B}_f f(k) \\
y(k) &= \bar{C}x(k)
\end{aligned}
\tag{1.9}
$$

which can be further written as [LI 07b; YE 06b; YE 06a]

$$
x(k+1) = \bar{A}x(k) + \bar{B}u(k) + g(k) + \bar{B}_d d(k) + \bar{B}_f f(k)
\tag{1.10}
$$

where

$$
g(k) = -\bar{\Gamma}_1 \Delta u_k, \quad \Delta u_k = u(k) - u(k-1).
\tag{1.11}
$$

Different from the sampled-data system without taking into account the network-induced delay, there exists a time-varying term $g(k)$ in the state evolution equation of system (1.10) and (1.11). When τ_k is random, $g(k)$ can be regarded as a random disturbance in (1.10). Therefore, it's natural to adopt a low-pass filter to reduce the impact of $g(k)$ on the residual signal. However, the idea could not be done by only designing a traditional optimal residual generator first and then adding a low-pass filter to its output. The reason is that the optimization of the traditional residual generator does not mean that NCS consisting of it and a post-filter is still optimal. So it is necessary to consider both of the two parts of the new system (i.e. the residual generator and the low-pass filter) when designing the fault detection system [YE 04a].

As an extension of the results in [YE 04b], a fault detection approach based on parity space and Stationary Wavelet Transform (SWT) for NCS with random network-induced delay has been introduced in [YE 04a], which is briefly introduced as follows.

Let

$$\star_{s,k} = \left[\ \star^T(k-s)\quad \star^T(k-s+1)\quad \cdots\quad \star^T(k)\ \right]^T \tag{1.12}$$

where $\star$ may represent u, y, d, f, respectively.

Let

$$H_{u,s} = \begin{bmatrix} 0 & 0 & \cdots & 0 \\ \bar{C}\bar{B} & 0 & \ddots & \vdots \\ \vdots & \ddots & \ddots & 0 \\ \bar{C}\bar{A}^{s-1}\bar{B} & \cdots & \bar{C}\bar{B} & 0 \end{bmatrix} \tag{1.13}$$

and define $H_{s,k}$, $H_{f,s}$, $H_{g,s}$ as the matrices obtained by replacing $\bar{B}$ in (1.13) with $\bar{B}_d$, $\bar{B}_f$ and identity matrix I, respectively.

Let

$$H_{o,s} = \left[\ \bar{C}^T\quad \bar{A}^T\bar{C}^T\quad \cdots\quad (\bar{A}^s)^T\bar{C}^T\ \right]^T$$

Then a parity space and SWT based residual generator is defined by

$$r_{s,k} = v_s(y_{s,k} - H_{u,s}u_{s,k}) \tag{1.14}$$

$$r_{s,k}^{WT} = WT_{r_s}^a(j_m, k) \tag{1.15}$$

whose dynamics is governed by

$$r_{s,k} = v_s(H_{d,s}d_{s,k} + H_{f,s}f_{s,k} + H_{g,s}g_{s,k}) \tag{1.16}$$

$$r_{s,k}^{WT} = WT_{r_s}^a(j_m, k) \tag{1.17}$$

where v_s is the parity vector to be designed which should be selected from the parity spaces P_s defined by $P_s = \{v_s | v_s H_{o,s} = 0\}$, and $WT_{r_s}^a(j_m, k)$ denotes the approximation coefficients of the SWT of $r_{s,k}$, under scale j_m, which can be considered as a low-pass filtering of $r_{s,k}$. The dynamics can be written in the following explicit form [YE 04a]

$$r_{s,k}^{WT} = v_s(H_{d,s}N_{l,j_m}^d\, d_{s+i_{set},k} + H_{f,s}N_{l,j_m}^f\, f_{s+i_{set},k} + H_{g,s}N_{l,j_m}^q\, q_{s+i_{set},k})$$

where N_{l,j_m}^d, N_{l,j_m}^f, N_{l,j_m}^q are known matrices, whose definitions can be found in [YE 04a].

1.5.1.2. *Structure matrix of network-induced time delay*

According to (1.10) and (1.11), [WAN 06a; YE 06a; YE 04a; LIU 05] have proposed a so-called structure matrix of τ_k to address the fault diagnosis for NCS. The main procedures are a) transforming $g(k)$ into a form of (known part)×(unknown part), where the known part can extract the known information (such as A, B, Δu_k) from $g(k)$ as much as possible, and the unknown part includes the unknown information related to τ_k, b) using traditional robust fault detection methods to make the robustness to τ_k. These results are further summarized as Taylor approximation [YE 04a], eigenvalue-decomposition and Padé approximation [YE 06b], accurate structure matrix of τ_k and PCA [YE 06a].

Consider a simpler NCS model such as:

$$x(k+1) = \bar{A}x(k) + \bar{B}u(k) + g(k) + f(k)$$
$$y(k) = \bar{C}x(k) \tag{1.18}$$

When the sampling period h is small enough, using the Taylor approximation of e^{Ah}, $g(k)$ will approximate to

$$g(k) \approx \bar{E}_{\tau,k}\tau_k$$
$$\bar{E}_{\tau,k} = -B\Delta u_k \tag{1.19}$$

A time-varying parity space based residual generator is defined as

$$r_{s,k} = v_{s,k}(y_{s,k} - H_{u,s}u_{s,k}) \tag{1.20}$$

with

$$r_{s,k} = v_{s,k}(H_{\tau,s,k}\tau_{s,k} + H_{f,s}f_{s,k}) \tag{1.21}$$

when $v_{s,k} \in P_s$, where

$$H_{\tau,s,k} = \begin{bmatrix} 0 & 0 & \cdots & 0 & 0 \\ \bar{C}\bar{E}_{\tau,k-s} & 0 & 0 & 0 & 0 \\ \bar{C}\bar{A}\bar{E}_{\tau,k-s} & \bar{C}\bar{E}_{\tau,k-s+1} & \ddots & 0 & 0 \\ \vdots & \vdots & \ddots & \vdots & \vdots \\ \bar{C}\bar{A}^{s-1}\bar{E}_{\tau,k-s} & \bar{C}\bar{A}^{s-2}\bar{E}_{\tau,k-s+1} & \cdots & \bar{C}\bar{E}_{\tau,k-1} & 0 \end{bmatrix} \tag{1.22}$$

To meet $v_{s,k} \in P_s$ and to decouple the residual signal from the network-induced delays, the parity vector is determined by solving

$$v_{s,k}H_{o,s} = 0 \text{ and } v_{s,k}H_{\tau,s,k} = 0 \tag{1.23}$$

This approach has good robustness to network-induced delay only if both h and τ_k are small enough.

1.5.1.3. *Robust deadbeat fault filter*

In [LI 07b], the authors assume that the statistical behavior of network-induced delay τ_k is random and governed by the Markov chain

$$\theta_k \in \mathcal{S} = \{1, 2, \ldots, s\}, \ \forall k \in \mathbb{Z}_+ \tag{1.24}$$

with the transition probabilities λ_{ij} denoting as $\lambda_{ij} = \text{pr}[\theta_{k+1} = j | \theta_k = i]$, $\lambda_{ij} \geq 0$ and $\sum_{j=1}^{s} \lambda_{ij} = 1$ for any $i \in \mathcal{S}$. For sake of simplifying notations, B_{1,τ_k} is denoted as B_{1,θ_k} and Δu_k as w_k. Then, the model of NCS is by replaced the state space system (1.10) with the following particular Markov jump linear system:

$$\begin{cases} x_{k+1} &= A x_k + B u_k + F f_k + B_{1,\theta_k} w_k, \\ y_k &= C x_k, \end{cases} \tag{1.25}$$

The following filter is presented as the residual generator of NCS (1.25):

$$\begin{cases} \hat{x}_{k+1} &= A \hat{x}_k + B u_k + K(y_k - C \hat{x}_k), \\ \alpha_k &= L(y_k - C \hat{x}_k), \end{cases} \tag{1.26}$$

where $\hat{x}_k$ is the state of the filter, α_k the residual generator or the fault indicator. Filter gain $K \in \text{rr}^{n \times m}$ and projector $L \in \mathbb{R}^{q \times m}$ are unknown matrices to be found for the solution of the fault detection and isolation problem.

From (1.25) and (1.26), the state estimation error $e_k = x_k - \hat{x}_k$ and the output of the filter α_k propagate as

$$\begin{cases} e_{k+1} = (A - KC)e_k + F f_k + B_{1,\theta_k} w_k \\ \alpha_k = LC e_k \end{cases}, k \in \mathbb{Z}_+, \theta_k \in \mathcal{S} \tag{1.27}$$

Let $G_{n\alpha}(z)$ be the transfer function from f_k to the output residual α_k. Then the following theorem is presented to design K and L such that

$$\begin{aligned} G_{n\alpha}(z) &= LC(zI - (A - KC))^{-1} F \\ &= \text{diag}\{z^{-\rho_1}, \ldots, z^{-\rho_q}\}, \end{aligned} \tag{1.28}$$

which ensures the isolation of multiple faults.

THEOREM.– *Under the condition* $\text{rank}(\Psi) = q$, *the solutions of (1.28) can be parametrized as* $K = \omega\Pi + \bar{K}_{\theta_k}\Sigma$, $L = \Pi$, *with* $\Sigma = \beta(I - \Psi\Pi)$, $\Pi = \Psi^+$, $\omega = AD$ *and* $\Psi = CD$, *where* $\bar{K}_{\theta_k} \in \text{rr}^{n \times m - q}$ *is the free parameter to be designed,* Ψ^+ *is the pseudo-inverse of* Ψ *and* β *is an arbitrary matrix chosen so that* $\text{rank}(\Sigma) = m - q$.

From the theorem above, the FIF (1.26) is rewritten from the free parameter $\bar{K}_{\theta_k}$ as:

$$\begin{cases} \hat{x}_{k+1} &= A \hat{x}_k + B u_k + \omega\alpha_k + \bar{K}_{\theta_k}\Sigma(y_k - C\hat{x}), \\ \alpha_k &= \Pi(y_k - C\hat{x}_k), \end{cases} \tag{1.29}$$

where α_k is a deadbeat filter of fault n_k and given by:

$$\alpha_k = \breve{\alpha}_k + \left[\; n^1_{k-\rho_1} \quad \cdots \quad n^i_{k-\rho_i} \quad \cdots \quad n^q_{k-\rho_q} \;\right]^T \tag{1.30}$$

where $\breve{\alpha}_k$ is the fault indicator signal without faults and propagates from the fault-free state estimation error $\bar{e}_k = \tilde{x}_k - \hat{x}_k$ as:

$$\begin{cases} \bar{e}_{k+1} & = \; (\bar{A} - \bar{K}_{\theta_k}\bar{C})\bar{e}_k + B_{1,\theta_k} w_k, \\ \breve{\alpha}_k & = \; \Pi C \bar{e}_k, \end{cases} \tag{1.31}$$

where $\bar{A} = A - \omega\Pi C$, $\bar{C} = \Sigma C$ and $\tilde{x}_k$ is the fault-free state. The transfer function from w_k to $\breve{\alpha}_k$ is then given by:

$$G_{w\breve{\alpha}}(z) = \Pi C(zI - (\bar{A} - \bar{K}_{\theta_k}\bar{C}))^{-1} B_{1,\theta_k}. \tag{1.32}$$

Let $\hat{\alpha}_k$ be the faults indicator signal without disturbances. From Equation (1.28), the transfer function $G_{f\hat{\alpha}}(z)$ from fault f to fault indicator $\hat{\alpha}_k$ is a pure delay and

$$\operatorname{norm} G_{f\hat{\alpha}}(z)_\infty := \sup_{\theta_0 \in \mathcal{S}} \sup_{0 \neq f \in \ell_2} \frac{\operatorname{norm} \hat{\alpha}_2}{\operatorname{norm} f_2} = 1, \tag{1.33}$$

where $\operatorname{norm} s_{\ell_2} = (\sum_{k=0}^\infty \operatorname{norm} s_k)^{1/2}$ is the ℓ_2 norm of the signal s_k.

Then the free parameters $\bar{K}_{\theta_k}$ are designed in order to

C1: ensure that the energy ratio between useful and disturbance signal w_k defined on the fault indicators is maximized,

C2: locate the closed-loop poles within a prescribed region in the complex plane in order that the residual dynamics has the given transient properties,

which can be formulated in the following theorem:

THEOREM.– *For given discs $D_i(\xi_i, \delta_i)$, if there exist matrices $P_i = P_i^T > 0$, G_i and Y_i for prescribed scalars $\gamma > 0$, $-1 < -\xi_i + \delta_i < 1$, $\forall i = \theta_k \in \mathcal{S}$ such that*

$$\begin{bmatrix} -P_i & 0 & \bar{A}^T G_i^T - \bar{C}^T Y_i^T & C^T \Pi^T \\ 0 & -\gamma^2 I & B_{1,i}^T G_i^T & 0 \\ G_i \bar{A} - Y_i \bar{C} & G_i B_{1,i} & \bar{P}_i - G_i - G_i^T & 0 \\ \Pi C & 0 & 0 & -I \end{bmatrix} < 0, \tag{1.34}$$

$$\begin{bmatrix} -\delta_i^2 P_i & \bar{A}^T G_i^T - \bar{C}^T Y_i^T - \xi_i G_i^T \\ G_i \bar{A} - Y_i \bar{C} - \xi_i G_i & P_i - G_i - G_i^T \end{bmatrix} < 0, \tag{1.35}$$

where $\bar{A} = A - \omega\Pi C$, $\bar{C} = \Sigma C$, then the free parameters are designed as $\bar{K}_i = G_i^{-1}Y_i$ ensuring the SMS of the error system (1.31) and the constraints C1 and C2. At the minimal possible value of γ leading to a solution $P_i = P_i^T > 0$, G_i and Y_i, the energy ratio between useful and disturbance signal defined on the fault indicators will be maximized.

Given discs $D_i(\xi_i, \delta_i)$, $i = \theta_k \in \mathcal{S}$, the search problem of the lowest possible value of γ can be formulated as the following convex optimization problem:

$$\mathcal{OP} : \min_{P_i = P_i^T > 0, G_i, Y_i} \gamma \tag{1.36}$$

1.5.1.4. *Other work*

In practice, the delay may be more than one sampling period. In some cases, this long time delay may distort the timing order of the message arriving at the receiver [HU 03; LIN 00; LI 04b; HAL 88].

In this way, the integrity of the information transmission is guaranteed. The discrete state model of the system with network-induced delay can be described as

$$\begin{aligned} x(k+1) &= \bar{A}x(k) + \bar{B}_0 u(k-1) + \bar{B}_1 u(k-l+1) + \bar{B}_d d(k) + \bar{B}_f f_a(k) \\ y(k) &= \bar{C}x(k) + f_s(k) \end{aligned} \tag{1.37}$$

which is a familiar discrete time system with input time delays utilizing a reduced-order memory-less state observer with a γ-stability margin, an observer-based fault detection method was presented for system (1.37) by comparing the output of the observer with the actual output of the practical system [ZHE 03]. The residual function for this approach is

$$\begin{aligned} r(z) &= Q\bar{C}P^{-1}\bar{B}_d d(z) + Q\bar{C}P^{-1}\bar{B}_f f_a(z) \\ &+ [Q - Q\bar{C}P^{-1}(zI - \bar{A})V(zI_n - \Lambda_r)^{-1}L]f_s(z) \end{aligned} \tag{1.38}$$

where $P = (zI_n - \bar{A})[I_n + V(zI_r - \Lambda_r)^{-1}L\bar{C}]$.

To remove the effect of the disturbance, it is required that

$$Q\bar{C}P^{-1}\bar{B}_d = H(zI_n - P^T)^{-1}\bar{B}_d = 0.$$

The simulation results demonstrating the feasibility of this approach can be found in [ZHE 03].

A method for fault detection in NCS with unknown network-induced delay, which may be greater than h, is proposed in [WAN 06b]. In this method, an NCS model for unknown network-induced delay which may be greater than h [HAL 88; HU 03] has been developed.

1.5.2. *Fault diagnosis of NCS with packet losses*

Packet losses happen when packets are dropped due to link failure or packets are dropped on purpose in order to avoid congestion or guarantee the most recent data to receiver. Although a single packet loss neither deteriorates system performance nor destabilizes the system, the consecutive packet losses have an impact on overall performance.

1.5.2.1. *Deterministic packet losses*

The deterministic packet losses have also been discussed, either in terms of switching systems [SEI 05; TIP 03; ZHA 01b; WAN 06b] or in terms of delayed differential equations [YUE 05; YU 05]. [YAN 98] firstly addressed the fault diagnosis for a class of state-delayed dynamic systems, in which the actuator and sensor faults as well as other effects such as disturbances and non-linearities were considered as unknown inputs. More recently, [KOE 05] dealt with the problem of full-order observer design for linear continuous delayed state and input systems with unknown input and time-varying delays. A method to design an unknown input observer (UIO) for such systems was proposed based on delay-dependent stability conditions of the state estimation error system.

[DIN 00] developed a weighting transfer function matrix to describe the desired behavior of residual respect to fault. The observer-based fault detection filter for a class of linear systems with time-varying delays was designed such that the error between the generated residual and fault is as small as possible in the sense of H_∞-norm.

1.5.2.2. *Stochastic packet losses*

In [ZHA 04], the fault detection problem of systems with stochastic packet losses is developed. The structure of standard model based residual generator is modified and dynamic network resource allocation is represented as

$$e(k+1) = \begin{cases} (A - LC)e(k) + (E_f - LF_f)f(k) + L\theta(k), & \gamma(k) = 0 \\ (A - LC)e(k) + (E_f - LF_f)f(k), & \gamma(k) = 1. \end{cases} \quad (1.39)$$

$$r(k) = \begin{cases} WCe(k) + WF_f f(k) - W\theta(k), & \gamma(k) = 0 \\ WCe(k) + WF_f f(k), & \gamma(k) = 1. \end{cases} \quad (1.40)$$

where $\theta(k)$ is the difference between real value of the measurement $y(k)$ and the used value $y^a(k)$, namely $\theta(k) := y(k) - y^a(k)$. $\gamma(k)$ is a stochastic variable representing data communication status. $\gamma(k) = 1$ means that the measurement at time point k arrives correctly, while $\gamma(k) = 0$ means that this measurement is lost.

To reduce the false alarm rate caused by missing measurements, a residual evaluation scheme is then developed as:

$$r_{\text{eval}} > J_{th}, \text{ a fault is detected}$$

$$r_{\text{eval}} \leq J_{th}, \text{ no fault is detected}$$

where $r_{\text{eval}} = \left(\sum_{j=0}^{\infty} r^T(j)r(j) \right)^{1/2}$. To compute the threshold J_{th}, a convex optimization problem is then developed to find the minimum of $\frac{\mathbf{E}[\|r\|_2]}{\|\theta\|_2}$, which is formulated as a disturbance attenuation problem of Markovian jump linear systems.

[MAO 07] designed a robust fault detection filter for networked control systems with large transfer delays. The multi-rate sampling method is combined with the augmented state matrix method to model the long random delay networked control systems as Markovian jump systems. The H_∞ fault detection filter is designed based on the obtained model.

Some works take into account simultaneous time-delays and packet losses, see e.g. [YUE 05; ZHA 05; YU 05].

1.5.3. *Fault diagnosis of NCS with limited communication*

The capacity of the communication network and its ability to carry a reasonable amount of information per unit of time plays an important role in the stability of NCS. When introducing the network into the control loop, issues like the channel/network capacity, encoding/decoding schemes and quantization naturally arise.

There is increasing attention to define the minimum bit rate which is needed to stabilize NCS through feedback, see e.g. [SAH 00; TAT 00; SAV 03] and the references therein. In order to describe the quantization effects on the performance NCS, some research effort has been devoted to develop new quantization scheme to achieve lower bit-rates, see e.g. [BRO 00; DEL 89; ELI 00; ISH 02; WON 97; HOK 04].

In [ZHA 06], the fault detection problem for networked control systems with limited data transmission rate is considered. In order to avoid the uncertainty caused by transmission delays and packet loss, a periodic communication sequence is proposed as:

$$y(k) = N_k y_p(k) \tag{1.41}$$

$$u_p(k) = M_k u(k) \tag{1.42}$$

where $y \in R^{\omega_m}$ represents the sensor signals transmitted from the sensors to the central station through the network, $N_k \in R^{\omega_m \times m}$ is a θ-periodic matrix formed by selecting ω_m rows of the identity matrix. $u \in R^p$ represents the signal generated by the controller, $M_k \in R^{p \times p}$ is a θ-periodic diagonal matrix with a number of ω_p non-zero element 1 on the diagonal.

1.5.4. *Fault-tolerant control of NCS*

Based on the fault diagnosis algorithm for NCS at section 1.5.2, the fault-tolerant control of NCS can be obtained. The existing methods of fault tolerant control (FTC) techniques against actuator faults can be categorized into two groups: passive [SEO 96; CHE 04] and active approaches [ZHA 02; ZHA 03]. [ZHE 03] proposed a passive controller for NCS by considering random time-delays. If a failure outside those considered in the design occurs, the stability and performance of the closed-loop system cannot be guaranteed. Such potential limitations of passive approaches motivate the research of the active FTC (AFTC).

With regard to AFTC, it is meant an on-line and real-time fault diagnosis process and a controller reconfiguration mechanism. Since the AFTC approaches propose a flexibility to select different controllers according to different component failures, better performance of the closed-loop system is expected. However, the above case holds true only if the fault diagnosis process does not make an incorrect or delayed decision. Some results have been obtained on AFTC which is immune to imperfect fault diagnosis process, see [MAH 03; WU 97]. [MAK 04] further developed the above issue by using the cost control approach and on-line controller switching in order to guarantee the stability of a closed-loop system. However, the NCS case is not considered in this work.

[LI 07a] addressed the stability guaranteed active fault tolerant control of NCS. The design of the procedures are summarized down below:

i) design a passive fault-tolerant controller so that the closed-loop system stability is guaranteed for all actuator failure modes;

ii) under the assumption that a particular actuator is fault free, the controller is repeatedly redesigned using only this actuator so that the robust performance is improved without jeopardizing the stability property of the design in i).

1.6. Summary

When control systems are implemented on top of a RTOS, scheduling policies are used to allocate scheduling parameters, i.e. usually priorities, to the control tasks. Traditional policies, such as RM and EDF, are designed only with computational constraints in mind, thus they are not necessarily suited when a control performance is the main objective to achieve. To better cope with control problem, flexible scheduling policies must be used, where the scheduling parameters are chosen according to control constraints. This is for example the case of the (m, k)-firm selective dropping policy described in section 1.2.2. This capability is further exploited to set up accelerable tasks in section 2.3 and to co-design control loops and real-time scheduling in section 4.3 and in Chapter 5.

Another way to implement control aware scheduling policies consists in continuously adapting the scheduling parameters in a closed loop controller. In this way feedback schedulers, whose basic design and examples are provided in section 1.4, actively control the execution resources usage, such as CPU load, network bandwidth or processor speed and voltage. They can be further used as stand-alone components to control computing related devices, as sketched in section 1.4.2.1 and surveyed in [LU 02] and [HEL 04]. Beyond pure execution resource control, feedback schedulers can be used as building blocks to implement integrated control of the plant and computing/networking resources as designed in Chapter 4.

Finally, when the capabilities of control loops are exceeded, e.g. due to uncertainties overshooting or to the appearance of faults, diagnosis and FTC techniques must supplement pure control robustness and adaptivity. Some basic concepts of model-based fault diagnosis, including observer-based parity space methods and FTC for NCS, are discussed in section 1.5. Other basic concepts in FDI/FTC, together with specific issues related to the processing of faults in networked control systems, are discussed in detail in Chapter 6 and assessed with the test-bed in Chapter 7.

1.7. Bibliography

[ABD 97] ABDELZAHER T., ATKINS E., SHIN K., QoS Negotiation in Real-Time Systems and Its Application to Automated Flight Control, *IEEE Real-Time Technology and Applications Symposium*, Montreal, Canada, June 1997.

[ABE 98] ABENI L., BUTTAZZO G., Integrating Multimedia Applications in Hard Real-Time Systems, *Proceedings of the 19th Real-Time Systems Symposium*, Madrid, Spain, 1998.

[ÅRZ 99] ÅRZÉN K.-E., A Simple Event-Based PID Controller, *14th IFAC World Congress*, Beijing, China, July 1999.

[ÅST 97] ÅSTRÖM K. J., WITTENMARK B., *Computer-Controlled Systems*, Information and System Sciences Series, Prentice Hall, third edition, 1997.

[BAR 90] BARUAH S., MOK A., ROSIER L., Preemptively scheduling hard-realtime sporadic tasks on one processor, *Real-Time Systems Symposium*, Lake Buena Vista, USA, p. 182–190, December 1990.

[BEN 91] BENVENISTE A., BERRY G., The synchronous approach to reactive and real-time systems, *Proceedings of the IEEE*, vol. 79, num. 9, p. 1270–1282, 1991.

[BEN 06] BEN GAID M., ÇELA A., HAMAM Y., IONETE C., Optimal Scheduling of Control Tasks with State Feedback Resource Allocation, *American Control Conference ACC'06*, Minneapolis, USA, June 2006.

[BRA 00] BRANICKY M. S., PHILLIPS S. M., ZHANG W., Stability of Networked Control Systems: Explicit Analysis of Delay, *American Control Conference ACC'00*, vol. 4, Chicago, USA, June 2000.

[BRO 00] BROCKETT R. W., LIBERZON D., Quantized Feedback Stabilization of Linear Systems, *IEEE Transactions on Automatic Control*, vol. 45, num. 7, p. 1279–1289, 2000.

[BUT 00] BUTTAZZO G., ABENI L., Adaptive Rate Control through Elastic Scheduling, *39th Conference on Decision and Control*, Sydney, Australia, December 2000.

[CAC 00] CACCAMO M., BUTTAZZO G., SHA L., Elastic Feedback Control, *12th Euromicro Conference on Real-Time Systems*, Stockholm, Sweden, June 2000.

[CER 00] CERVIN A., EKER J., Feedback Scheduling of Control Tasks, *39th IEEE Conference on Decision and Control*, Sydney, Australia, December 2000.

[CER 02] CERVIN A., EKER J., BERNHARDSSON B., ARZEN K.-E., Feedback-Feedforward Scheduling of Control Tasks, *Real-Time Systems*, vol. 23, num. 1-2, p. 25–53, July 2002.

[CER 03a] CERVIN A., Integrated Control and Real-Time Scheduling, PhD thesis, Department of Automatic Control, Lund Institute of Technology, Sweden, April 2003.

[CER 03b] CERVIN A., EKER J., The Control Server: A Computational Model for Real-Time Control Tasks, *15th Euromicro Conference on Real-Time Systems*, Porto, Portugal, p. 113-120, July 2003.

[CER 05] CERVIN A., Analysis of Overrun Strategies in Periodic Control Tasks, *16th IFAC World Congress*, Prague, Czech Republic, July 2005.

[CHE 04] CHENG C., ZHAO Q., Reliable control of uncertain delayed systems with integral quadratic constraints, *IEE Proceedings Control Theory Applications*, vol. 151, num. 6, p. 790–796, 2004.

[DEL 89] DELCHAMPS D. F., Extracting state information from a quantized output record, *System and Control Letters*, vol. 13, num. 5, p. 365–372, 1989.

[DIN 00] DING S. X., DING E. L., JEINSCH T., A new optimization approach to the design of fault detection filters, *SafeProcess'00*, Budapest, Hungary, p. 250–255, June 2000.

[DOR 62] DORF R., FARREN M., PHILLIPS C., Adaptive sampling frequency for sampled-data control systems, *IEEE Transactions on Automatic Control*, vol. 7, num. 1, p. 38–47, 1962.

[DUR 09] DURAND S., MARCHAND N., Further Results on Event-Based PID Controller, *European Control Conference ECC'09*, Budapest, Hungary, August 2009.

[EKE 99] EKER J., CERVIN A., A Matlab Toolbox for Real-Time and Control Systems Co-Design, *6th International Conference on Real-Time Computing Systems and Applications*, Hong-Kong, China, December 1999.

[EKE 00] EKER J., HAGANDER P., ARZEN K.-E., A feedback scheduler for real-time controller tasks, *Control Engineering Practice*, vol. 8, num. 12, p. 1369–1378, 2000.

[ELI 00] ELIA N., MITTER S. K., *Quantized linear systems In System Theory: Modeling, Analysis, and Control*, Kluwer, 2000.

[FAN 07] FANG H., YE H., ZHONG M., Fault diagnosis of networked control systems, *Annual Reviews in Control*, vol. 31, num. 1, p. 55–68, 2007.

[FRA 94] FRANK P., Enhancement of robustness in observer-based fault detection, *International Journal of Control*, vol. 59, num. 4, p. 955–981, 1994.

[GEO 95] GEORGE L., MUHLETAHLER P., RIVIERRE N., Optimality and nonpreemptive realtime scheduling revisited, *Report* num. RR2516, INRIA, 1995.

[GER 98] GERTLER J., *Fault Dectection and Diagnosis in Engineering Systems*, Marcel Dekker, USA, 1998.

[HAL 88] HALEVI Y., RAY A., Integrated Communication and Control Systems: Part I- Analysis, *ASME Journal of Dynamic Systems, Measurement and Control*, vol. 110, num. 4, p. 367–373, 1988.

[HAM 95] HAMDAOUI M., RAMANATHAN P., A dynamic priority assignment technique for streams with (m, k)-firm deadlines, *IEEE Transactions on Computers*, vol. 44, num. 4, p. 1443–1451, December 1995.

[HEL 04] HELLERSTEIN J., DIAO Y., PAREKH S., TILBURY D., *Feedback Control of Computing Systems*, Wiley-IEEE Press, New York, 2004.

[HOK 04] HOKAYEM P. F., ABDALLAH C. T., Inherent issues in networked control systems: a survey, *American Control Conference ACC'04*, Boston, USA, p. 4897–4902, June 2004.

[HRI 99] HRISTU D., Optimal control with limited communication, PhD thesis, Harvard University, 1999.

[HSI 74] HSIA T. C., Analytic design of adaptive sampling control law in sampled data systems, *IEEE Transactions on Automatic Control*, vol. 19, num. 1, p. 39–42, 1974.

[HU 03] HU S.-S., ZHU Q.-X., Stochastic optimal control and analysis of stability of networked control systems with long delay, *Automatica*, vol. 39, num. 11, p. 1877–1884, 2003.

[ISH 02] ISHII H., FRANCIS B., Stabilization with control networks, *Automatica*, vol. 38, num. 10, p. 1745–1751, 2002.

[JEF 91] JEFFAY K., STANAT D. F., MARTEL C. U., On non-preemptive scheduling of periodic and sporadic tasks, *Real-Time Systems Symposium*, San Antonio, USA, p. 129–139, December 1991.

[JIA 07] JIA N., SONG Y.-Q., SIMONOT-LION F., Graceful Degradation of the Quality of Control through Data Drop Policy, *European Control Conference ECC'07*, Kos, Greece, July 2007.

[JOS 86] JOSEPH M., PANDYA P., Finding Response Time in a Real-Time System, *BCS Computer journal*, vol. 29, num. 5, p. 390–395, October 1986.

[KOE 05] KOENIG D., BEDJAOUI N., LITRICO X., Unknown input observers design for time-delay systems application to an open-channel, *44th IEEE Conference on Decision and Control and the European Control Conference*, Sevilla, Spain, p. 5794–5799, December 2005.

[KOP 03] KOPETZ H., BAUER G., The time-triggered architecture, *Proceedings of the IEEE*, vol. 91, num. 1, p. 112–126, 2003.

[LEU 04] LEUNG J., *Handbook of scheduling: algorithms, models, and performance analysis*, Chapman Hall/CRC, 2004.

[LI 04a] LI J., SONG Y.-Q., SIMONOT-LION F., Schedulability analysis for systems under (m,k)-firm constraints, *IEEE WFCS2004*, Vienna, Austria, September 2004.

[LI 04b] LI S., WANG Z., SUN Y., Observer-based compensator design for networked control systems with long time delays, *30th Annual Conference of IEEE Industrial Electronics Society*, Busan, Korea, p. 678–683, November 2004.

[LI 05] LI S., YU L., WANG Z., SUN Y., Approach to Guaranteed Cost Control for Networked Control Systems, *Developments in Chemical Engineering and Mineral Processing*, vol. 13, num. 3, p. 351–361, 2005.

[LI 06] LI S., SAUTER D., AUBRUN C., Robust Fault Isolation Filter Design for Networked Control systems, *11th IEEE International Conference on Emerging Technologies and Factory Automation*, Prague, Czech Republic, p. 681–688, September 2006.

[LI 07a] LI S., SAUTER D., AUBRUN C., YAMÉ J.-J., Stability guaranteed active fault tolerant control of networked control systems, *European Control Conference, ECC'07*, Kos, Greece, p. 180–186, July 2007.

[LI 07b] LI S., WANG Y., XIA F., SUN Y., Guaranteed cost control of networked control systems with time-delays and packet losses, *International Journal of wavelets, multiresolution and information processing*, vol. 4, num. 4, p. 691–706, 2007.

[LI 09] LI J., SONG Y.-Q., DLB: a novel real-time QoS control mechanism for multimedia transmission, *International Journal High Performance Computing and Networking*, vol. 6, num. 1/2, p. 4–14, 2009.

[LIN 00] LINCOLN B., BERNHARDSSON B., Optimal Control over Networks with Long Random Delays, *International Symposium on Mathematical Theory of Networks and Systems*, Perpignan, France, p. 84-90, July 2000.

[LIN 02a] LINCOLN B., BERNHARDSSON B., LQR Optimization of Linear System Switching, *IEEE Transactions on Automatic Control*, vol. 47, num. 10, p. 1701–1705, October 2002.

[LIN 02b] LING Q., LEMMON M. D., Robust performance of soft real-time networked control systems with data dropouts, *41st IEEE Conference on Decision and Control*, Las Vegas, USA, p. 1225–1230, December 2002.

[LIU 73] LIU C., LAYLAND J., Scheduling Algorithms for Multiprogramming in Hard Real-Time Environment, *Journal of the ACM*, vol. 20, num. 1, p. 40–61, February 1973.

[LIU 00] LIU J., *Real-time systems*, Prentice-Hall, Inc., 2000.

[LIU 05] LIU H., CHENG Y., YE H., A combinative method for fault detection of networked control systems, *20th IAR/ACD Annual Meeting*, Mulhouse , France, p. 59–63, November 2005.

[LIU 06] LIU D., HU X., LEMMON M., LING Q., Firm real-time system scheduling based on a novel QoS constraint, *IEEE Transactions on Computers*, vol. 55, num. 3, p. 320–333, March 2006.

[LU 00] LU C., STANKOVIC J., ABDELZAHER T., TAO G., SON S., MARLEY M., Performance Specifications and Metrics for Adaptive Real-Time Systems, *Real-Time Systems Symposium*, Orlando, USA, December 2000.

[LU 01] LU C., ABDELZAHER T. F., STANKOVIC J. A., SON S. H., A Feedback Control Approach for Guaranteeing Relative Delays in Web Servers, *IEEE Real-Time Technology and Applications Symposium*, Taipei, Taiwan, June 2001.

[LU 02] LU C., STANKOVIC J.-A., TAO G., SON S.-H., Feedback Control Real-Time Scheduling: Framework, Modeling, and Algorithms, *Real Time Systems*, vol. 23, num. 1, p. 85-126, 2002.

[MAH 03] MAHMOUD M., JIANG J., ZHANG Y., Stabilization of active fault tolerant control systems with imperfect fault detection and diagnosis, *Stochastic Analysis and Applications*, vol. 21, num. 3, p. 673–701, 2003.

[MAK 04] MAKI M., JIANG J., HAGINO K., A stability guaranteed active fault-tolerant control against actuator failures, *International Journal of Robust and Nonlinear Control*, vol. 14, num. 12, p. 1061–1077, 2004.

[MAN 00] MANGOUBI R. S., EDELMAYER A. M., Model-based fault detection: the optimal past, the robust present and a few thoughts on the future, *SafeProcess'00*, Budapest, Hungary, p. 64–75, June 2000.

[MAO 07] MAO Z., JIANG B., SHI P., H_∞ fault detection filter design for networked control systems modelled by discrete Markovian jump systems, *Control Theory & Applications, IET*, vol. 1, num. 5, p. 1336–1343, 2007.

[NAI 97] NAIR G. N., EVANS R. J., State estimation via a capacity-limited communication channel, *36th Conference on Decision and Control*, San Diego, USA, p. 866–871, December 1997.

[NIL 98] NILSSON J., Real-time control systems with delays, PhD thesis, Lund University, Sweden, 1998.

[OHL 07] OHLIN M., HENRIKSSON D., CERVIN A., TrueTime 1.5–Reference Manual, January 2007.

[PAR 02] PAREKH S., GANDHI N., HELLERSTEIN J., TILBURY D., JAYRAM T., BIGUS J., Using Control Theory to Achieve Service Level Objectives in Performance Management, *Real-Time Systems Journal*, vol. 23, num. 1-2, p. 127–141, 2002.

[PAT 00] PATTON R., CHEN J., On eigenstructure assignment for robust fault diagnosis, *International Journal of Robust and Non Linear Control*, vol. 10, num. 14, p. 1193–1208, 2000.

[POG 03] POGGI E.-M., SONG Y.-Q., KOUBAA A., WANG Z., Matrix-DBP for (m,k)-firm real-time guarantee, *Conference of Real Time Systems*, Paris, France, p. 457–482, April 2003.

[QIN 01] QIN S., LI W., Detection, identification and reconstruction of faulty sensors with maximized sensitivity, *A.I. Ch. E. Journal*, vol. 34, num. 39, p. 1963–1976, 2001.

[QUA 00] QUAN G., HU X., Enhanced Fixed-priority Scheduling with (m, k)-firm Guarantee, *Real-Time Systems Symposium*, Orlando, USA, p. 79–88, Nov. 2000.

[RAM 99] RAMANATHAN P., Overload Management in Real-Time Control Applications Using (m, k)-Firm Guarantee, *IEEE Transactions on Parallel and Distributed Systems*, vol. 10, num. 6, p. 549–559, June 1999.

[REH 04] REHBINDER H., SANFRIDSON M., Scheduling of a limited communication channel for optimal control, *Automatica*, vol. 40, num. 3, p. 491–500, March 2004.

[RYU 97] RYU M., HONG S., SAKSENA M., Streamlining real-time controller design: from performance specifications to end-to-end timing constraints, *IEEE Real Time Systems Symposium*, San Francisco, USA, December 1997.

[SAH 00] SAHAI A., Evaluating channels for control capacity reconsidered, *American Control Conference*, Chicago, USA, p. 2358–2362, June 2000.

[SAK 98] SAKSENA M., PTAK A., FREEDMAN P., RODZIEWICZ P., Schedulability Analysis for Automated Implementations of Real-Time Object-Oriented Models, *IEEE Real-Time Systems Symposium*, Madrid, Spain, December 1998.

[SAN 00] SANFRIDSON M., Problem Formulations for QoS Management in Automatic Control, *Report* num. TRITA-MMK 2000:3, ISSN 1400-1179, ISRN KTH/MMK-00/3-SE, KTH, Stockholm, Sweden, 2000.

[SAN 02] SANDSTRÖM K., NORSTRÖM C., Managing Complex Temporal Requirements in Real-Time Control Systems, *9th IEEE International Conference and Workshop on the Engineering of Computer-Based Systems (ECBS'02)*, Lund, Sweden, April 2002.

[SAV 03] SAVKIN A. V., PETERSEN I. R., Set-valued state estimation via a limited capacity communication channel, *IEEE Transactions on Automatomatic Control*, vol. 48, num. 4, p. 676–680, April 2003.

[SCH 02] SCHINKEL M., CHEN W.-H., RANTZER A., Optimal control for systems with varying sampling rate, *American Control Conference ACC'02*, Anchorage, USA, May 2002.

[SEI 05] SEILER P., SENGUPTA R., An H_∞ approach to networked control, *IEEE Transactions on Automatic Control*, vol. 50, num. 3, p. 356–364, 2005.

[SEO 96] SEO C., KIM B., Robust and reliable H_∞ control for linear systems with parameter uncertainty and actuator failure, *Automatica*, vol. 32, num. 3, p. 465–467, 1996.

[SET 96] SETO D., LEHOCZKY J. P., SHA L., SHIN K. G., On Task Schedulability in Real-Time Control Systems, *17th IEEE Real-Time Systems Symposium*, New York, USA, December 1996.

[SIM 98] SIMON D., CASTILLO E., FREEDMAN P., Design and Analysis of Synchronization for Real-time Closed-loop Control in Robotics, *IEEE Transactions on Control Systems Technology*, vol. 6, num. 4, p. 445–461, July 1998.

[SIM 03] SIMON D., SENAME O., ROBERT D., TESTA O., Real-time and delay-dependent control co-design through feedback scheduling, *CERTS'03 Workshop on Co-design in Embedded Real-time Systems*, Porto, Portugal, July 2003.

[SIM 05a] SIMON D., BENATTAR F., Design of real-time periodic control systems through synchronisation and fixed priorities, *International Journal of Systems Science*, vol. 36, num. 2, p. 57–6, 2005.

[SIM 05b] SIMON D., ROBERT D., SENAME O., Robust control / scheduling co-design: application to robot control, *11th IEEE Real-Time and Embedded Technology and Applications Symposium*, San Francisco, USA, March 2005.

[SPU 96] SPURI M., Analysis of deadline scheduled real-time systems, Report num. RR2772, INRIA, January 1996.

[TAT 00] TATIKONDA S., Control under communication constraints, PhD thesis, Massachusetts Institute of Technology, 2000.

[TIN 94] TINDELL K., Fixed Priority Scheduling of Hard Real-Time Systems, PhD thesis, Department of Computer Science, University of York, UK, 1994.

[TIP 03] TIPSUWAN Y., CHOW M.-Y., Control Methodologies in Networked Control Systems, *Control Engineering Practice*, vol. 11, num. 10, p. 1099–1111, 2003.

[WAN 06a] WANG Y. Q., YE H., CHENG Y., WANG G. Z., Fault detection of ncs based on eigendecomposition and Pade approximation, *SafeProcess'06*, Beijing, China, p. 937–941, June 2006.

[WAN 06b] WANG Y. Q., YE H., WANG G. Z., A new method for fault detection of networked control systems, *1st IEEE Conference on Industrial Electronics and Applications*, Singapore, China, p. 1–4, May 2006.

[WES 99] WEST R., SCHAWN K., Dynamic window-constrained scheduling for multimedia applications, *6th IEEE International Conference On Multimedia Computing and Systems, ICMCS'99*, Florence, Italy, June 1999.

[WES 04] WEST R., ZHANG Y., SCHWAN K., POELLABAUER C., Dynamic Window-Constrained Scheduling of Real-Time Streams in Media Servers, *IEEE Transactions on Computers*, vol. 53, num. 6, p. 744–759, June 2004.

[WON 97] WONG W. S., BROCKETT R. W., Systems with finite communication bandwidth constraints-Part I: state estimation problems, *IEEE Transactions Automatic Control*, vol. 42, num. 9, p. 1294–1299, 1997.

[WU 97] WU N. E., Robust feedback design with optimized diagnostic performance, *IEEE Transactions on Automatic Control*, vol. 42, num. 9, p. 1264–1268, 1997.

[XIA 08] XIA F., SUN Y., *Control and scheduling Codesign: Flexible resource management in real-time control systems*, Springer, 2008.

[YAN 98] YANG H. L., SAIF M., Observer design and fault diagnosis for state-retarded dynamical systems, *Automatica*, vol. 34, num. 2, p. 217–227, 1998.

[YAN 06] YANG T. C., Networked control system: a brief survey, *IEE Proc.-Control Theory Applications*, vol. 153, num. 4, p. 403–412, 2006.

[YE 04a] YE H., DING S. X., Fault detection of networked control systems with network-induced delay, *8th International Conference on Control, Automation, Robotics and Vision*, Kunming, China, p. 294–297, December 2004.

[YE 04b] YE H., WANG G. Z., DING S. X., A new parity space approach for fault detection based on stationary wavelet transform, *IEEE Transactions Automatic Control*, vol. 49, num. 2, p. 281–287, 2004.

[YE 06a] YE H., LIU R. H. H., WANG G. Z., A new approach for fault detection of networked control systems, *14th Symposium on System Identification SYSID'06*, Newcastle, Australia, p. 654–659, March 2006.

[YE 06b] YE H., WANG Y. Q., Application of parity relation and statinary wavelet transform to fault detection of networked control systems, *Proceedings of 1th IEEE Conference on Industrial Electronics and Applications*, Singapore, China, May 2006.

[YU 05] YU M., WANG L., CHU T., HAO F., Stabilization of networked control systems with packet dropout and transimission delays:continuoust-time case, *European Journal of Control*, vol. 11, num. 1, p. 40–49, 2005.

[YUE 05] YUE D., HAN Q.-L., LAM J., Network-based robust H_∞ control of systems with uncertainty, *Automatica*, vol. 41, num. 6, p. 999–1007, 2005.

[ZHA 01a] ZHANG W., Stability analysis of networked control systems, PhD thesis, Case Western Reserve University, Cleveland, Ohio, USA 2001.

[ZHA 01b] ZHANG W., BRANICKY M. S., PHILLIPS S. M., Stability of networked control systems, *IEEE control systems Magazine*, vol. 21, num. 1, p. 84–99, 2001.

[ZHA 02] ZHANG Y. M., JIANG J., An active fault-tolerant control system against partial actuator failures, *IEE Proceedings Control Theory and Applications*, vol. 149, num. 1, p. 95–104, 2002.

[ZHA 03] ZHANG Y. M., JIANG J., Bibliographical review on reconfigurable fault-tolerant control systems, *5th SafeProcess'03*, Washington, USA, p. 265–276, June 2003.

[ZHA 04] ZHANG P., DING S. X., FRANK P. M., DADER M., Fault detection of networked control systems with missing measurements, *5th Asian Control Conference*, Melbourne, Australia, p. 1258–1263, July 2004.

[ZHA 05] ZHANG L., CHEN Y. S. T., HUANG B., A new method for stabilization of networked control systems with random delays, *IEEE Transactions on Automatic Control*, vol. 50, num. 8, p. 1177–1181, 2005.

[ZHA 06] ZHANG P., DING S. X., Fault detection of networked control systems with limited communication, *SafeProcess'06*, Beijing, China, p. 1135–1140, June 2006.

[ZHE 03] ZHENG Y., Fault diagnosis and fault tolerant control of networked control systems, PhD thesis, Huazhong University of Science and Technology, China, 2003.

[ZHO 96] ZHOU K., DOYLE J. C., GLOVER K., *Robust and optimal control*, Prentice-Hall Inc., 1996.

Chapter 2

Computing-aware Control

2.1. Overview

To implement a controller, the basic idea consists of running the whole set of control equations in a unique periodic real-time task whose clock gives the controller sampling rate. In fact, all parts of the control algorithm do not have an equal weight and urgency w.r.t. the control performance. To minimize the latency, a control law can be basically implemented as two real-time blocks; the urgent one sends the control signal directly computed from the sampled measures, while updating the state estimation or parameters can be delayed or even more computed less frequently [ÅST 97].

In fact, a complex system involves sub-systems with different dynamics which must be further coordinated [TÖR 98]. Assigning different periods and priorities to different blocks according to their relative weight allows for a better control of critical latencies and for a more efficient use of the computing resource [SIM 98]. However, in such cases finding adequate periods for each block is out of the scope of current control theory and must be done through case studies, simulation and experiments.

Latencies have several sources: the first one comes from the computation duration itself, and worst-case-execution times (WCET) are difficult to get. In multi-tasking systems, they come from pre-emption due to concurrent tasks with higher priority, from precedence constraints and from synchronization. Another source of delays is the communication medium and protocols when the control system is distributed on a network of connected devices. In particular, it has been observed that in synchronous

Chapter written by Mongi BEN GAID, David ROBERT, Olivier SENAME, Alexandre SEURET and Daniel SIMON.

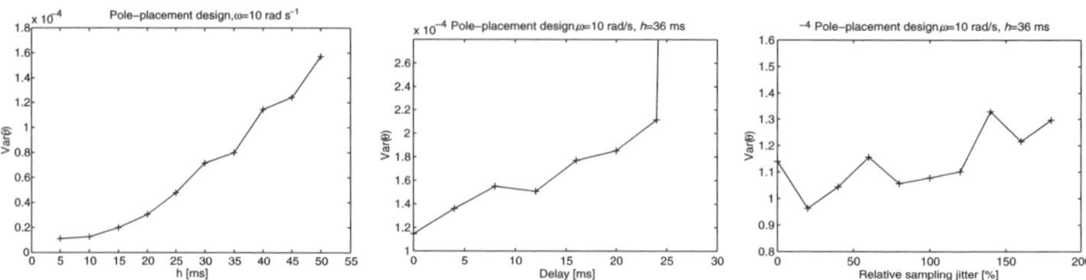

Figure 2.1. *Performance loss w.r.t. timing deviations*

multi-rate systems the value of sampling-induced delays shows complex patterns and can be surprisingly long [CHE 88; WIT 01].

Control systems are often cited as examples of "hard real-time systems" where jitter and deadline violations are strictly forbidden. In fact, experiments show that this assumption may be false for closed-loop control. Any practical feedback system is designed to obtain some stability margin and robustness w.r.t. the plant parameters uncertainty. This also provides robustness w.r.t. timing uncertainties: closed-loop systems are able to tolerate some amount of sampling period and computing delays deviations, jitter and occasional data loss without loss of stability or integrity. For example in [CER 03], the loss of control performance has been checked experimentally using an inverted pendulum, for which a linear quadratic (LQ) controller has been designed according to a nominal sampling period and null delay and jitter. Figure 2.1 (borrowed from [CER 03] with permission of the author) shows the output performance (position error variance) when, respectively, the period, the *I/O* latency and the output jitter are increased: the controller behavior can still be considered correct as long as the sample-induced disturbances stay inside the performance specification bounds.

The following sections deal with computing-aware robust control, where several control methods dealing with robustness and/or adaptation w.r.t. implementation induces timing uncertainties are successively exposed.

A general consequence of the execution of control algorithms on digital distributed platforms is inducing delays from different sources in the control loops, which should be taken into account in the control algorithms tuning. A first idea in the design of dependable control systems consists of using robust controllers, i.e. controllers which are slightly sensitive to both process model and execution resource uncertainties. Therefore, section 2.2 provides a survey of the main existing results concerning the control of systems with delays. Then the "weakly hard" constrained nature of control loops w.r.t. timing deviations is revisited and formalized in section 2.3, using the accelerability property for control tasks when the control periods are allowed to vary along a slotted timescale. Finally, as the variations of the control intervals can be both a consequence of network induced delays and a control variable to manage the CPU and /or network load, a LPV/H_∞-based robust variable sampling control design is described in section 2.4.

2.2. Robust control w.r.t. computing and networking-induced latencies

2.2.1. *Introduction*

Delays appear naturally in the modeling of several physical processes. In general, the delays come from transportation of materials or the transmission of information. Stability analysis of time-delay systems is thus an important topic in many disciplines of science and engineering [GU 03; NIC 01; RIC 03]. Motivating applications are found in diverse areas, such as biology, chemistry, telecommunication control engineering, economics, and population dynamics [KOL 99a]. There has been an increased interest in the area of time-delay systems over the last two decades due to the emerging area of networked embedded systems, which are systems where sensor and actuator devices communicate with control nodes over a communication network. In such systems, processing time and pre-emption in the network nodes, together with propagation delays in the inter-node communication, necessarily leads to time delays affecting the overall closed-loop control system. Various phenomena related to delays in networked controlled systems have recently been considered, e.g. packet losses [HES 07; NAG 08] and robust sampling [FRI 04b]. The locations where delays appear in a control loop are summarized in Figure 2.2.

1) The actuators and the sensors are generally subsystems which have their own dynamics. A first source of delay is the time taken to achieve the computation of the control algorithm itself. This duration is directly related to the algorithm complexity and to the hardware capabilities. It is known that evaluating the WCET of a given program is a very long (and anyway imprecise) duty, especially with modern processors

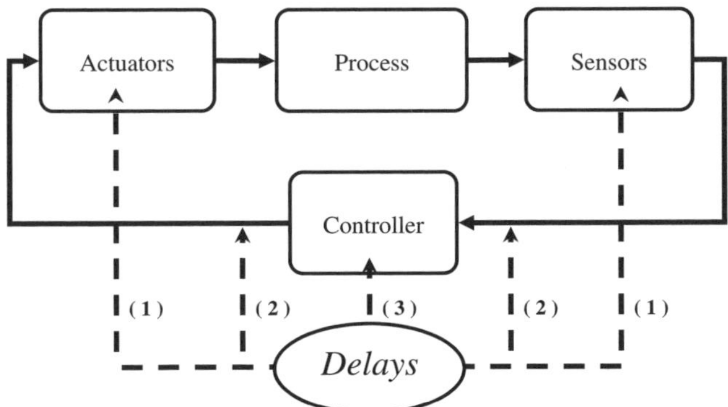

Figure 2.2. *Location of delays in a networked control loop*

using caches and pipelines [DEV 05]. Even for quite simple algorithms using a constant number of statements this duration is usually not constant, due to jitter coming from the hardware architecture and from the operating system's overheads. Moreover many algorithms used in engineering applications and involved in control loops have a variable complexity, depending on input data and operating conditions. This is, for instance, the case in video processing where the computational complexity may strongly vary according to the observed scene, e.g. the number of basic visual features to be extracted and processed. Other algorithms, such as in optimization, have a variable and badly known rate of convergence, so that the number of iterations needed to reach a predefined accuracy may vary considerably. A second source of delays comes from the operating system and scheduling policies. In a complex control system several computing activities share the computing resources under control of a real-time scheduler. Tasks are scheduled according to their importance and/or urgency, stated by their priority. Static schedulers are quite predictable but lead to inflexible implementations. A more flexible and adaptive sharing of computing resources is provided by dynamic and pre-emptive schedulers, where high-priority tasks can pre-empt lower priority ones. However, the complexity of real-life control systems, e.g. where computing activities are triggered by data dependences or asynchronous events, lead to very complex scheduling patterns and increase even more the complexity and unreliability of precise timing analysis of the real-time control system, e.g. [WIT 01].

To avoid increasing the complexity in the model, it is possible to gather computation and pre-emption induced latencies in a single delay in the forward control path. Note that in a well-designed control implementation this "local" delay is usually smaller or equal to the control interval. However, it is not measurable or known when the computation begins, thus it can be handled by the control algorithm only by estimations of its lower and upper bounds.

2) In a basic control loop, data is exchanged from one entity to another (for instance, from the controller to actuator or from the sensors to the controller). Depending on the system this communication might not be instantaneous. The latency between the time a data is sent and the time it is received increase considerably when the controller and the process are remotely installed. This is particularly the case in the so-called networked control systems (NCS) where the communication is achieved through a communication network. Such systems are attracting a lot of attention nowadays. The main problem and interest, at least for the field of time-delay systems, is that the delays become time varying with a high amplitude. Depending on the communication link (wire or wireless communication) and on the communication protocol (TCP, UDP, ZIGBEE, etc.), the quality of the communication can be very low so that exchange data can be lost during their transfer and leads to additional delays. This constitutes a large scale of problems which are exposed in [HES 07; ZAM 08]

In order to cope with the problems connected to delays, one has to understand the difficulties and the complexity that arrive together with delays. In the sequel, a first section briefly points out basic problems which appear when it comes to time-delay systems. This presentation is based on a very simple example. A second section presents the model of delay functions that are usually used in the literature. More especially it will present the usual assumptions that are required to design stability conditions. Finally, the last section exposes the two extensions of the Lyapunov theory to deal with the stability analysis of time-delay systems based on time-domain methods.

2.2.2. *What happens when delays appear?*

This section introduces the problem of time-delay systems in terms of mathematical considerations. More particularly, this section exposes some reasons for which researchers are still investigating some the topic. To have a better understanding and reading of this section, simple examples are now studied. Consider the following simple example. Through this example, several aspects of time-delay systems are presented. This helps the reader to have a practical approach to understand the relevant point. Let $x \in R$ be a variable whose evolution is governed by

$$\forall t > t_0, \qquad \dot{x}(t) = -x(t - \tau), \tag{2.1}$$

where $\tau > 0$ represents the constant delay. If one considers the non-delay case, i.e. $\tau = 0$, it is well known that the solutions of the system are stable and are of the form $x(t) = x(t_0)e^{t_0 - t}$. In the following, particular aspects of this equation with delay will allow us to point out the major difficulties of time delay systems and the difference between the non-delay case.

2.2.2.1. *Initial conditions*

Consider the case where $\tau = -\pi/2$. The two functions $x_1(t) = \sin(t)$ and $x_2(t) = \cos(t)$ are trivial solutions of 2.1. The solutions are shown in Figure 2.3.

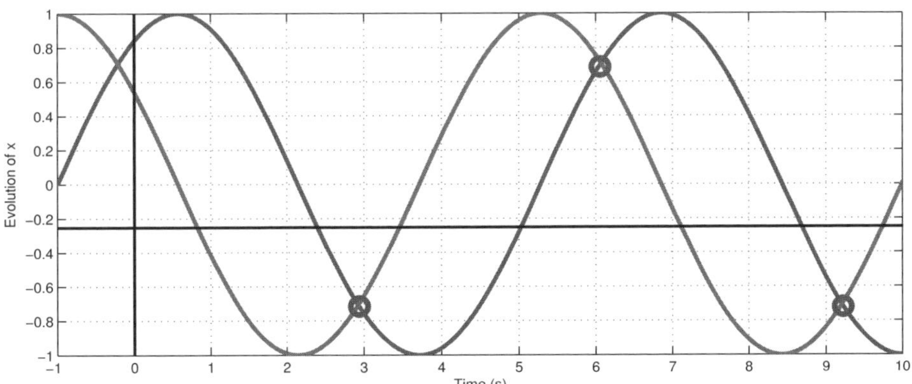

Figure 2.3. *Possible solution for $\tau = \pi/2$*

In this figure we can find a contradiction with the Cauchy theorem. In the non-delay case, if two solutions of the same linear differential equation cross, then the two solutions are the same. In this simple example, it is clear that the two solutions x_1 and x_2 cross an infinite number of times but are, by definition not equal. This problem comes from the fact that the state of a time-delay system is not only a vector considered at an instant t as non-delay case is not, but is a function taken over an interval (or a window) of the form $[t - \tau, \ t]$. Consequently, it is not sufficient to initialize the state of the system by only including the initial position of the state at time t_0. It is required to define a vector function $\phi : [t_0 - \tau, \quad t_0] \rightarrow R$ such that $x(\theta) = \phi(\theta)$ for all θ lying in the delay interval $[t_0 - \tau, \quad t_0]$.

Note that the Cauchy theorem still holds. It is rewritten as follows: if two solutions are equal over an interval of length τ, then the solutions are equal over the whole simulation time.

2.2.2.2. *Infinite dimensional systems*

Consider $\tau = 1$ and the initial conditions $\phi(\theta) = 1$, for all θ in $[t_0 - \tau, \quad t_0]$. The solutions are shown in Figure 2.4.

As expected, in the non-delay case, the solution is an exponential decreasing function. In the delay case, the solution is not of this form anymore. First the solution has an oscillatory behavior around 0. These oscillations are the usual and expected effect of the introduction of delay in a system. For small values of the delay, these oscillations can be of of very low amplitude and thus negligible. However, for greater values of τ (for instance $\tau = 2$), the oscillations have a larger amplitude and the solution is unstable.

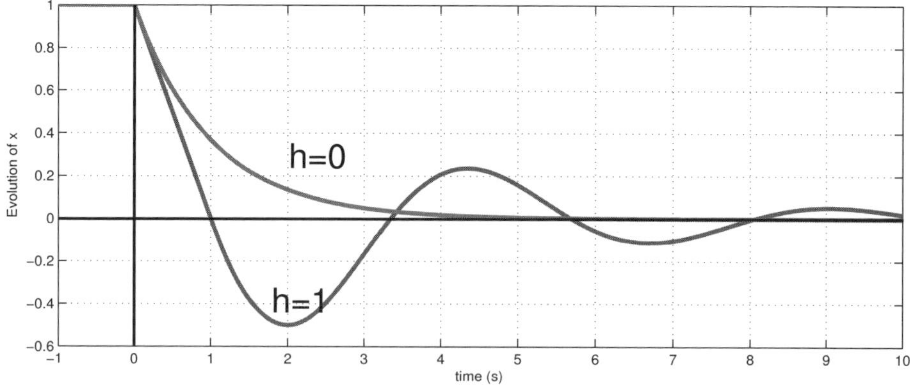

Figure 2.4. *Solution for $\tau = 0, 1$ and constant initial conditions*

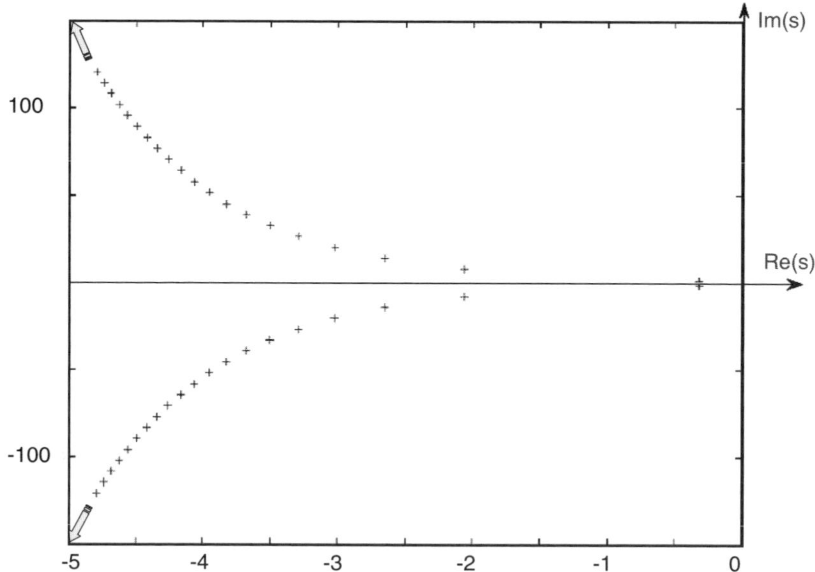

Figure 2.5. *Solution for $\tau = 0, 1$ and constant and zero initial conditions*

Considering $\tau = 1$ and the same initial conditions, it is possible to construct the solution of the system by integrating interval by interval:

$$
\begin{aligned}
t &\in [-1,\ 0], & x(t) &= 1, \\
t &\in [\,0,\ 1], & x(t) &= 1 - t, \\
t &\in [\,1,\ 2], & x(t) &= 1/2 - t + t^2,
\end{aligned}
$$

$$\ldots$$

Thus, the solution of the system is a polynomial function whose degree increases with time. One can then see the time-delay system as an infinite dimensional system since its solution is a polynomial of infinite dimension.

Another property of time-delay systems to understand this type of systems is of infinite dimension, is to consider the Laplace transform of equation 2.1. The characteristic equation is

$$s + e^{-\tau s} = 0$$

This characteristic equation has an infinite number of complex solutions as shown in Figure 2.5. It is clear that this also implies that a time-delay system is of infinite dimension.

Remark. The stability conditions from roots still holds, i.e. the stability is ensured if the roots of the characteristic equation have a negative real part (see [GU 03] or [NIC 01] for more detailed explanations).

2.2.3. *Delay models*

In this section, a brief overview of the delay functions is provided in the current literature has been given.

a) Constant delays: The first studies about the stability of time-delay systems mainly concerned this type of delays together with linear time-invariant systems. A lot of stability criteria were developed by on frequency approach [DAM 94], LMI [GU 03], [NIC 01]. They variously deal with known or unknown, bounded and unbounded delays. Since the middle of the 1990's, several conditions were also expressed in terms of linear matrix inequalities and were able to deal with more complex problems such as linear systems with norm-bounded uncertainties (see [KOL 99b], [LI 97] and [NIC 01]).

b) Bounded time-varying delays: The choice of constant time-delay restriction becomes less relevant when it comes to practical problems as NCS where the delays are induced by networked type of communications [LOP 06; HES 07; ZAM 08] or as in a more practical fluid transportation pipe [ANT 07], to give just two examples. The case of (known or unknown) variable delays have also been the topic of numerous researchers ([RIC 03; KAO 07] and the reference there in). In such a case, the delays appear as a positive scalar functions of the time or of the state of the process. A first type of conditions on the functions is to consider that the delay is bounded, i.e. there exists a positive and known scalar $\tau_2 > 0$ such that [HAL 97]

$$0 \leq \tau(t) \leq \tau_2.$$

c) Interval time-varying delays or non-small delays: A large majority of the existing results on time-varying delays only deal with delays functions which vary between 0 and an upper bound. However, in transportation problem or in networked control systems, the delay functions only vary in an interval excluding zero. The case of considering a delay which is sporadically equal to 0 indeed means that, for instance, the transport of a fluid or of information is done instantaneously, which is not relevant in practice. Thus, consider that delay functions which can only vary in a non-zero interval are relevant. One can define the conditions on this type of time-varying delays as follows. There exist two scalars $0 \leq \tau_1 < \tau_2$ so that

$$0 < \tau_1 \leq \tau(t) \leq \tau_2.$$

Only recent articles deal with this problem [FRI 04a; JIA 05].

d) Delays with constraints on their first derivative: Numerous stability conditions require that the delay functions can vary arbitrarily. The following constraints is often included. Consider a positive scalar d so that

$$\dot{\tau}(t) \leq d. \tag{2.2}$$

More specifically, the conditions require the constraint that d is strictly less than 1. To understand this condition, consider the function $f(t) = t - \tau(t)$ which corresponds to the delayed value considered at time t. The condition $d < 1$, means that this function is strictly increasing. This can be interpreted as the fact that the transport of the fluid or the delayed information is considered in chronological order.

e) Piece-wise time-varying delays: In practice and more especially in networked control systems, the delay function is not a continuous function. Consider, for instance, the case of sampling over communication networks. Sampling a signal as an increasing sequence of time $\{t_k\}_k$ can be seen as a delayed signal with the delay $\tau(t) = t - t_k$ (see [FRI 04b],[NAG 08] and the reference therein). These cases are very interesting and relevant because of the scientific and practical aspects. The main difficulty comes from the fact that the delay function is discontinuous at the sampling instant and from the fact that the delay derivative is equal to 1 almost all the time, which is critical regarding condition (2.2).

In general, the stability conditions that we can find in the literature included a combination of the previous constraints. More importantly and more interestingly, the stability of the system strongly depends on the type of delay function which is considered. For instance, the stability a system characterized by given constant delay τ_2, an unknown time-varying delay bounded by τ_2 or a sampling delay (with a constant sampling period) are different (see for instance article [NAG 08; RIC 03] and the references therein). Thus investigating in more accurate tools to cope with each of the previous type of delays is still an active topic and even if time-delay systems have already been paid a lot of attention, there are still a large number of open problems which need to be solved.

2.2.4. *Stability analysis of TDS using Lyapunov theory*

This section gives a reminder of the theoretical background on the stability analysis of time-delay systems, based only on a time-domain approach and the second method of Lyapunov.

2.2.4.1. *The second method*

Consider the following time-delay system:

$$\dot{x}(t) = f(t, x(t), x_t), \quad x_{t_0}(\theta) = \phi(\theta), \text{ for } \theta \in [-\tau, 0]. \tag{2.3}$$

It is assumed that this system has a unique solution and a steady state $x_t = 0$ (is the steady state is non-zero, a change of coordinate can allow it).

The second method of Lyapunov is based on the existence of a function V of the state x_t which is positive definite so that its derivative along the trajectories of (2.3),

$\frac{dV}{dt}$ is definite negative, if $x_t \neq 0$. This "direct" method is only valid for a reduced class of time-delay systems since the derivative function $\frac{dV}{dt}$ depends on past values of x (included in x_t). It is thus restrictive and sometimes complex to cope with time-delay systems. However, two extensions of the second method of Lyapunov have been provided especially for the case of functional differential equations. In the case of ordinary differential equations (i.e. without delays), a candidate for a Lyapunov function is of the form $V = V(t, x(t))$. It only depends on the current "position", the current state of the system. In the retarded case, this function is not sufficient to analyze the stability since it does not contain the full state of the system, i.e. the current "position" and also past values of it. The stability analysis requires more assumption on the Lyapunov functions.

Two approaches have been provided to cope with functional differential equations: The first one, called the Lyapunov–Razumikhin approach, is based on the same type of Lyapunov function $V = V(t, x(t))$ but leads to difficulties since the derivative also depends on past values of x. The second one, named the Lyapunov–Krasovskii approach, is based on a functional, and not a function, of the form $V = V(t, x_t)$ also leads to several difficulties since it depends on the state x_t of the time-delay system. These two approaches are briefly introduced in the sequel.

2.2.4.2. *The Lyapunov–Razumikhin approach*

Consider a Lyapunov function of the form $V(t, x(t))$ which only requires to consider vector in $\mathbb{R}^n$ as in the ordinary case. However, the following theorem shows that it only requires that the derivative of $V(t, x(t)) \leq 0$ along the trajectories of the system to decrease over the delay interval, i.e. for all $s \in [t - \tau, \tau]$. This test can be restricted to the solutions which tend to leave a close set of the form $V(t, x(t)) \leq c$ around the equilibrium.

THEOREM.–*[[KOL 99a]] Consider increasing functions u, v and $w : \mathbb{R}_+ \to \mathbb{R}_+$ so that $u(\theta)$ and $v(\theta)$ are strictly positive for all $\theta > 0$. Assume that the vector field f of (2.3) is bounded for bounded values of its arguments. Then if there exists a continuous function $V : \mathbb{R} \times \mathbb{R}^n \to \mathbb{R}_+$ such that*

– $u(\|\phi(0)\|) \leq V(t, \phi) \leq v(\|\phi\|)$,

– $\dot{V}(t, \phi) \leq -w(\|\phi(0)\|)$ for all trajectories of (2.3) satisfying:

$$V(t + \theta, \phi(t + \theta)) \leq V(t, \phi(t)), \quad \forall \theta \in [-\tau, 0], \tag{2.4}$$

then the solution $x_t = 0$ of (2.3) is uniformly stable.

Moreover, if $w(\theta) > 0$ for all $\theta > 0$ and if there exists a strictly increasing function $p : \mathbb{R}_+ \to \mathbb{R}_+$ satisfying $p(\theta) > \theta$ for all $\theta > 0$ so that

– $u(\|\phi(0)\|) \leq V(t, \phi) \leq v(\|\phi\|)$,

$- \dot{V}(t, \phi) \leq -w(\|\phi(0)\|)$, *for all trajectories of (2.3) satisfying:*

$$V(t + \theta, x(t + \theta)) \leq p(V(t, x(t))), \quad \forall \theta \in [-\tau, 0], \tag{2.5}$$

then V is called the Lyapunov–Razumikhin function and the solution $x_t = 0$ of (2.3) is uniformly asymptotically stable.

Remark. In practice, the more common functions p are of the form $p = q\theta$ where q is a constant strictly greater than 1 and the Lyapunov–Razumikhin functions to design are generally chosen as simple quadratic functions of the form $V(t) = x^T(t)Px(t)$ where P is a positive definite matrix. Condition (2.5) thus becomes

$$x^T(t + \theta)Px(t + \theta) \leq qx^T(t)Px(t), \quad \forall \theta \in [-\tau, 0], \quad \text{and} \quad q > 1.$$

Note that it was proved, in [DRI 77], that the two approaches are equivalent in the case of constant delays. In the time-varying delay case, the Lyapunov–Razumikhin approach has the particularity to easily take into account variable delay without considering additional constraints on the derivative of the delay function of the type (2.2). However, in such a situation, it leads to more conservative stability conditions than the Lyapunov–Krasovskii approach described in what follows.

2.2.4.3. *The Lyapunov–Krasovskii approach*

This method is an extension of the Lyapunov theorem to the case of functional differential equations. It consists of the research of a *functional* of the form $\mathcal{V}(t, x_t)$ which decreases along the trajectories of (2.3). Similarly, V is called a functional because it depends on the state of the time-delay system x_t which a vector function considers the delay interval $[t - \tau, \tau]$.

THEOREM.–*[[KOL 99a]] Consider continuous and increasing functions u, v, w: $\mathbb{R}_+ \to \mathbb{R}_+$ so that $u(\theta)$ and $v(\theta)$ are strictly positive for all $\theta > 0$ and $u(0) = v(0) = 0$. Assume that the vector field f of (2.3) is bounded for bounded values of its arguments. Then if there exists a functional $\mathcal{V} : \mathbb{R} \times C \to \mathbb{R}_+$ so that*

$- u(\|\phi(0)\|) \leq \mathcal{V}(t, \phi) \leq v(\|\phi\|)$,

$- \dot{\mathcal{V}}(t, \phi) \leq -w(\|\phi(0)\|)$ *for all $t \geq t_0$ along the trajectories of (2.3),*

then the solution $x_t = 0$ of (2.3) is uniformly stable. Moreover if $w(\theta) > 0$ for all $\theta > 0$, then the solution x_t is uniformly asymptotically stable.

If $\mathcal{V}$ satisfies the following conditions:

$- u(\|\phi(0)\|) \leq \mathcal{V}(t, \phi) \leq v(\|\phi\|)$,

$- \dot{\mathcal{V}}(t, \phi) \leq -w(\|\phi(0)\|)$, *for $t \geq t_0$ and $w(\theta) > 0$ for all $\theta > 0$,*

$- \mathcal{V}$ *is Lipschitz with respect to its second argument,*

Then the solution x_t of (2.3) is exponentially stable and the functional $\mathcal{V}$ is called Lyapunov–Krasovskii functional.

Remark. In the previous theorem, the notation $\dot{\mathcal{V}}(t,\phi)$ refers to the derivative in the sense of Dini, i.e. $\dot{\mathcal{V}}(t,\phi) = \lim_{\epsilon \to 0^+} \sup \frac{\mathcal{V}(t+\epsilon,x_{t+\epsilon})-\mathcal{V}(t,x_t)}{\epsilon}$.

The main idea of the theorem is to determine a positive definite functional $\mathcal{V}$ whose derivative along the trajectories of (2.3) is negative definite. Of course the main difficulties come from the design of such a functional if it exists. The classical type of functionals are of the following form ([KOL 96], section 2.2.2):

$$
\begin{aligned}
\mathcal{V}(t,\phi) = \ & \phi^T(0)P(t)\phi(0) + 2\phi^T(0)\left(\int_{-\tau}^{0} Q(t,\sigma)\phi(\sigma)d\sigma\right) \\
& + \int_{-\tau}^{0}\int_{-\tau}^{0}\phi^T(\sigma R(t,\sigma,\rho)\phi(\rho)d\sigma d\rho + \int_{-\tau}^{0}\phi^T(\zeta)S(\zeta)\phi(\zeta)d\zeta,
\end{aligned}
$$

where P, Q, R, and S are square matrix functions of dimension $n \times n$. $P(t)$ and $S(\zeta)$ are symmetric positive definite and R satisfies $R(t,\sigma,\rho) = R^T(t,\rho,\sigma)$. It is assumed that each element of these matrices are bounded and their derivatives are piece-wise continuous functions.

In practice, the research of such functionals leads to complex problems to solve. Often, researchers limit themselves to the cases where the matrix functions P, Q, R, and S are constant. In such a situation, the previous Lyapunov–Krasovskii functional becomes:

$$
\begin{aligned}
\mathcal{V}(t,\phi) = \ & \phi^T(0)P\phi(0) + \int_{-\tau}^{0}\phi^T(\sigma)S\phi(\sigma)d\sigma + \phi^T(0)\int_{-\tau}^{0}Q\phi(\sigma)d\sigma \\
& + \int_{-\tau}^{0}\int_{-\tau}^{0}\phi^T(\sigma)R\phi(\rho)d\sigma d\rho,
\end{aligned}
$$

This type of Lyapunov–Krasovskii function leads to the positive property that it proves the stability only for the delay τ and not lower values of the delay. This is positive for some systems where the delay has a stabilizing effect (see for instance [SEU 09c; MIC 04; GU 97]). However, it is possible to design another type Lyapunov–Krasovskii functionals which allows stability to be assured for constant or time-varying bounded delays. This is done by considering another integral functional expressed with the derivative of the state $\dot{x}$ of the form

$$
V'(t,\dot{\phi}) = \int_{-\tau}^{0}\int_{t-\theta}^{t}\dot{x}^T(s)U\dot{x}(s)ds d\theta
$$

However, we have to be careful with the use of this functional since it might introduce some additional dynamics or constraints, since the functional now depends on the derivative $\dot{\phi}$ and not ϕ itself (see [GU 03] and [NIC 01] for more details).

Note that recent stability conditions have been established using matrix functions for the parameters Q, R, and S. The discretization method introduced in [GU 97; GU 03], allows us to construct piece-wise linear functions as the matrix parameters by dividing the delay interval into several smaller intervals on which the parameters of the functionals are linearly varying. The stability analysis leads to less conservative conditions than in the case of constant parameters but extensions to the time-varying delay case are not straightforward. In [PAP 07], a method to build functionals with varying parameters based on the sum of squares tools is introduced. Another recent approach method is based on Integral Quadratic Constraints [KAO 07]. It provides a new interpretation of the Lyapunov–Krasovskii functionals and the potential conservatism (see [ARI 07]). Another method has been provided based on the solutions of an arbitrary differential equations to build exponential [SEU 09b] and polynomial [SEU 09a] parameters.

Remark. One has to understand that those two theorems only lead to sufficient conditions of stability. In [KHA 01], a complete Lyapunov–Krasovskii functional (i.e. which corresponds to necessary and sufficient conditions of stability) is constructed by solving a functional differential equation. This approach is useful to derive robustness conditions with respect to delay variations [FRI 08a] or parameter uncertainties [KOL 03], [MON 05].

In general, the two previous theorems applied to the case of linear systems with delay is expressed in terms of linear matrix inequalities. Based on semi-definite programming (see [BOY 94; GHA 00]), it becomes easy to find solutions of this type of problems using for instance the LMI Toolbox from Matlab (http://www. mathworks.com) or Scilab (http://www.scilab.org). These two approaches are not only reduced to cope with the stability of time-delay systems. These methods have been extended to numerous problems such as stability and stabilization of linear parameter-varying (LPV) systems (see [BRI 08], section 2.4 and the references therein), H_∞ control (for instance [FRI 02; LI 97]), stabilization finite time [MOU 06], input-to-output or input-to-state stability ([FRI 08b],[GU 03]).

2.2.5. *Summary: time-delay systems and networking*

This section proposed a brief overview on the problems which appear in the context of time-delay systems. Even if delay requires careful attention, there already exist a large number of stability conditions including robustness issues with respect to the parameter uncertainties and delay variations. Networked control systems are characterized by a complex interaction between heterogenous components and timing incertitudes. It is unlikely that the values of computing, scheduling and networking latencies can be accurately modeled and predicted, therefore robust control approaches are particularly attractive and convincing to cope with the various sources of induced latencies in networked control systems, as it is easier (and real-time cheaper) to evaluate an upper bound of aggregated delays rather than an actual measurement or accurate prediction. Note that, even if being of old concern, research on time-delay systems is still a very active field where many recent results are of prime interest in NCS.

2.3. Weakly hard constraints

In this section, a particular model of uncertainty is addressed. This uncertainty is rather related to the ability or the inability of the control task to update the control, at given instances, which are known at (control and real-time scheduling) design time. This uncertainty model may originate from the application of the weekly-hard real-time framework [BER 01]. There are two main motivations behind the application of this framework as follows:

– The need to minimize computational resource economic costs, which is a strong requirement in the development process of embedded systems, especially in the context of their mass production. In real applications, and due to task execution time variations, there is an important gap between average-case and worst-case processor utilization. Due to this gap, the application of state-of-the-art worst-case design methodology leads to an over-dimensioning of the required resources and consequently to their under-utilization. The reduction of computational resources cost may be achieved if methods allowing the needed resources to be dimensioned according to average utilization needs, and not to worst-case utilization scenarios, are employed. In both cases, achieving the stability and ensuring a minimum performance level are required.

– The need to define application-level-degraded modes in distributed control systems. These degraded modes may be activated, for example, in the case of a processor failure. In order to ensure a graceful degradation of the application performance, and to avoid a sudden breakdown of the whole system, the application has to continue running in a degraded mode, where all the tasks of faulty processors are dispatched into the other running processors. This may cause an overload situation.

The framework of weakly hard real-time scheduling was introduced in [BER 01]. This framework encapsulates many previously introduced task models and schedulability constraints such as (m, k)-firm [RAM 95] and skipover [KOR 95]. It allows handling tasks that can tolerate a clearly specified number of missed deadlines during a window of time. For those reasons, this framework appears to be a suitable approach allowing the design and implementation of control tasks based on their average execution times. In order to exploit this framework in a rigorous way, appropriate control design approaches have to be proposed. The design of such strategies was addressed in [BEN 08], and will be summarized throughout this section.

Although the weakly hard real-time scheduling paradigm received significant interest, few works were dedicated to the problem of the control design under weakly hard scheduling constraints. [RAM 99] proposed the use of the (m, k)-firm scheduling concept to achieve a graceful degradation in a situation of processor overload. In this approach, the different jobs of a control task are classified into mandatory and optional. Mandatory jobs are guaranteed to be completed before their deadlines. Optional jobs may either meet or miss their deadlines. The proposed associated control design methodology, which relies on the optimal linear quadratic periodic control theory, modifies the gains of the mandatory jobs in order to minimize the degradation which may result from executing the original control law in an overloaded processor. However, no control design methodology was proposed to compute the control law of the optional jobs (which simply maintain the previous controls constant). For this reason, this design method does not allow the successful executions of the optional instances to be exploited.

This section describes a control design method allowing the advantages of the weakly hard real-time scheduling concept to be exploited. It first provides a summary and then extends the contributions of [BEN 08].

2.3.1. *Problem definition*

Consider the control system described by the following difference equation:

$$x(k + 1) = f(k, x(k), u(k)). \tag{2.6}$$

For any discrete instants k_a and k_b such that $k_a \leq k_b$, notation $u(k_a, k_b)$ represents the control input sequence $(u(k_a), \ldots, u(k_b))$.

A cost functional J is associated with system (2.6), and is defined by

$$J(x(0), u(0, \infty)) \triangleq \sum_{k=0}^{\infty} q(k, x(k), u(k)). \tag{2.7}$$

It is assumed that q is chosen to ensure that when $J(x(0), u(0, \infty))$ is finite, then the controlled system (2.6) is uniformly asymptotically stable. $J(x(k_a), u(k_a, k_b))$ denotes the cost function corresponding to an evolution starting from state $x(k_a)$ at instant k_a to instant k_b, and where the control sequence $u(k_a, k_b)$ is applied.

System (2.6) is controlled by a periodic control task τ, whose execution period is assigned according to average utilization considerations. To simplify the discussions, it is assumed that this period is identical to the time period separating two consecutive discrete instants. However, due to the variations of its own execution time (up to its Worst Case Execution Time, or WCET), and also to the variations of the processor load, the jobs of task τ are not ensured to be completed by their deadlines; the deadline of a job is defined as the release time of its subsequent job. Therefore, in order to characterize the jobs that complete by their deadlines, and the others that will miss their deadlines, and will be consequently aborted, the notion of execution sequence is introduced.

DEFINITION.– *An execution sequence σ is an infinite sequence of elements of $\{0,1\}$.*

According to this definition, execution sequences are elements of $\{0,1\}^{\mathbb{N}}$. An execution sequence σ is associated with each completion of task τ, and defined by

$$\begin{cases} \sigma(k) = 1 & \text{if the job activated at instant } k \text{ completes} \\ & \text{its execution by its deadline,} \\ \sigma(k) = 0 & \text{otherwise.} \end{cases}$$

Let $\mathcal{E}$ be the set valued function that associates to each execution sequence the invocation count (i.e. the discrete instant of activation) of the jobs that are finished before their deadlines. Formally, $\mathcal{E}$ is defined by

$$\mathcal{E}(\sigma) \triangleq \{k \in \mathbb{N} \text{ such that } \sigma(k) = 1\} .$$

Fortunately, using a priority-based scheduling, and knowing the WCET of all the tasks that have priority over τ, it is possible to guarantee that selected jobs of τ will always meet their deadlines. This may be ensured using weakly hard schedulability analysis techniques developed in [RAM 99] and [BER 01]. In the following, it is assumed that task τ guarantees a (μ, κ)-constraint where μ and κ are two integers so that $\mu \leq \kappa$ (i.e. the deadlines of μ out of any κ consecutive jobs of τ are met). It is also assumed that this constraint is met by guaranteeing that the jobs whose invocation count k verifies

$$k = \left\lfloor \left\lceil \frac{k\mu}{\kappa} \right\rceil \frac{\kappa}{\mu} \right\rfloor \tag{2.8}$$

will always meet their deadlines. These jobs are called *mandatory* jobs. The other jobs, which are not guaranteed to be completed by their deadlines, are called *optional* jobs. It has been proved in [RAM 99] that when the jobs of a given task are classified according to (2.8), then the pattern of mandatory jobs will be κ-periodic. This means that if the job activated at instant k is mandatory, than any job activated at instant $k + i\kappa$, $(i \in \mathbb{N})$ is also mandatory. Consequently, when only these mandatory jobs are guaranteed to be completed before their deadlines, the worst-case execution sequence γ that may be associated with τ is defined by

$$\begin{cases} \gamma(k) = 1 & \text{if } k = \left\lfloor \left\lceil \frac{k\mu}{\kappa} \right\rceil \frac{\kappa}{\mu} \right\rfloor, \\ \gamma(k) = 0 & \text{otherwise.} \end{cases} \tag{2.9}$$

When relation (2.8) is applied to impose the worst-case execution pattern of task τ, the corresponding worst-case execution sequence γ is guaranteed to be a κ-periodic execution sequence, verifying $\gamma(k) = \gamma(k + \kappa)$.

2.3.2. *Notion of accelerable control*

For any given control law $\xi(k)$ defined over $\mathbb{N}$, $\xi_\sigma(k)$ denotes the control input to the plant, taking into account execution constraints. Therefore, $\xi_\sigma(k)$ is defined by

$$\begin{cases} \xi_\sigma(k) = \xi(k) & \text{if } \sigma(k) = 1, \\ \xi_\sigma(k) = \xi_\sigma(k-1) & \text{otherwise.} \end{cases}$$

Now the notion of accelerable control can be introduced.

DEFINITION.– *[BEN 08] Assume that a control law $\xi(k)$ was defined, so that for all $x_0 \in \mathbb{R}^n$, the cost function $J(x_0, \xi_\gamma(0, \infty))$ corresponding to the worst-case execution sequence γ is finite. The control law $\xi(k)$ is called accelerable according to performance index (2.7) and worst-case execution sequence γ, if for all execution sequences σ_1 and σ_2 so that $\mathcal{E}(\gamma) \subseteq \mathcal{E}(\sigma_2) \subseteq \mathcal{E}(\sigma_1)$, for all $x_0 \in \mathbb{R}^n$, the associated cost functions satisfy $J(x_0, \xi_{\sigma_1}(0, \infty)) \leq J(x_0, \xi_{\sigma_2}(0, \infty))$.*

A control task executing an accelerable control law will be called accelerable control task. An accelerable control task has the property that the more executions are performed, the better the control performance. When used in conjunction with weakly hard real-time scheduling design, an accelerable control task allows us to take advantage of the extra computational resources that may be allocated to it (the successful execution of optional task instances), and to improve the control performance with respect to worst-case design methods. In practice, however, control laws designed using standard sampled-data control design methods are not necessarily accelerable. This fact was demonstrated in [BEN 08].

2.3.3. *Design of accelerable controllers*

Let σ be an execution sequence and ℓ a time instant. $\sigma \bullet \ell$ denotes the execution sequence defined by

$$\begin{cases} (\sigma \bullet \ell)(k) = \sigma(k) & \text{if } k \in \mathbb{N} \text{ and } k \neq \ell, \\ (\sigma \bullet \ell)(\ell) = 1 & \text{otherwise.} \end{cases}$$

In the following, a general method for constructing accelerable control laws is presented. Let γ be a worst-case execution sequence determined according to (2.8). Let k_a and k_b two discrete instants such that $k_a \leq k_b$. Let $\mathcal{U}_\rho(k_a, k_b)$ be the set of admissible control inputs, defined between instants k_a and k_b, taking into account the resource constraints that are modeled by the execution sequence ρ. The set $\mathcal{U}_\rho(k_a, k_b)$ is formally defined as $\{u(k_a, k_b)$, so that $u(k) = u(k-1)$ if $\rho(k) = 0\}$.

In the rest, the following assumption are made.

ASSUMPTION.– *For all $k \in \mathbb{N}$ and $x \in \mathbb{R}^n$,*

$$\arg \min_{u(k,\infty) \in \mathcal{U}_{(\gamma \bullet k)}(k,\infty)} J(x, u(k,\infty)) \ exist.$$

For any given $k \in \mathbb{N}$ and $x(k) \in \mathbb{R}^n$, an optimal control sequence corresponding to an evolution over an infinite horizon, starting at instant k from state $x(k)$, and taking into account the computation constraints defined by the execution sequence $\gamma \bullet k$, will be denoted as

$$u^*_{(\gamma \bullet k)}(k, \infty) \triangleq \arg \min_{u(k,\infty) \in \mathcal{U}_{(\gamma \bullet k)}(k,\infty)} J(x(k), u(k,\infty)). \tag{2.10}$$

An optimal solution of problem (2.10) is a control sequence $u^*_{(\gamma \bullet k)}(k, \infty) = (u^*_{(\gamma \bullet k)}(k), u^*_{(\gamma \bullet k)}(k+1), u^*_{(\gamma \bullet k)}(k+2), \ldots)$ that minimizes the cost function J, corresponding to an evolution over an infinite horizon, starting from state $x(k)$ at instant k, and assuming that

– the job (mandatory or optional) activated at instant k will meet its deadline and update the plant,

– the subsequent computation constraints (from instant $k + 1$ to ∞) are described following the worst-case execution sequence.

Based on the solutions of optimization problems (2.10), it is possible to construct, in a simple way, an accelerable control law. Let $u^\bullet(k)$ be the first element of the optimal control sequence $u^*_{(\gamma \bullet k)}(k, \infty)$:

$$u^\bullet(k) = u^*_{(\gamma \bullet k)}(k). \tag{2.11}$$

Strategy $u^\bullet(k)$ may be seen as a "robust control" approach against execution uncertainties satisfying the introduced weakly hard model. It allows us to be minimized the cost function J for the "worst-case uncertainty" from the implementation. Under the Assumption in section 2.3.3, strategy (2.11) provides a general method for constructing accelerable control laws. The following theorem, borrowed form [BEN 08], states the accelerability properties of strategy (2.11).

THEOREM.– *Let γ be a worst-case execution sequence. Under Assumption 2.3.3, control law $u^\bullet(k)$, as defined in (2.11), is accelerable in accordance to (2.7) and γ.*

2.3.4. *Accelerable LQR design for LTI systems*

In this subsection, only Linear Time Invariant (LTI) plants are considered. Under the Assumption in section 2.3.3, which will be met if reachability properties (as defined for linear time-varying systems) are fulfilled and when the cost functions are appropriately chosen, it becomes possible to compute off-line a closed form of $u^\bullet$, which will be a time-varying state feedback. In the rest of this section, the following assumption is made.

ASSUMPTION.–

$$f(k, x(k), u(k)) = Ax(k) + Bu(k),$$

$$q(k, x(k), u(k)) = \left[\begin{array}{c} x(k) \\ u(k) \end{array} \right]^T Q \left[\begin{array}{c} x(k) \\ u(k) \end{array} \right],$$

where $Q = \left[\begin{array}{cc} Q_1 & Q_{12} \\ Q_{12}^T & Q_2 \end{array} \right] \geq 0$ and $Q_2 > 0$.

In the following, the methodology underlying the design of the accelerable control law $u^\bullet$ is developed. The following definitions are first introduced. Let γ be a worst-case execution sequence defined according to (2.9). The application of the theorem in section 2.3.3 to LTI systems is described by the following corollary.

COROLLARY.– *[[BEN 08]] Let γ be a worst-case execution sequence defined according to (2.9). Under the assumptions in sections 2.3.3 and 2.3.4, the state feedback control law defined by*

$$u^\bullet(k) = -L_\gamma(k)x(k) \tag{2.12}$$

is accelerable in accordance to (2.7) and γ, where

$$
\begin{aligned}
L_\gamma(k) = & \left(\tilde{Q}_{2_\gamma}(k) + \tilde{B}_\gamma^T(k)\tilde{S}_\gamma(d_\gamma(k))\tilde{B}_\gamma(k) \right)^{-1} \\
& \times \left(\tilde{B}_\gamma^T(k)\tilde{S}_\gamma(d_\gamma(k))\tilde{A}_\gamma(k) + \tilde{Q}_{12_\gamma}^T(k) \right).
\end{aligned}
\tag{2.13}
$$

Matrices $\tilde{S}_\gamma(k)$ are defined for as the steady state solutions of the following Riccati equation

$$
\begin{aligned}
\tilde{S}_\gamma(k) = & \tilde{A}_\gamma^T(k)\tilde{S}_\gamma(d_\gamma(k))\tilde{A}_\gamma(k) + \tilde{Q}_{1_\gamma}(k) \\
& - \left(\tilde{A}_\gamma^T(k)\tilde{S}_\gamma(d_\gamma(k))\tilde{B}_\gamma(k) + \tilde{Q}_{12_\gamma}(k) \right) \\
& \times \left(\tilde{B}_\gamma^T(k)\tilde{S}_\gamma(d_\gamma(k))\tilde{B}_\gamma(k) + \tilde{Q}_{2_\gamma}(k) \right)^{-1} \\
& \times \left(\tilde{B}_\gamma^T(k)\tilde{S}_\gamma(d_\gamma(k))\tilde{A}_\gamma(k) + \tilde{Q}_{12_\gamma}^T(k) \right),
\end{aligned}
\tag{2.14}
$$

where

$$d_\gamma(k) \triangleq \min_{k_m} \left\{ k_m, k_m > k \text{ and } \gamma(k_m) = 1 \right\}, \tag{2.15}$$

$$\Phi(i, k) \triangleq \left[\begin{array}{cc} A^{i-k} & \sum_{j=0}^{i-k-1} A^j B \\ 0_{m,n} & I_m \end{array} \right],$$

$$\tilde{A}_\gamma(k) \triangleq A^{d_\gamma(k)-k}, \tag{2.16}$$

$$\tilde{B}_\gamma(k) \triangleq \sum_{i=0}^{d_\gamma(k)-k-1} A^i B, \tag{2.17}$$

$$\tilde{Q}_\gamma(k) \triangleq \begin{cases} Q & \text{if } d_\gamma(k) - k = 1, \\ Q + \sum_{i=k+1}^{d_\gamma(k)-1} \Phi(i,k)^T Q \Phi(i,k) & \text{if } d_\gamma(k) - k > 1, \end{cases} \tag{2.18}$$

and $\tilde{Q}_\gamma(k)$ is partitioned as

$$\tilde{Q}_\gamma(k) = \begin{bmatrix} \tilde{Q}_{1_\gamma}(k) & \tilde{Q}_{12_\gamma}(k) \\ \tilde{Q}_{12_\gamma}^T(k) & \tilde{Q}_{2_\gamma}(k) \end{bmatrix}.$$

2.3.5. *Kalman filtering and accelerability*

Consider the following linear discrete-time system

$$x(k+1) = Ax(k) + w(k) \tag{2.19}$$

$$y(k) = Cx(k) + v(k) \tag{2.20}$$

where $y(k) \in \mathbb{R}^p$ is the output, $w(k) \in \mathbb{R}^n$ and $v(k) \in \mathbb{R}^p$ are Gaussian random vectors with zero mean and respective co-variance matrices $W \geq 0$ and $V > 0$. The pair (A, W) is supposed stabilizable and that the pair (A, C) is detectable. $w(k)$ is independent of $w(l)$ for $l < k$. The initial state is assumed to be a Gaussian random vector with zero mean and co-variance X_0. To simplify the further developments, v and w are assumed mutually independent.

The control task τ has to determine an estimate of the state of system (2.3.5), which minimizes the co-variance of the *a posteriori* estimation error, and under the computation constraints defined by a given execution sequence σ (which is unknown at design time). The *a priori* and the *a posteriori* estimates of state $x(k)$ are respectively denoted by $\hat{x}(k|k-1)$ and $\hat{x}(k|k)$. Matrices $P(k|k-1)$ and $P(k|k)$ will denote respectively the co-variance matrices of the *a priori* (i.e $x(k) - \hat{x}(k|k-1)$) and the *a posteriori* (i.e $x(k) - \hat{x}(k|k)$) estimation errors. Let $\mathbf{y}_k = [y(0), \ldots, y(k)]^T$ and $\sigma_k = [\sigma(0), \ldots, \sigma(k)]^T$. $\hat{x}(k|k-1)$, $\hat{x}(k|k)$, $P(k|k-1)$ and $P(k|k)$ are defined by the following equations

$$\hat{x}(k|k-1) = \mathbb{E}\left[x(k)|\mathbf{y}_{k-1}, \sigma_{k-1}\right] \tag{2.21}$$

$$P(k|k-1) = \mathbb{E}\left[(x(k) - \hat{x}(k))(x(k) - \hat{x}(k))^T |\mathbf{y}_{k-1}, \sigma_{k-1}\right] \tag{2.22}$$

$$\hat{x}(k|k) = \mathbb{E}\left[x(k)|\mathbf{y}_k, \sigma_k\right] \tag{2.23}$$

$$P(k|k) = \mathbb{E}\left[(x(k) - \hat{x}(k))(x(k) - \hat{x}(k))^T |\mathbf{y}_k, \sigma_k\right] \tag{2.24}$$

The computational constraints, which prevent task τ from computing an estimate of the state at instants k where $\sigma(k) = 0$ may be modeled using the approach of [SIN 04]. In this approach, the output noise model is modified so that

$$p(v(k)|\sigma(k)) = \begin{cases} \mathcal{N}(0, V) & \text{if } \sigma(k) = 1 \\ \mathcal{N}(0, \beta^2 I) & \text{if } \sigma(k) = 0, \text{ with } \beta \longrightarrow \infty \end{cases} \qquad (2.25)$$

In [SIN 04], it has been shown that the equations of the Kalman filter with intermittent observations are defined as follows.

$$\hat{x}(k|k-1) = A\hat{x}(k-1|k-1) \qquad (2.26)$$

$$P(k|k-1) = AP(k-1|k-1)A^T + W \qquad (2.27)$$

$$L(k) = P(k|k-1)C^T \left(CP(k|k-1)C^T + V\right)^{-1} \qquad (2.28)$$

$$\hat{x}(k|k) = \hat{x}(k|k-1) + \sigma(k)L(k)\left(y(k) - C\hat{x}(k|k-1)\right) \qquad (2.29)$$

$$P(k|k) = P(k|k-1) - \sigma(k)L(k)CP(k|k-1) \qquad (2.30)$$

In particular, the iterations of the co-variance matrices $P_k^- = P(k|k-1)$ and $P_k^+ = P(k|k)$ are defined by

$$P_{k+1}^- = AP_k^- A^T + W - \sigma(k)\left[AP_k^- C^T \left(CP_k^- C^T + V\right)^{-1} CP_k^- A^T\right] \qquad (2.31)$$

and

$$P_k^+ = AP_{k-1}^+ A^T + W - \sigma(k)\left[P_k^- C^T \left(CP_k^- C^T + V\right)^{-1} CP_k^-\right] \qquad (2.32)$$

These last equations show that the Kalman filter is accelerable. In fact, the terms

$$AP_k^- C^T \left(CP_k^- C^T + V\right)^{-1} CP_k^- A^T$$

and

$$P_k^- C^T \left(CP_k^- C^T + V\right)^{-1} CP_k^-$$

are definite positive (because $V > 0$ and $P_k^- \geq 0$). Consequently, when $\sigma(k) = 1$, the co-variance matrices values P_{k+1}^- and P_k^+ (which represent the quality of the state estimation) will be better than respectively $(AP_k^- A^T + W)$ and $(AP_{k-1}^- A^T + W)$ (which represent their respective values if $\sigma(k) = 0$). This shows the intrinsic accelerability properties of the Kalman filter (2.3.5).

2.3.6. *Robustifying feedback scheduling using weakly hard scheduling concepts*

Feedback scheduling is a control theoretical approach to real-time scheduling of systems with variable workload. The feedback scheduler may be seen as a "scheduling controller" that receives filtered measures of tasks execution times and acts on tasks periods in order to optimize a quality of control criterion. In practice, the feedback scheduler is implemented as a particular real-time task, which is triggered periodically (at a period which is sufficiently greater than other task periods), in order to adjust control tasks periods according to their average execution time estimates, while optimizing a control performance criterion. However, when achieving high processor utilization rates is desired, these basic implementations of feedback scheduling algorithms do not provide any guarantees on the number of deadline misses, input output latencies and effective sampling periods that control tasks may experience during the sampling period of the feedback scheduler. In fact, between two consecutive executions of the feedback scheduler, the execution time of each task may vary in a random fashion up to a maximal value. There may exist some situations when the execution times of some tasks exceed their estimated average value. Consequently, the effective CPU utilization may also exceed the estimated average CPU utilization, according to which task periods were adjusted. In this overload situation, lower priority tasks may experience several deadline misses, long input output latency and effective sampling period, leading to a severe degradation of the control performance. This problem is illustrated by the following example.

EXAMPLE.– Consider two control tasks τ_1 and τ_2. The execution times of these tasks vary respectively between [1, 3] and [0.5, 1.5]. Their measured average execution times are, respectively, equal to 2 and 1. The utilization set points for each task are respectively equal to 66% and 25%. Consequently, the assigned periods are, respectively, equal to 3 and 4. Figure 2.6(a) depicts a situation when three consecutive executions of control task τ_1 exceed their average estimated execution time. In this situation, the second task experiences an input output latency of about three consecutive sampling periods, which may significantly degrade its performance.

Fortunately, using a weakly hard scheduling approach allows dealing with these situations of transient overload. Instead of causing the "starvation" of lower priority tasks, these approaches allow performance to be gradually degraded, when overload

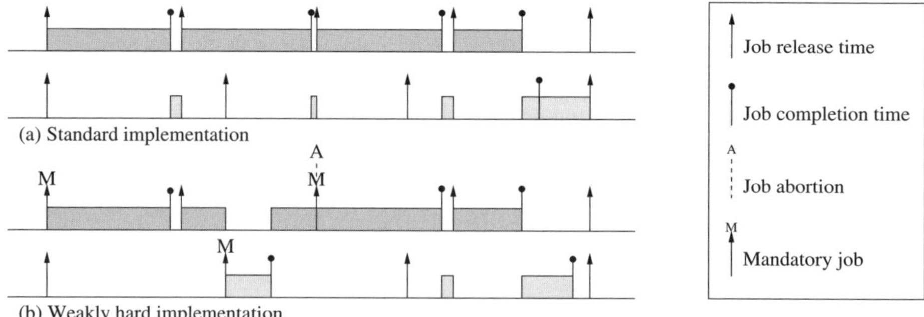

Figure 2.6. *Illustration of overload handling using weakly hard scheduling concepts*

occurs, as illustrated in Figure 2.6(b). In this situation, the marking of the jobs of tasks τ_1 and τ_2 was performed according to the (m, k)-pattern (1,2). This ensures that in the most severe overload condition, no more than two consecutive deadlines of tasks τ_1 and τ_2 are missed.

2.3.7. *Application to the attitude control of a quadrotor*

Consider the problem of the attitude control of a quadrotor. A quadrotor is a rotorcraft that is lifted and propelled by four rotors. Chapter 7 provides a detailed description of the used quadrotor model as well as a review of the mathematical concepts used to describe it. The attitude (i.e. orientation) of the quadrotor is completely described using a unitary quaternion $q = [q_0 \ \vec{q}^T]^T$, with $\|q\|_2 = 1$, where q_0 and $\vec{q} = [q_1 \ q_2 \ q_3]^T$ are, respectively, the scalar and vector parts of the quaternion. The unit quaternion represents the rotation from an inertial frame to the body frame attached to the quadrotor. The dynamic evolution of the attitude quaternion is given by

$$\begin{pmatrix} \dot{q}_0 \\ \dot{\vec{q}} \end{pmatrix} = \frac{1}{2} \begin{pmatrix} -\vec{q}^T \\ I_3 q_0 + [\vec{q}^\times] \end{pmatrix} \omega, \tag{2.33}$$

where $\omega \in \mathbb{R}^3$ is the angular velocity of the quadrotor in the body frame and $[\vec{q}^\times]$ is the skew symmetric tensor associated with $\vec{q}$

$$[\vec{q}^\times] = \begin{pmatrix} 0 & -q_3 & q_2 \\ q_3 & 0 & -q_1 \\ -q_2 & q_1 & 0 \end{pmatrix}.$$

The rotational motion of the quadrotor, neglecting the gyroscopic torques, is given by

$$I_f \dot{\omega} = -[\omega^\times] I_f \omega + \tau_a + \tau_{\text{dist}}, \tag{2.34}$$

where $I_f \in \mathbb{R}^{3 \times 3}$ is the inertia matrix of the quadrotor with respect to the body frame. I_f is constant. $\tau_a \in \mathbb{R}^3$ represents the torques resulting from the differences of the relative speeds of the four rotors, and may be written as

$$\tau_a = \begin{bmatrix} \tau_{\text{roll}} \\ \tau_{\text{pitch}} \\ \tau_{\text{yaw}} \end{bmatrix},$$

and $\tau_{\text{dist}} \in \mathbb{R}^3$ describe the aerodynamic disturbances acting on the quadrotor.

The objective of the attitude controller is to drive the quadrotor attitude to a desired value, which is specified by a unitary quaternion. In practice, this attitude may also be specified by Euler angles (ϕ, θ, ψ), which are more intuitive than quaternions.

To design the attitude controller, consider the linearized model of equations (2.33) and (2.34), and where Euler angles are used instead of quaternions for the description of the attitude. This linearized model is described by

$$\dot{x}_c(t) = A_c x_c(t) + B_c u_c(t) + B_c \tau_{\mathrm{dist}}(t),\qquad(2.35)$$

with $A_c = \begin{bmatrix} 0_3 & I_3 \\ 0_3 & 0_3 \end{bmatrix}$, $B_c = \begin{bmatrix} 0_3 \\ I_f^{-1} \end{bmatrix}$, $x_c = \begin{bmatrix} e \\ w \end{bmatrix}$, $e = \begin{bmatrix} \phi \\ \theta \\ \psi \end{bmatrix}$ and $u_c = \tau_a$.

The attitude controller acts on τ_a, based on e and ω. It is assumed that e and ω are measured by an Inertial Measurement Unit (IMU). The attitude controller is implemented as periodic task τ, with period $T_s = 50$ ms. Let $x(k) = x_c(kT_s)$. Assume that the estimated WCET of task τ is three times its average execution time. According to the weakly hard scheduling philosophy, a (1,3)-firm constraint is associated with task τ. The corresponding worst-case execution sequence is described by the periodic execution sequence $\gamma = (1,0,0,1,0,0,\ldots)$. An accelerable control design, based on the worst-case execution sequence γ was performed (according to corollary 1). In the following simulations, the assigned set point for the attitude controller is $x_{c_{sp}} = 0$. The accelerable control law $u^{\bullet}$ is defined by

$$u^{\bullet}(k) = -L_\gamma(k)x(k),$$

where

$$
\begin{cases}
L_\gamma(k) = \begin{bmatrix} 0.1783 & 0 & 0 & 0.0824 & 0 & 0 \\ 0 & 0.1370 & 0 & 0 & 0.0555 & 0 \\ 0 & 0 & 0.0911 & 0 & 0 & 0.0555 \end{bmatrix} & \text{if } k \bmod 3 = 0 \\[12pt]
L_\gamma(k) = \begin{bmatrix} 0.2169 & 0 & 0 & 0.0935 & 0 & 0 \\ 0 & 0.1797 & 0 & 0 & 0.0667 & 0 \\ 0 & 0 & 0.1197 & 0 & 0 & 0.0689 \end{bmatrix} & \text{if } k \bmod 3 = 1 \\[12pt]
L_\gamma(k) = \begin{bmatrix} 0.2681 & 0 & 0 & 0.1075 & 0 & 0 \\ 0 & 0.2474 & 0 & 0 & 0.0838 & 0 \\ 0 & 0 & 0.1686 & 0 & 0 & 0.0915 \end{bmatrix} & \text{if } k \bmod 3 = 2
\end{cases}
$$

In this particular example, the influence of sampling period reduction manifests itself on different control performance attributes, in particular on the disturbance rejection abilities. Figure 2.7 illustrates this point, and compares the controlled attitude of quadrotor, described by Euler angles (ϕ, θ, ψ), in three different situations as follows.

– The situation where all the optional jobs are not executed (0% hit); this corresponds to the state of the practice hard real-time scheduling design, where resources are dimensioned according to worst-case utilization situations.

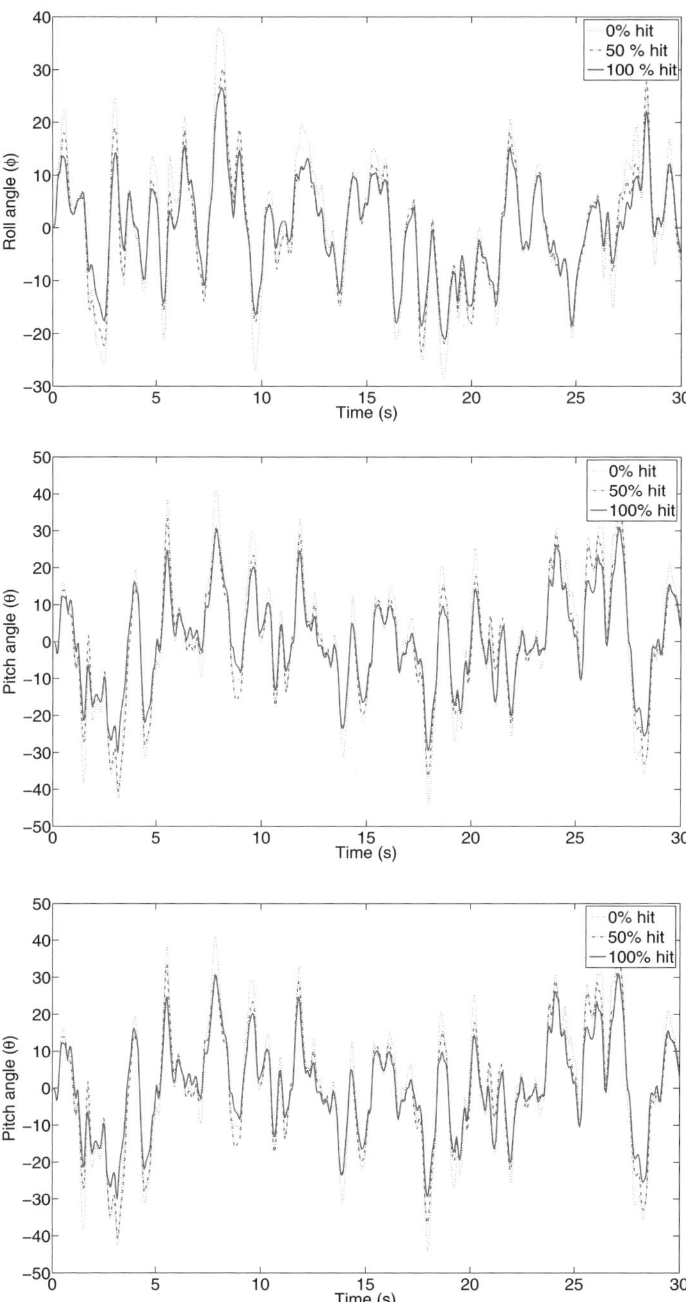

Figure 2.7. *Quadrotor roll ϕ, pitch θ, and yaw ψ for different optional jobs hit ratios*

– The situation where all optional jobs are triggered for execution (according to the weakly hard real-time scheduling philosophy) and when 50% of them meet their deadlines and update the control, according to a random Bernoulli probability distribution, with success probability 1/2 (50% hit).

– The best-case situation where all the optional jobs are triggered and meet their deadlines (100% hit).

The aerodynamic disturbances $\tau_{\mathrm{dist}}(t)$ are modeled by band-limited white noises, with noise power 10^{-3} and period 10^{-1}, and which are produced using different seeds. Figure 2.7 shows that significant improvements in control performance result from the weakly hard real-time design, with respect to the worst-case real-time design. The band-limited white noise disturbances are better rejected. These improvements are due to the fact that in the accelerable weakly hard design, optional instances are executed when possible, with conveniently computed and compensated control gains.

Finally, Figure 2.8 depicts the cumulative cost functions that are associated with the previous simulations. They illustrate the intuitive notion behind accelerability: the more optional instances are executed, the better the control performance. It is worth noticing that although the design was performed on the linearized model, these significant improvements are observed on the nonlinear model. This suggests the robustness of the used control design methodology.

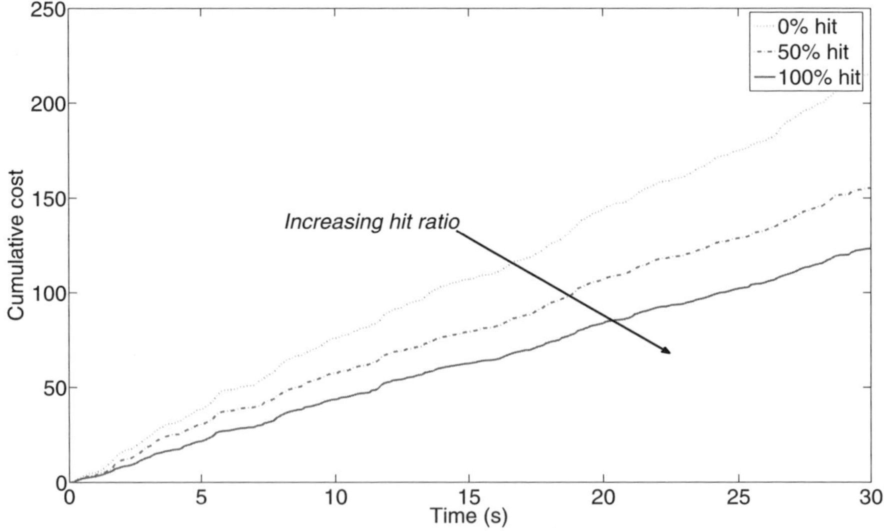

Figure 2.8. *Cumulative costs for different optional jobs hit ratios*

2.4. LPV adaptive variable sampling

As seen in section 1.4, the key actuator to be used for CPU utilization or network bandwidth control is the control interval. A first idea is to design a bank of controllers, each of them being designed and tuned for a specific sampling frequency, and to switch between them according to the decisions of the feedback scheduler. However, it has been observed that switching without caution between such controllers may lead to instability although each controller in isolation is stable [SCH 02].

To ensure the system's stability for a measured variable input delay, [SAL 05] proposes for example a gain scheduled approach based on time-varying observers and state feedback controllers, synthesized using linear matrix inequalities (LMI) and quadratic Lyapunov functions. Indeed, gain scheduling is a popular approach to control nonlinear plants, allowing to some extent to re-use the well-known linear control theory and tools [LEI 00]. The model of the plant is linearized around a set of operating points, and the control responsibility is switched to the controller whose specification is the closest to the actual operating conditions. LPV approaches were then developed to enforce the overall control system stability during switching.

Based on these tools, this section presents a varying sampling rate control algorithm based on LPV gain scheduling design: the gain scheduled controller is able to guarantee the system's stability whatever the instants and speed of variation of the control task intervals. A specified performance level must also be preserved in the allowed interval range. However, this approach is up to now only designed for linear systems. The first point is the problem formulation so that it can be solved following the LPV design of [APK 95]. Remember that this design ensures the stability and performance robustness of the closed-loop parameter-varying system whatever the variations of the parameters inside their predefined allowed range.

Here a parameterized discretization of the continuous time plant and of the weighting functions leads to a discrete-time, sampling period-dependent, augmented plant. In particular, the plant discretization approximates the matrix exponentials appearing in the discretized model by a Taylor series of order N. The original LPV design builds a discrete-time sampling period dependent controller through the convex combination of 2^N controllers, which may be conservative and complex to implement. In the particular case where the control interval is in only varying parameter, the dependency between the variable parameters, which are the successive powers of the sampling period $h, h^2, ..., h^N$, is used to reduce the number of controllers to be combined down to $N + 1$. This reduction of the polytopic set drastically decreases the conservatism of the original design and makes the solution easier to implement. A summary of this approach, which is described in details in [ROB 07a], [ROB 09], and [ROB 07b], is given in the following sections.

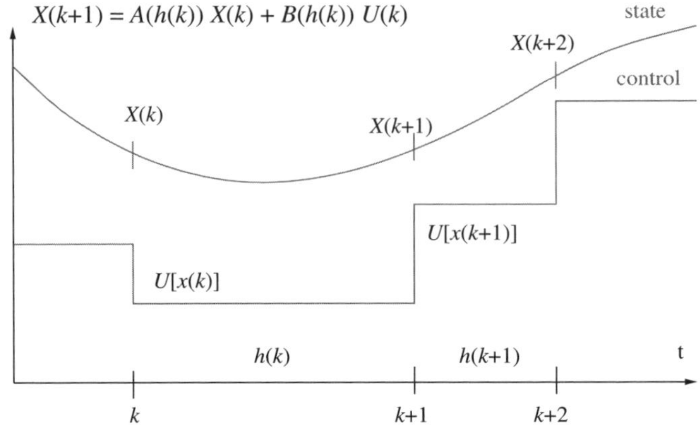

Figure 2.9. *Scheduled varying sampling scheme*

2.4.1. *A polytopic discrete-plant model*

A state space representation of a continuous time plant is

$$G : \begin{cases} \dot{x} &=& Ax + Bu \\ y &=& Cx + Du. \end{cases} \tag{2.36}$$

The exact discretization of this system with a zero-order hold at the sampling period h can be computed, e.g. see [ÅST 97], leading to the discrete-time LPV system 2.37

$$G_d : \begin{cases} x_{k+1} &=& A_d(h)\,x_k + B_d(h)\,u_k \\ y_k &=& C_d(h)\,x_k + D_d(h)\,u_k, \end{cases} \tag{2.37}$$

with h ranging in $[h_{\min}; h_{\max}]$. The corresponding sampling and hold scheme are depicted in Figure 2.9: the control signal $U[t(k)]$ computed at the kth instant from measure $X[t(k)]$ is hold until instant $t(k+1)$, which is known and given by a controlled scheduler, e.g. by a feedback scheduler as described in section 1.4.2.3 or a (m,k)-firm policy as used in section 5.2.5.

However, computing A_d and B_d involve matrix exponentials of the original A and B matrices and thus are not affine on h. To get a polytopic model and then apply an LPV design, the exponential is approximated by a Taylor series of order N. Since h is assumed to belong to the interval $[h_{\min},\ h_{\max}]$ with $h_{\min} > 0$, the sampling period is approximated around the nominal value h_0 of the sampling period, as

$$h = h_0 + \delta \qquad \text{with} \qquad h_{\min} - h_0 \leq \delta \leq h_{\max} - h_0. \tag{2.38}$$

It can be written that

$$\begin{pmatrix} A_d(h) & B_d(h) \\ 0 & I \end{pmatrix} = \begin{pmatrix} A_{h_0} & B_{h_0} \\ 0 & I \end{pmatrix} \begin{pmatrix} A_\delta & B_\delta \\ 0 & I \end{pmatrix}, \tag{2.39}$$

where

$$\left(\begin{array}{cc} A_{h_0} & B_{h_0} \\ 0 & I \end{array} \right) := \exp\left(\left(\begin{array}{cc} A & B \\ 0 & 0 \end{array} \right) h_0 \right) \tag{2.40}$$

and

$$\left(\begin{array}{cc} A_\delta & B_\delta \\ 0 & I \end{array} \right) := \exp\left(\left(\begin{array}{cc} A & B \\ 0 & 0 \end{array} \right) \delta \right). \tag{2.41}$$

In order to get a polytopic model, a Taylor series of order N is used to approximate the matrix exponential in 2.41, and allows us to get

$$A_\delta \approx I + \sum_{i=1}^{N} \frac{A^i}{i!} \delta^i \quad \text{and} \quad B_\delta \approx \sum_{i=1}^{N} \frac{A^{i-1}B}{i!} \delta^i. \tag{2.42}$$

This leads to

$$A_d(h) = A_{h_0} A_\delta \quad \text{and} \quad B_d(h) = B_{h_0} + A_{h_0} B_\delta \tag{2.43}$$

Let us define $H = [\delta, \delta^2, \ldots, \delta^N]$ the vector of parameters that belong to a convex polytope (hyper-polygon) $\mathcal{H}$ with 2^N vertices.

$$\mathcal{H} = \left\{ \sum_{i=1}^{2^N} \alpha_i(\delta)\omega_i : \alpha_i(\delta) \geq 0, \sum_{i=1}^{2^N} \alpha_i(\delta) = 1 \right\}$$

$$\{\delta, \delta^2, \ldots, \delta^N\}, \quad \delta^i \in \{\delta^i_{\min}, \delta^i_{\max}\}.$$

Each vertex is defined by a vector $\omega_i = [\nu_{i_1}, \nu_{i_2}, \ldots, \nu_{i_N}]$, where ν_{i_j} can take the extrema values $\{\delta^j_{\min}, \delta^j_{\max}\}$ with $\delta_{\min} = h_{\min} - h_0$ and $\delta_{\max} = h_{\max} - h_0$.

The matrices $A_d(\delta)$ and $B_d(\delta)$ are therefore affine in H and given by the polytopic forms

$$A_d(H) = \sum_{i=1}^{2^N} \alpha_i(\delta)A_{d_i}, \quad B_d(H) = \sum_{i=1}^{2^N} \alpha_i(\delta)B_{d_i},$$

where the matrices at the vertices, i.e. A_{d_i} and B_{d_i}, are obtained by the calculation of $A_d(\delta)$ and $B_d(\delta)$ at each vertex of the polytope $\mathcal{H}$. The polytopic coordinates α_i which represent the position of a particular parameter vector $H(\delta)$ in the polytope $\mathcal{H}$ are given solving

$$H(\delta) = \sum_{i=1}^{2^N} \alpha_i(\delta)\omega_i , \quad \alpha_i(\delta) \geq 0 , \quad \sum_{i=1}^{2^N} \alpha_i(\delta) = 1. \tag{2.44}$$

As the gain-scheduled controller will be a convex combination of 2^N "vertex" controllers, the choice of the series order N gives a trade-off between the approximation accuracy and the controller complexity.

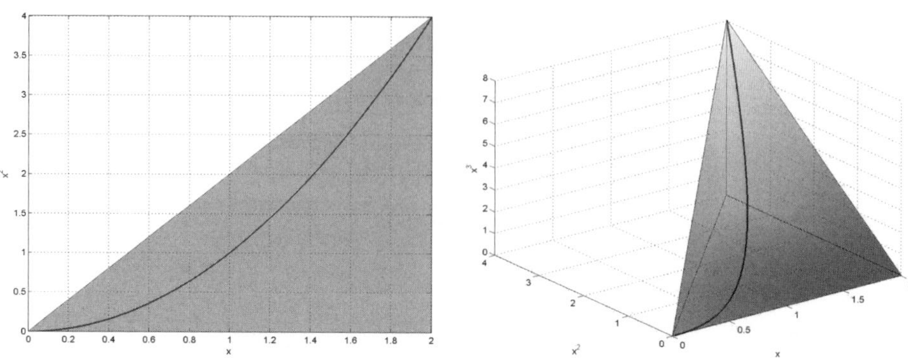

Figure 2.10. *Polytope reduction for N = 2 and 3*

To decrease the volume and number of vertices of the matrice polytope, the dependency between the successive powers of the parameter h is exploited. Remember that the vertices ω_i of $\mathcal{H}$ are defined by $h, h^2, \ldots, h^N$, with $h^i \in \{h^i_{\min}, h^i_{\max}\}$. Indeed, the representative point of the parameters set is constrained to be on a one-dimensional curve, so that the polytope of interest can be reduced to the "lower" $N + 1$ vertices, as illustrated in Figure 2.10 for the cases $N = 2$ and 3.

2.4.2. *Performance specification*

In the H_∞ framework, the general control configuration of Figure 2.11 is considered, where W_i and W_o are weighting functions specifying closed-loop performances (see [SKO 96]). The objective here is to find a controller K so that internal stability is achieved and $\|\tilde{z}\|_2 < \gamma \|\tilde{w}\|_2$, where γ represents the H_∞ attenuation level.

Classic control design assumes constant performance objectives and produces a controller with a unique sampling period. This sampling period is chosen according to the controller bandwidth, the noise sensitivity and the availability of computation resources. When the sampling period varies, the usable controller bandwidth also varies

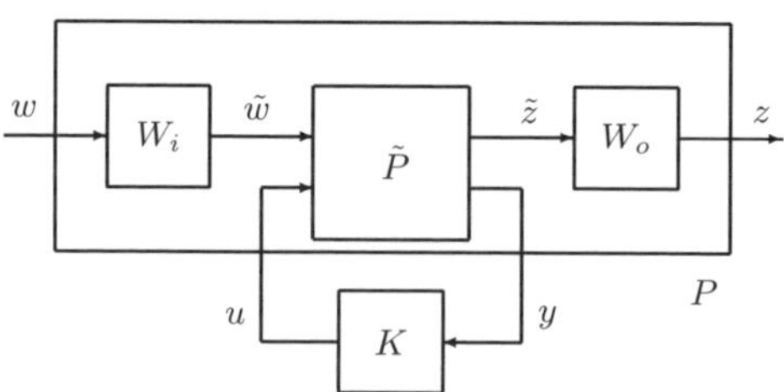

Figure 2.11. *Focused interconnection*

and the closed-loop objectives should logically be adapted; therefore, the bandwidth of the weighting functions is adapted. In this aim, W_i and W_o are split into two parts:

– a constant part with constant poles and zeros. This allows, for instance, to compensate for oscillations or flexible modes which are, by definition, independent of the sampling period. This part is merged with the plant before its discretization.

– the variable part contains poles and zeros whose pulsations are expressed as an affine function of the frequency $f = 1/h$, which allow the bandwidth of the weighting functions to be adapted. These poles and zeros are here constrained to be *real* by the discretization step. Finally, opportune cancelations make the *discretized* templates independent of h, facilitating further interconnections.

Indeed, preliminary experiences with varying sampling control [ROB 05] pointed out the advantages of performance adaptation w.r.t. to the sampling rate to preserve stability margins and to keep the control size inside wise bounds.

2.4.3. *LPV/H_∞ control design*

The interconnection between the discrete-time polytopic model of the plant $\tilde{P}$ (now including the constant part of the weighting functions) and the variable weighting functions W_i and W_o leads to the discrete-time LPV-augmented plant $P(H)$ is depicted in Figure 2.11.

The H_∞ control design for linear parameter-varying systems detailed in [APK 95] is used here. The method states that under some mild conditions, there exists a gain-scheduled controller:

$$\begin{cases} x_{K_{k+1}} & = & A_K(H)x_{K_k} + B_K(H)y_k \\ u_k & = & C_K(H)x_{K_k} + D_K(H)y_k, \end{cases} \tag{2.45}$$

where $x_K \in \mathbb{R}^n$, ensuring overall parameter trajectories, for the closed-loop system:

– closed-loop quadratic stability

– $\mathscr{L}_2$-induced norm of the operator mapping w into z bounded by γ, i.e. $\|z\|_2 < \gamma\|w\|_2$.

$N + 1$ controllers are reconstructed at each vertex of the parameter polytope (corresponding with the extreme values of the parameters). The gain-scheduled controller $K(H)$ is then the convex combination of these controllers

$$K(H) : \begin{pmatrix} A_K(H) & B_K(H) \\ C_K(H) & D_K(H) \end{pmatrix} = \sum_{i=1}^r \alpha_i(h) \begin{pmatrix} A_{K_i} & B_{K_i} \\ C_{K_i} & D_{K_i} \end{pmatrix}$$

with $\alpha_i(h)$ so that $H = \sum_{i=1}^r \alpha_i(h)\omega_i$. Note that on-line scheduling of the controller needs the computation of $\alpha_i(h)$ knowing h. Considering a Taylor's expansion around

h_0 with

$$\delta_{\min} = h_{\min} - h_0 \text{ and } \delta_{\max} = h_{\max} - h_0,$$

and the case of the reduced polytope, explicit solutions are easily recursively computed using

$$\begin{cases} \alpha_1 = \frac{\delta_{\max} - \delta}{\delta_{\max} - \delta_{\min}} \\ \alpha_n = \frac{\delta_{\max}^n - \delta^n}{\delta_{\max}^n - \delta_{\min}^n} - \sum_1^{n-1} \alpha_i \quad, \quad n = [2, ..., N] \\ \alpha_{N+1} = 1 - \sum_1^N \alpha_i. \end{cases} \qquad (2.46)$$

2.4.4. *Experimental assessment*

The latter approach has been experimentally assessed using a "*T*"-inverted pendulum, as extensively described in [ROB 07b].

As such a T-pendulum system is difficult to be controlled, our main objective here is to get a closed-loop stable system, to emphasis the practical feasibility of the proposed methodology for real-time control.

The sampling interval is assumed to be in the interval [1,3] ms. Note that the sampling rate seems to be very fast compared with the closed-loop-desired performance: it appears from previous studies and experiments that such fast sampling is necessary to achieve closed-loop stability for this nonlinear device, whatever the control algorithm [NAT 04].

After some trials and comparisons [ROB 09], the control synthesis has been implemented using the reduced polytope model and a Taylor's expansion truncated at the order 2. Hence, three vertex controllers must be combined for every new value of the control interval.

The plant is controlled through Matlab/Simulink using the Real-time Workshop and xPC Target. Two cases are displayed. First, in Figure, 2.12 the sampling period variation is continuous and follows a sinusoidal signal of frequency 0.15 rad s^{-1}. The left plot represents simulation results and the right one a real experiment. As the control interval varies continuously, the controller is adapted at each sample. Therefore, the polytopic coordinates computation 2.46 and convex combination 2.4.3 must be computed previously to the control signal calculation using the state feedback controller 2.45. Anyway, the overall on-line computation remains bounded and simple enough to be easily implemented in real time.

The position reference for the pendulum is a square pattern. Note that the settling time varies with the sampling frequency, accordingly to the definition of the variable weighting functions used for the performance specification. On the right part,

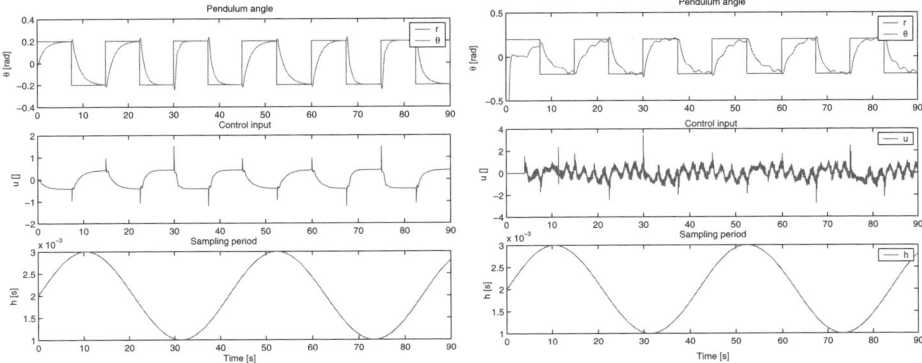

Figure 2.12. *Motions of the T pendulum under a sinusoidal sampling period*

the experimental data exhibit additional spikes and noise which are a consequence of dry friction and elasticity (or stick-slip) in the pendulum actuation. However, the experimental plant is still stable and the control signals remain bounded, despite these unmodeled mechanical defects whose effect is known to be difficult to compensate for [OLS 98].

Then, in Figure 2.13, step changes of the sampling rate are experimented. Note that now the controller's gains computation only needs to be performed at the control interval switches, i.e. in this case, the on-line overhead of the method compared with a classical H_∞ controller becomes negligible.

As expected from the sampling-dependent performance objectives, the settling time is minimal when the sampling period is maximal, and conversely. There are no abrupt changes in the control signal, even when the sampling period suddenly varies

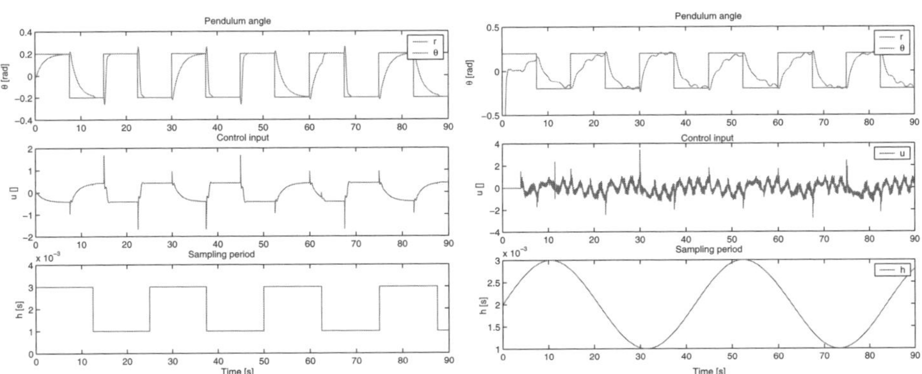

Figure 2.13. *Motions of the T pendulum under a square sampling period*

from 1 to 3 ms. Finally, similar results are obtained in simulation and experimental tests, which illustrates the inherent robustness property of the H_∞ design.

Indeed, few assumptions about sampling have been made for this control design. The main point is that the control interval is known and lies between the predefined bounds $[h_{\min}; h_{\max}]$, whatever the origin of the control interval variations, its speed and its frequency. Two cases may be considered:

– the control interval is a control variable which can be used by a feedback scheduler to manage the CPU load share, e.g. as in section 1.4: in that case the desired control interval is computed by a scheduling controller and sent to the real-time scheduler which manages the control tasks execution;

– the control interval variations may be due to sensor scheduling, e.g. induced by a communication channel between the sensor and controller nodes: in that case the interval between the successive expected appearance of data at the controller input are delivered by the scheduling policy controller, e.g. a (m, k)-firm scheduler as defined in section 1.2.2.

Some further simulations, still using the same case study with the "T"-inverted pendulum and associated LPV/H_∞-based controller, have been made to illustrate the capabilities and robustness of the method. In the simulation, depicted in Figure 2.14, the control interval has been randomly varied at every sample, with values in the set $\{1, 2, 3\}$ ms. Indeed, this case mimics data dropping between the sensor and control sites, where the network interface is fed by the sensor at a 1 ms rate, and randomly drops up to two packets out of three with a maximum interval of 3 ms between successful transmission. This transmission pattern may be the result of a feedback scheduler using a (1,3)-firm data-dropping policy to manage the network bandwidth.

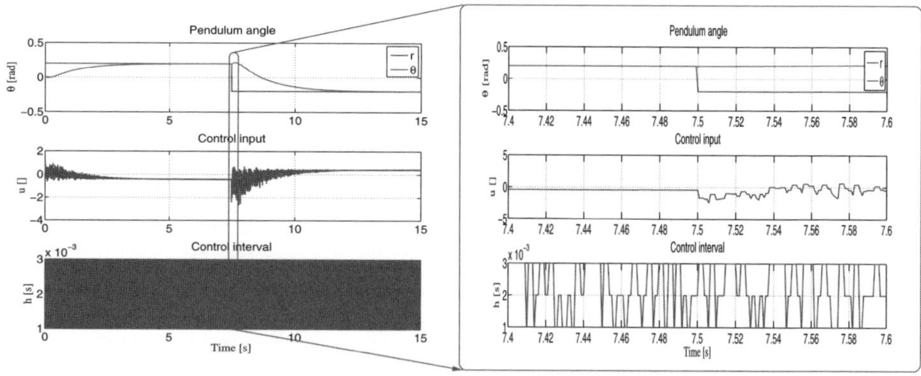

Figure 2.14. *LPV with (1-3)-firm input scheduling*

Despite the fast and random changes of the control interval, the system remains stable. The settling time conforms with the performance specification, its actual value lies between the performance templates defined with the variable weighting functions for $h_{\min}$ and $h_{\max}$. The control size remains bounded and reasonably small even with the appearance of chattering. Similar simulation results have been obtained when the control interval varies at every sample with random float values in the specified interval.

This LPV/H_∞-based variable sampling control synthesis assumes that the variable parameter, i.e. the control interval in this particular case, is known before the synthesis is performed. Indeed, this is true for the desired interval sent to the operating system by a feedback scheduler, or for the scheduled packets incoming interval on the control node, if jitter is small enough and negligible. However, during control processing and communications between nodes, delays appear due to both the control computation on a real CPU and to pre-emption due to higher priority tasks sharing the control node, and to networking between the sensor, control and actuator's node if any. These added delays are not known in real time when the controller's gains computation starts, as they cannot be accounted in the polytopic parameters computation and in the elementary vertex controllers convex combination.

A possible way to cope with these unmodeled and unmeasured delays is to compute at each controller execution a discrete set of control signals (corresponding with the range of expected latencies), send all the set to the actuator's node, and apply the control signal which best fit the actual measured interval at the actuator node. This is for example the approach of [SAL 09]. A drawback of such an approach is overloading the CPU and network with unused control signals computation and transmission.

In a safely designed and schedulable real-time control system running in nominal conditions, the computing and pre-emption induced latencies are typically smaller than a control period. It is expected that the effect of these quite small latencies can be absorbed by the robustness of the on-board controller. Indeed, the fact that the control experiments depicted in Figures 2.12 and 2.13 are successful, despite nothing was done to model, identify and compensate for the computation and scheduling latencies, indicates that the proposed control design provides such capabilities.

The simulation plot in Figure 2.15 further explores the already used case study robustness w.r.t. unmodeled control delays. The control interval is constant and set to 2 ms, and latencies of increasing values up to 12 ms are added in the control path to simulate unmodeled compound computation and scheduling latencies. It can be observed that the control stability is kept despite quite large added latencies (unstability appears beyond 15 ms delays), while the control quality decreases with the appearance of a classical oscillatory behavior.

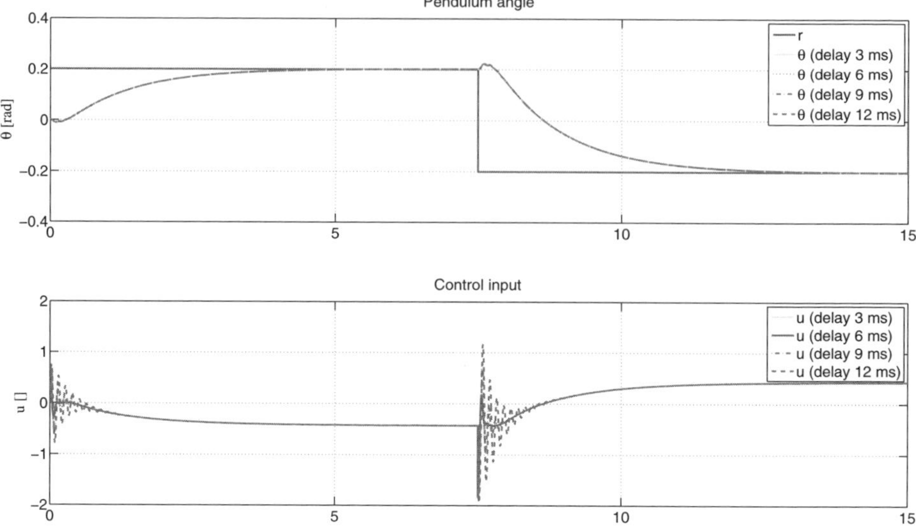

Figure 2.15. *LPV with unmodeled output delay*

Beyond this case study, the robustness of this LPV/H_∞ variable sampling control method deserves to be further formally explored, for example using the results exposed in section 2.2. Anyway, this design method already appears to be effective to preserve the plant's stability and performance objectives during arbitrarily fast control interval variations. Therefore, the method can further be used to cope with varying computing and networking resources availability, for example, as sketched in the state-based feedback scheduler described in section 4.2.

2.5. Summary

Networked control systems are characterized by a complex interaction between heterogenous components and uncertain behaviors. Compared with classical centralized systems, this complex interaction induces disturbances such as latencies due to communication links and protocols, computing durations and pre-emption between concurrent activities. The complexity and combined incertitudes of the components involved in an NCS make untractable an accurate modeling and precise prediction of the timing patterns in control loops. Robust control techniques, which rely on modeling incertitudes bounds rather than on perfect knowledge, are an effective answer to the control of uncertain systems. To this end, section 2.2 provides a summary of results on the control of time-delay systems. Beyond the basic models and theorems of the domain, recent results dealing with time-varying delays are of prime interest for handling network induced delays in control systems.

Section 2.3 deals with a particular model of uncertainty and timing weakly hard constraints. The method works on a slotted timescale and assumes that the control tasks are accelerable, i.e. more executions are performed, better is the control performance. A method to design accelerable controllers for LTI systems is provided, estimation and robustness problems are also addressed. Combining such accelerable controllers with (m, k)-firm policies allows for a smart degradation of the control performance in case of transient overloads of the execution platform.

With the robustness approach control tasks are designed to be tolerant to the timing disturbances. A more active approach is used in section 2.4 to make controllers adaptive to varying known control intervals. An LPV gain scheduling methodology is used, where the variable parameter is the control interval. It is associated with a H_∞ design to guarantee both stability and requested performance over all the predefined range of intervals, and provides real-time compliant controllers.

These control designs, which are either robust or adaptive w.r.t. timing deviations, can be further used as building blocks in integrated controllers combining plant control and execution resources constraints, as described later in Chapters 4 and 5.

Note that, while these designs are mainly suited for linear time invariant plants, approaches based on model predictive control, and able to deal with plants nonlinearity, are briefly described in section 4.3.

2.6. Bibliography

[ANT 07] ANTHONIS J., SEURET A., RICHARD J.-P., AND RAMON H., Design of a pressure control system with dead band and time delay, *IEEE Transactions on Control Systems Technology*, vol. 15, p. 1103–1111, 2007.

[APK 95] APKARIAN P., GAHINET P., AND BECKER G., Self-scheduled H_∞ control of linear parameter-varying systems: a design example, *Automatica*, vol. 31, p. 1251–1262, 1995.

[ARI 07] ARIBA Y., AND GOUAISBAUT F., Robust stability of time-delay systems with interval delays, *46th IEEE Conference on Decision and Control*, New Orleans, USA, December 2007.

[ÅST 97] ÅSTRÖM K. J., AND WITTENMARK B., *Computer-Controlled Systems*, Information and System Sciences Series, Prentice Hall, Englewood Cliffs, NJ, 3rd edition, 1997.

[BEN 08] BEN GAID M.-M., SIMON D., AND SENAME O., A sesign methodology for weakly-hard real-time control, *IFAC World Congress*, Seoul, Korea, July 2008.

[BER 01] BERNAT G., BURNS A., AND LLAMOSÍ A., Weakly hard real-time systems, *IEEE Transactions on Computers*, vol. 50, p. 308–321, 2001.

[BOY 94] BOYD S., ELGHAOUI L., AND FERON E., *Linear Matrix Inequalities in System and Control Theory*, SIAM, Philadelphia, 1994

[BRI 08] BRIAT C., Robust control and observation of LPV time-delay systems, PhD thesis, Grenoble-INP, France, 2008.

[CER 03] CERVIN A., Integrated control and real-time scheduling, PhD thesis, Department of Automatic Control, Lund Institute of Technology, Sweden, April 2003.

[CHE 88] CHEN J., ARMSTRONG B., FEARING R., AND BURDICK J., Satyr and the Nymph: software archetype for real time robotics, *IEEE-ACM Joint Computer Conference*, Dallas, USA, November 1988.

[DAM 94] DAMBRINE M., Contribution à l'étude de la stabilité des systèmes à retards, PhD thesis, Lille University of Science and Technology, France, 1994.

[DEV 05] DEVERGE J., AND PUAUT I., Safe measurement-based WCET estimation, *5th International Workshop on worst-case execution time analysis*, Palma de Mallorca, Spain, July 2005.

[DRI 77] DRIVER D., *Ordinary and Delay Differential Equations*, Springer-Verlag, New York, 1977.

[FRI 02] FRIDMAN E., AND SHAKED U., A descriptor system approach to H_∞ control of linear time-delay systems, *IEEE Transactions on Automatic Control*, vol. 47, p. 253–270, 2002.

[FRI 04a] FRIDMAN E., Stability of linear functional differential equations: a new Lyapunov technique, *Proceedings of Mathematical Theory of Networks and Systems*, September 2004.

[FRI 04b] FRIDMAN E., SEURET A., AND RICHARD J.-P., Robust sampled-data stabilization of linear systems: an input delay approach, *Automatica*, vol. 40, p. 1141–1446, 2004.

[FRI 08a] FRIDMAN E., AND NICULESCU S.-I., On complete Lyapunov–Krasovskii functional techniques for uncertain systems with fast-varying delays, *International Journal of Robust and Nonlinear Control*, vol. 8, p. 364–374, 2008.

[FRI 08b] FRIDMAN E., YEGANEFAR N., AND DAMBRINE M., On input-to-state stability of systems with time-delay: a matrix inequalities approach, *Automatica*, vol. 44, p. 2364–2369, 2008.

[GHA 00] GHAOUI L. E., AND NICULESCU S.-I., *Advances in Linear Matrix Inequality Methods in Control*, Advances in Design and Control Series, SIAM, Philadelphia, 2000.

[GU 97] GU K., Discretized LMI set in the stability problem of linear uncertain time-delay systems, *International Journal of Control*, vol. 68, p. 923–934, 1997.

[GU 03] GU K., KHARITONOV V.-L., AND CHEN J., *Stability of Time-delay Systems*, Birkhauser, Basle, 2003.

[HAL 97] HALE J., *Theory of Functional Differential Equations*, Springer-Verlag, New York, 1997.

[HES 07] HESPANHA J., NAGHSHTABRIZI P., AND XU Y., A survey of recent results in networked control systems, *Proceedings of the IEEE*, vol. 95, p. 138–162, 2007.

[JIA 05] JIANG X., AND HAN Q.-L., On H_∞ control for linear systems with interval time-varying delay, *Automatica*, vol. 41, p. 2099–2106, 2005.

[KAO 07] KAO C. Y., AND RANTZER A., Stability analysis of systems with uncertain time-varying delays, *Automatica*, vol. 43, p. 959–970, 2007.

[KHA 01] KHARITONOV V., AND ZHABKO A., Lyapunov--Krasovskii approach to robust stability of time delay systems, *1st IFAC/IEEE symposium on system structure and control*, Prague, Czech Republic, August 2001.

[KOL 96] KOLMANOVSKII V., AND SHAÏKHET L., *Control of Systems with Aftereffect*, American Mathematical Society, Providence, RI, 1996.

[KOL 99a] KOLMANOVSKII V., AND MYSHKIS A., *Applied Theory of Functional Differential Equations*, Kluwer, Dordrecht, 1999.

[KOL 99b] KOLMANOVSKII V., NICULESCU S.-I., AND RICHARD J.-P., On the Lyapunov–Krasovskii functionals for stability analysis of linear delay systems, *International Journal of Control*, vol. 72, p. 374–384, 1999.

[KOL 03] KOLMANOVSKII V., AND ZHABKO A., Lyapunov–Krasovskii approach to the robust stability analysis of time-delay systems, *Automatica*, vol. 39, p. 15–20, 2003.

[KOR 95] KOREN G., AND SHASHA D., Skip-over: algorithms and complexity for overloaded systems that allow skips, *16th IEEE Real-Time Systems Symposium*, Pisa, Italy, December 1995.

[LEI 00] LEITH D. J., AND LEITHEAD W. E., Survey of gain-scheduling analysis and design, *International Journal of Control*, vol. 73, p. 1001–1025, 2000.

[LI 97] LI X., AND DE SOUZA C.-E., Criteria for robust stability and stabilization of uncertain linear systems with state delay, *Automatica*, vol. 33, p. 1657–1662, 1997.

[LOP 06] LOPEZ I., PIOVESAN J., ABDALLAH C., LEE D., PALAFOX O. M., SPONG M., AND SANDOVAL R., Pratical issues in networked control systems, *American Control Conference*, Minneapolis, USA, June 2006.

[MIC 04] MICHIELS W., NICULESCU S.-I., MOREAU L., Using delays and time-varying gains to improve the static output feedback stabilizability of linear systems: a comparison, *IMA Journal of Mathematical Control and Information*, vol. 21, p. 393–418, 2004.

[MON 05] MONDIE S., KHARITONOV V., SANTOS O., Complete Lyapunov-Krasovskii functional with a given cross term in the time derivative, 44th *IEEE Conference on Decision and Control*, Sevilla, Spain, December 2005.

[MOU 06] MOULAY E., DAMBRINE M., PERRUQUETTI W., YEGANEFAR N., Une première approche de la stabilité et de la stabilisation en temps fini des systèmes à retard, *CIFA'06*, Bordeaux, France, May 2006.

[NAG 08] NAGHSHTABRIZI P., HESPANHA J., TEEL A., Exponential stability of impulsive systems with application to uncertain sampled-data systems, *Systems and Control Letters*, vol. 57, p. 378–385, 2008.

[NAT 04] NATALE O., SENAME O., CANUDAS DE WIT C., Inverted pendulum stabilization through the ethernet network, performance analysis, *American Control Conference ACC'04*, Boston, USA, June 2004.

[NIC 01] NICULESCU S.-I., *Delay Effects on Stability. A Robust Control Approach*, Springer-Verlag, New York, 2001.

[OLS 98] OLSSON H., ASTRÖM K.-J., DE WIT C. C., GÄFVERT M., AND LISCHINSKY P., Friction models and friction compensation, *European Journal of Control*, vol. 4, p. 176–195, 1998.

[PAP 07] PAPACHRISTODOULOU A., PEET M. M., AND NICULESCU S.-I., Stability analysis of linear systems with time-varying delays: delay uncertainty and quenching, *46th IEEE Conference on Decision and Control*, New Orleans, USA, December 2007.

[RAM 95] RAMANATHAN P., AND HAMDAOUI M., A dynamic priority assignment technique for streams with (m, k)-firm deadlines, *IEEE Transactions on Computers*, vol. 44, p. 1443–1451, 1995.

[RAM 99] RAMANATHAN P., Overload management in real-time control applications using (m, k)-firm guarantee, *IEEE Transactions on Parallel and Distributed Systems*, vol. 10, p. 549–559, June 1999.

[RIC 03] RICHARD J.-P., Time delay systems: an overview of some recent advances and open problems, *Automatica*, vol. 39, p. 1667–1694, 2003.

[ROB 05] ROBERT D., SENAME O., AND SIMON D., Sampling period dependent RST controller used in control/scheduling co-design, *16th IFAC World Conference*, Prague, Czech Republic, July 2005.

[ROB 07a] ROBERT D., SENAME O., AND SIMON D., A reduced polytopic LPV synthesis for a sampling varying controller: experimentation with a T inverted pendulum, *European Control Conference ECC'07*, Kos, Greece, July 2007.

[ROB 07b] ROBERT D., Contribution à l'interaction commande/ordonnancement, PhD thesis, INP Grenoble, January 2007.

[ROB 09] ROBERT D., SENAME O., AND SIMON D., An H_∞ LPV design for sampling varying controllers: experimentation with a T inverted pendulum, *IEEE Transactions on Control Systems Technology*, in DOI10.1109/TCST.2009.2026179, 2009.

[SAL 05] SALA A., Computer control under time-varying sampling period: an LMI gridding approach, *Automatica*, vol. 41, p. 2077–2082, 2005.

[SAL 09] SALA A., CUENCA A., AND SALT J., A retunable PID multi-rate controller for a networked control system, *Information Sciences*, vol. 179, p. 2390–2402, 2009.

[SCH 02] SCHINKEL M., CHEN W.-H., AND RANTZER A., Optimal control for systems with varying sampling rate, *American Control Conference ACC'02*, Anchorage, USA, May 2002.

[SEU 09a] SEURET A., Lyapunov–Krasovskii functionals parameterized with polynomials, 6th *IFAC Symposium on Robust Control Design - ROCOND'09*, Haifa, Israel, June 2009.

[SEU 09b] SEURET A., Stabilization of time-delay systems through linear differential equations using a descriptor representation, *European Control Conference - ECC'09*, Budapest, Hungary, August 2009.

[SEU 09c] SEURET A., EDWARDS C., SPURGEON S., AND FRIDMAN E., Static output feedback sliding mode control design via an artificial stabilizing delay, *IEEE Transactions on Automatic Control*, vol. 54, p. 256–265, 2009.

[SIM 98] SIMON D., CASTILLO E., AND FREEDMAN P., Design and analysis of synchronization for real-time closed-loop control in robotics, *IEEE Transactions on Control Systems Technology*, vol. 6, p. 445–461, July 1998.

[SIN 04] SINOPOLI B., SCHENATO L., FRANCESCHETTI M., POOLLA K., JORDAN M. I., AND SASTRY S. S., Kalman filtering with intermittent observations, *IEEE Transactions on Automatic Control*, vol. 49, p. 1453–1464, 2004.

[SKO 96] SKOGESTAD S., AND POSTLETHWAITE I., *Multivariable Feedback Control: Analysis and Design*, Wiley, New York, 1996.

[TÖR 98] TÖRNGREN M., Fundamentals of implementing real-time control applications in distributed computer systems, *Real Time Systems*, vol. 14, p. 219–250, 1998.

[WIT 01] WITTENMARK B., Sample-induced delays in synchronous multirate systems, *European Control Conference*, Porto, Portugal, p. 3276–3281, September 2001.

[ZAM 08] ZAMPIERI S., A survey of recent results in networked control systems, *Proc. 17th IFAC World Congress*, Seoul, Korea, July 2008.

Chapter 3

QoC-aware Dynamic Network QoS Adaptation

3.1. Overview

When engineers are designing complex networked control systems (NCS), the main efforts are generally put towards dealing with the negative influences arising from the network and its interactions with the global system performance. Theories and methods are hence developed to adapt the control to network-induced delays, packet losses, jitter or even asynchronous sampling. These methods encompass the estimation or observation of the quality of service (QoS), and it is assumed that these parameters are non-controllable. However, it might be possible, in a particular situation, to improve the performances offered by the network rather than modifying the control parameters, and even degrading the global system performance. In this situation, a set of techniques able to adjust the QoS offered by a network are proposed in order to enhance the QoC. They aim at providing a certain level of performance to a network data flow, while achieving an efficient and balanced utilization of network resources as defined by [ZAM 08]. The field of the QoS control includes applications related to the call admission method, scheduling policy, routing protocol, flow control strategies, and various other resource allocation problems. In the NCS framework, the hot issue is to adapt the network according to the evolution of the QoC parameter and not just the network's behavior. This means that application constraints coming from one or even more distributed control systems should be taken into account and that a relation

Chapter written by Christophe AUBRUN, Belynda BRAHIMI, Jean-Philippe GEORGES, Guy JUANOLE, Gérard MOUNEY, Xuan Hung NGUYEN and Eric RONDEAU.

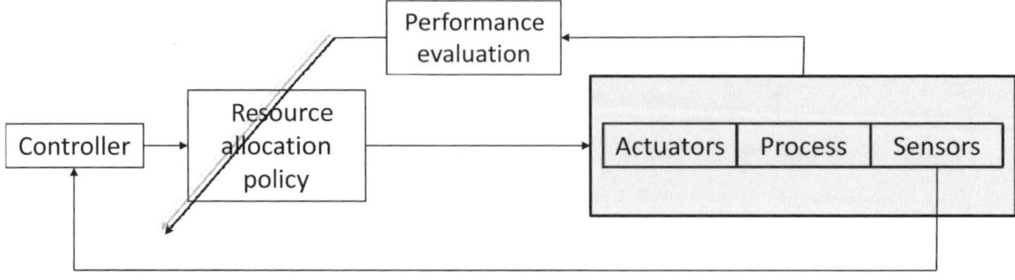

Figure 3.1. *General scheme for dynamic network QoS adaptation*

between QoC and QoS needs to be determined. For instance, QoC might be formulated in terms of overshoot or damping, whereas QoS is often expressed in terms of delays. The principle of the control of network strategy in the field of NCS is hence shown in Figure 3.1.

In Figure 3.1, the *performances evaluation* block enables the system to act on the network resources in order to adjust the QoS according to the application needs. It is responsible for identifying the influence arising from the network and evaluating the QoS improvements required. It is important to note here that a single network might be shared by different applications or different distributed control systems. Each of these applications operates with respect to different constraints. Different methods such as mean square error, stability analysis, etc. might be used here to express the QoC.

The *resource allocation policy* block in Figure 3.1 executes QoS adaptation. It is important to note here that the choice of QoS adaptation policy for a network depends on the protocols and standards defined by this network. Thus, the adaptation method dedicated to a given protocol will not necessarily give correct results with another network. Also, this chapter presents two network control strategies, each one related to a specific network.

In section 3.2, the CAN bus which is one of the most widely used protocols for industrial communication is considered. The CAN network was developed by the Bosch company for multiplexing issues in vehicles [BOS 91]. The dynamic network QoS adaptation proposed in section 3.2 for CAN consists of a dynamic message priority allocation mechanism based on control application needs. In this study, QoC is evaluated in terms of overshoot and phase margin, and control performance is associated with the mean square error. The adaptation mechanism is then related to the CAN hierarchical medium access method. Indeed, in CAN, a frame is labeled by an identifier which is used to resolve the bus contention and which hence determines the frame priority. Initially, the priority allocation is static which does not allow for taking into account the dynamics of the application. Section 3.2 proposes a hybrid priority scheme and an on-line priority allocation method.

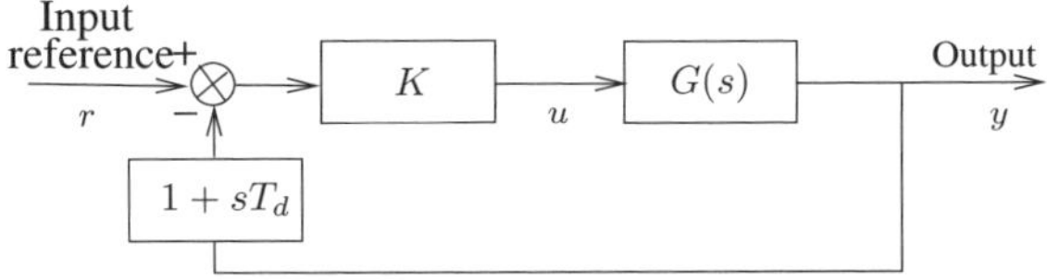

Figure 3.2. *Model*

Section 3.3 focuses on switched Ethernet architectures [IEE 02] which compared to the CAN bus were not initially defined for constrained communication, but are nevertheless used more and more to support real-time traffic. Here, the CSMA/CD medium access method does not include as with CAN, a hierarchical arbitration mechanism, which means that another QoS adaptation method is required. In switched Ethernet architectures, a simple FIFO scheduling policy is used to select the frames for output forwarding. By using Classification of Service (CoS), it is possible to replace the FIFO policy with a more sophisticated Weighted Round Robin (WRR). The approach proposed in section 3.3 consists of implementing an adaptive configuration of the scheduling policy parameters. By adjusting these parameters, it is possible to control the bandwidth offered to the different flows. The goal here is to provide a sufficient bandwidth according to the worst QoS level acceptable for global system control.

3.2. Dynamic CAN message priority allocation according to the control application needs

3.2.1. *Context of the study*

3.2.1.1. *The considered process control application*

The closed loop application is presented in Figure 3.2 by using the concept of continuous time transfer function based on the Laplace transform [ÅST 97]. The process for controlling is a DC-servo process described by the transfer function

$$G(s) = \frac{1000}{s(1+s)}.$$

The controller is a proportional derivative (PD) controller which considers the output derivation [ÅST 97]. The PD algorithm has the following form:

$$U(s) = K(R(s) - (1 + sT_d)Y(s)),$$

where $U(s)$, $R(s)$, and $Y(s)$ denote the Laplace transforms of the command signal u, input reference r, and output signal y. K is the proportional gain and T_d is the derivative time of the controller.

The closed loop transfer function $F(s)$ of this application is a second-order function

$$F(s) = \frac{\omega_n^2}{s^2 + 2\zeta\omega_n s + \omega_n^2}$$

characterized by the natural pulsation ω_n and the damping ζ ($\omega_n^2 = 1,000\,K$ and $2\zeta\omega_n = 1 + 1,000\,KT_d$). We want the following performances: overshoot $= 5\%$ and response time$= 100$ ms, which requires the following dynamic characteristics: $\zeta \simeq 0.7$ and $\omega_n \simeq 43$ rad s^{-1}; in these conditions, we have the rise time $t_r \approx 40$ ms. Then we need the following values for K and T_d: $K = 1.8$ rad s^{-1} and $T_d = 0.032$ s.

3.2.1.2. *Control performance evaluation*

In order to evaluate the quality of the control of the process control application we use the following cost function:

$$J = \int_0^T t.(r(t) - y(t))^2 dt.$$

The higher the cost function is, the worse is the control performance. We take 0.5 s for the value of T (at $T = 0.5$ s the transient behavior is finished and we are in the permanent behavior). This evaluator, applied to the application when it is implemented without the network, gives a cost J of 2.5385×10^{-4}.

3.2.1.3. *The implementation through a network*

3.2.1.3.1. Structure

We consider the implementation represented in Figure 3.3. The network operates both (i) between computer 1 (C1) in association with the numerical information provided by the AD conversion (this computer includes a task that we call the sensor task and which generates the sensor flow) and computer 2 (C2) where we have the reference and the controller (in C2 we have a task called controller task which generates

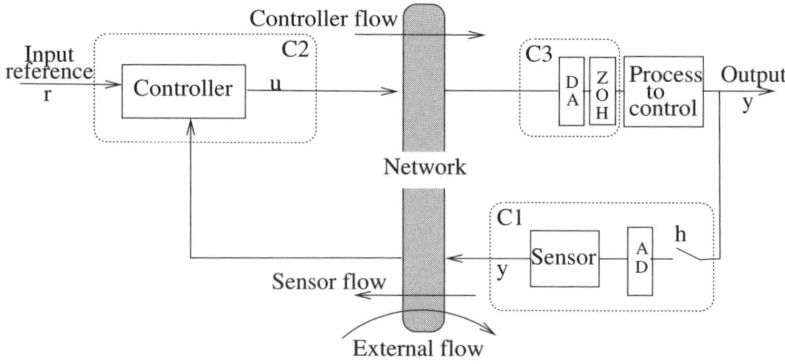

Figure 3.3. *Implementation through the network*

the controller flow), (ii) and between C2 and computer 3 (C3) which provides numerical information to the DA conversion in front of the zero-order hold (ZOH) which is connected to the actuator acting on the process to control.

The sensor flow which goes from the sensor to the controller will be noted as f_{sc}. The controller flow which goes from the controller to the actuator will be noted as f_{ca}. The task which generates the sensor flow is time-triggered (the sampling is based on a clock), whereas the task which generates the controller flow is event triggered (the controller waits for sensor sample reception before computing and generating its flow).

Generally, a network is not dedicated to only one application but shared between different applications. In order to make a general study of the process control application, when it is implemented through a network, we have to see, in particular, the influence of the flows of the other applications. It is why we have in Figure 3.3 what we call the external flow, noted as f_{ex}, which globally represents an abstraction of the flows of all the other applications. We also consider that this flow is periodic.

3.2.1.3.2. Choice of the sampling period

This choice is a basic action. The sampling period has, from the control point of view, an upper bound [JUR 58]. But from the network point of view a value that is too small gives a load that is too great. So, the choice results from a compromise. The relation $\frac{tr}{10} \leq h \leq \frac{tr}{4}$, which has been given in [ÅST 97], is generally used. We consider here the bound $\frac{tr}{4}$. As $tr \approx 40$ ms; we have $h = 10$ ms. The controller is discretized with this sampling period; the measured dynamic characteristics are an overshoot less than 5%, a rise time $t_r \simeq 34$ ms and a response time (at 5%) $res\_t \simeq 45$ ms. These characteristics will be our references to analyze the performances of the control application through the studied networks.

3.2.1.3.3. Considering the network CAN

As we want to emphasize message scheduling, we consider a CAN network limited to the MAC layer. The MAC layer determines the schedule for sending the frames of the controller, sensor flows, and external flows. For this study, we then have to specify the bit rate in the physical layer (we consider a bit rate of 125 Kbits s^{-1}) as well as the frame transmission rate requested by the MAC layer (that we call the *use request factor* (URF) and which represents the load imposed on the network by the applications).

By calling D_{sc}, D_{ca}, and D_{ex} the duration of the sensor flow frames, the controller flow frames, and the external flow frames, respectively, h the sampling period of the process control application (the period for the controller and sensor flow), and T_{ex} the period of the external flow, we have URF $= \frac{D_{sc}}{h} + \frac{D_{ca}}{h} + \frac{D_{ex}}{T_{ex}}$.

Concerning the numerical values, we consider that the frames of the controller flow and of the sensor flow have a length of 10 bytes, thus a duration of 640 µs. The frame of the external flow has a length of 16 bytes, thus a duration of 1,024 µs.

The component $\frac{D_{sc}}{h} + \frac{D_{ca}}{h}$ of the URF, which concerns the process control application and which represents the network capacity used by this application, has the value 12.8%. The use by the external frame of the network capacity will depend on its period T_{ex}. *It is this parameter that we will vary during our study in order to analyze the robustness of the scheduling of the process control application frames.* The frame scheduling in the MAC layer of CAN [BOS 91; CIA 02] is based on priorities (static priorities) which appear in the identifier field (ID field) of the frames. The scheduling is done by comparing the field ID bit by bit (we start from the most significant bit, MSB). In CAN, the bit 0 is a dominant bit, and the bit 1 is a recessive bit. The lower the numerical value of the CAN ID, the higher the priority. We consider here the standard length of 11 bits for the ID field.

3.2.1.4. Evaluation of the influence of the network on the behavior of the process control application

This work has been done by using the tool TrueTime [CER 03; OHL 07], a toolbox for simulating distributed real-time control systems.

3.2.1.4.1. Dedicated network CAN

This study shows the influence of the format of the serial frames on the process control application. The evaluator J, applied to the regulation application when it is implemented through the network CAN without the external flow (thus the network CAN is dedicated to the regulation application), gives a cost J of 2.565×10^{-4}. This value of J is called J_0. The value J_0 is very close to the value obtained without the network (see 3.2.1.2). *The value J_0 will be considered as the reference value to evaluate the performance of the regulation application, taking into account the influence of the external flow.*

In the condition of J_0, the time response to a unity input step has the following characteristics: overshoot $D\% = 5\%$ and damping $\zeta = 0.7$, response time about 45 ms. We can see that, with respect to the performance of the sampled process control application (section 3.2.1.3.2), the network has a very weak influence here. This is because the duration of each frame ($640\ \mu s$) is very small with respect to the sampling period (10 ms).

3.2.1.4.2. Shared network CAN: considering static priorities and showing their inadequacy

We got the following important results from previous studies [JUA 05]:

– The priority associated with the controller flow (P_{ca}) must be higher than the priority associated with the sensor flow (P_{sc}); this way, we get the best performances for the control application (the intuition is that the controller has to send its message as soon as it receives the messages of the sensor) .

URF(%)	J	$\frac{\Delta J}{J_0}$
30	2.773×10^{-4}	8.11%
80	3.281×10^{-4}	27.9%
90	3.915×10^{-4}	52.6%
99	7.370×10^{-4}	187.0%
100	$1,445 \times 10^{-3}$	463%

Table 3.1. J and $\frac{\Delta J}{J_0}$

– If the priority of the external flow (P_{ex}) is larger than P_{ca}, and if the use request factor of the external flow ($\mathrm{URF}_{\mathrm{ex}}$) becomes very high, the external flow will use (as it has the largest priority) the network (bus) more and more often and will prevent the flows of the process control application from using the bus. (Consequently, the process control application will have bad performances and thus cannot be implemented.)

Table 3.1 gives the results obtained by using fixed priorities with $P_{\mathrm{sc}} < P_{\mathrm{ca}} < P_{\mathrm{ex}}$. By considering increases of the global URF – first column – (from 30% to 100%) due to increase of $\mathrm{URF}_{\mathrm{ex}}$ (which increases from (30%–12.8%) to (100%–12.8%)), this table gives the evolution of the cost function J and the percentage of its variation with respect to J_0: $\frac{J-J_0}{J_0} = \frac{\Delta J}{J_0}$.

We have a degradation which increases with URF (when URF becomes too great we have a delay with a big jitter and making the performances so bad that we cannot implement the regulation application). The time response to a unity input step (when $\mathrm{URF} = 100\%$) is represented in Figure 3.4. We now have an overshoot of $D = 29\%$. The response time here is very long compared to the response time when there is no implementation through the network (100 ms), see section 3.2.1.1. These are the results which have prompted the work on the hybrid priorities.

3.2.1.5. *Idea of hybrid priority schemes: general considerations*

When we have static priorities, as we have seen in the previous example, when the loads are high, and if the flows of the process control application do not have the highest priority, we cannot get acceptable control performances. However, in general, it is not always possible to give the highest priority to a process control application: we can have more important applications and, furthermore, if we have at least two process control applications, one will obviously not have the highest priority.

The idea of hybrid priority results from this problem, as previously stated, and also from the following important observation: in the general case of a distributed system, we have a lot of applications which generate different classes of flows which have different needs in terms of transmission urgency (constant urgency or variable urgency (from weak to strong)). A class is a characterization *a priori*, and thus, specified off-line. A class is a set of flows.

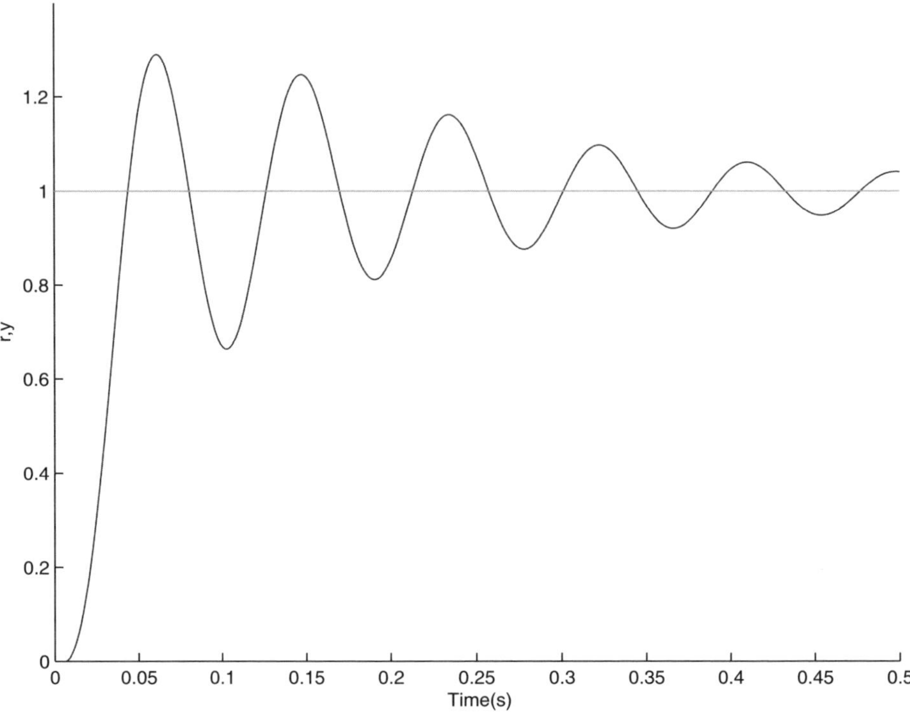

Figure 3.4. *Time response with* $URF = 100\%$ $((P_{\mathrm{ca}}, P_{\mathrm{sc}})< P_{\mathrm{ex}})$

The needs are an operational characteristic which depend on the behavior of the application concerned. The needs are specified off-line, if they are constant, and on-line if they are variable. In the latter case, we say they are "dynamic needs".

A process control application generates a class of two flows (controller flow and sensor flow) which have dynamic needs: strong urgency in a transient behavior after an input reference change (in order to follow the change) or after a disturbance (in order to make the regulation); small urgency in the permanent behavior.

3.2.1.5.1. The identifier (ID) field and the scheduling execution

The identifier field of a frame is divided into two levels (Figure 3.5): the first level represents the priority of a flow (it is a static priority specified off-line); the second level represents the priority of the transmission urgency (the urgency can be either constant or variable). The idea of structuring of the ID is present in the Mixed Traffic Scheduler [ZUB 97], [ZUB 00] which combines EDF (dynamic field) and FP (static field). In [WAL 01], the authors propose encoding the weighted absolute value of the

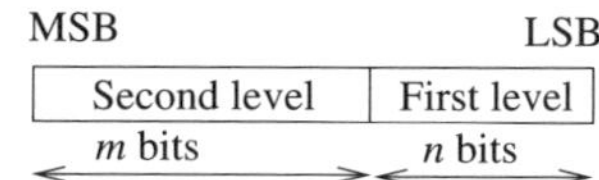

Figure 3.5. *Identifier field (hybrid priority)*

error in the dynamic field (this idea is also presented in [MAR 04]) and to resolve the collisions with the least significant bits (static field).

A constant transmission urgency is characterized by a static priority (one m bit combination) specified off-line. A variable transmission urgency is characterized by a dynamic priority (which can take, generally speaking, m bit combinations among a subset of the m bit combinations).

The frames of the flows f_{sc} and f_{ca} of a process control application have variable needs (strong urgency in a transient behavior after an input reference change (in order to follow the change quickly) or after a disturbance (in order to make the regulation quickly); weak urgency in a permanent behavior). That is why, in this study, we consider that the dynamic priority of the frames of the flows f_{sc} and f_{ca} of a process control application can take any m bit combination of the set of m bit combinations. The scheduling is executed by, first, comparing the second level (needs predominance), and, secondly, if the needs are identical, by comparing the first level (flow predominance).

3.2.1.5.2. Cohabitation of flows with constant needs and flows of process control applications (variable needs)

We have the objective of good performances for the process control applications in transient behavior. This means that the urgent needs of these flows can be satisfied very quickly. For that, we impose a maximum value to the flows with constant needs for the priority of these needs (concept of priority threshold (Pr_th) for the constant needs). In this way, a strong transmission urgency of a process control application flow (dynamic priority with a very high value i.e. higher than Pr_th) will be scheduled first.

3.2.1.5.3. Toward making dynamic priorities

The concept of the dynamic priorities requires specifying, at first, the characteristic of a process control application which gives information on the needs, and, secondly, how these needs can be translated into a dynamic priority (computation of a dynamic priority, instants of re-evaluation of a dynamic priority). We propose to express the needs with a signal which aptly characterizes the behavior of a process control application: it is the control signal u.

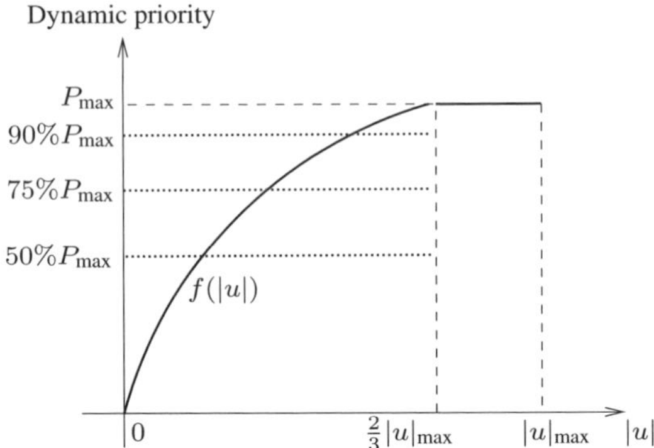

Figure 3.6. *The considered nonlinear function*

3.2.2. *Three hybrid priority schemes*

We have defined three schemes. The first is what we call the strict hybrid priority (hp) scheme (computation of the dynamic priority directly from a function of the control signal u; re-evaluation after each sampling instant). The second is the hp scheme extended with a static time strategy (STS) for the re-evaluation of the dynamic priority (re-evaluation not always after each sampling time). This scheme is noted *hp+sts*. The third is a scheme which does not compute the dynamic priority directly from the control signal u (definition of a timed dynamic priority reference profile and trip in this profile by means of an on-line temporal supervision based on a function of the control signal u). The dynamic priority is re-evaluated after each sampling instant. This third scheme, which implements a dynamic time strategy for the trip in the timed dynamic reference profile, is noted as *hp+dts*.

We will now detail these three schemes.

3.2.2.1. *hp scheme*

The needs are translated into a dynamic priority by considering an increasing function of $|u|$ (call it $f(|u|)$) characterized by a saturation for a value of $|u|$ less than the maximum of $|u|$ (noted $|u|_{\max}$). We do not want the dynamic priority to take its highest value only when $|u|$ is at its maximum but already for values before the maximum, in order to react quickly as soon as the needs begin to become important. So we decide (it is an arbitrary choice) to take $\frac{2}{3}|u|_{\max}$ as the value of $|u|$ where the dynamic priority reaches its highest value.

Several functions $f(|u|)$ have been studied [JUA 07b]. For this work, we consider the function $f(|u|)$ represented in Figure 3.6. This function is defined by

$$f(|u|) = \begin{cases} P_{\max} \sqrt{\dfrac{|u|}{\frac{2}{3}|u|_{\max}}}, & 0 \le |u| \le \frac{2}{3}|u|_{\max} \\ P_{\max}, & |u| > \frac{2}{3}|u|_{\max}. \end{cases}$$

The computation of the dynamic priority is done by the controller each time it receives a frame that the sensor sends after each sampling instant (dynamic priority re-evaluated after each sampling instant). Then, after the reception of a frame from the sensor, the controller sends a frame with the value of the new dynamic priority. This frame reaches all the sites (CAN is a bus) and as the sensor site knows the first level of the ID of f_{ca} (it is a constraint for our implementation), it will learn the dynamic priority that it will put in the next frame that it will send (the dynamic priority is then used by the two flows of a process control application). The implementation of the dynamic priority mechanism (calculation by the controller task and attribution by the sensor task) is represented in Figure 3.7.

Taking into account the task implementation (sensor task is time-triggered, controller task is event-triggered), note that it is the sensor task which transmits the first frame at the start of the application. For this first frame, the sensor site has no information about the dynamic priority and, thus, we consider that it uses the maximum priority. This way, the first f_{sc} frame reaches the controller site as quickly as possible.

3.2.2.2. *(hp+sts) scheme*

A criticism of the hp scheme is that we can have oscillatory behavior of the dynamic priority values (resulting from a damped sinusoidal transient behavior of u).

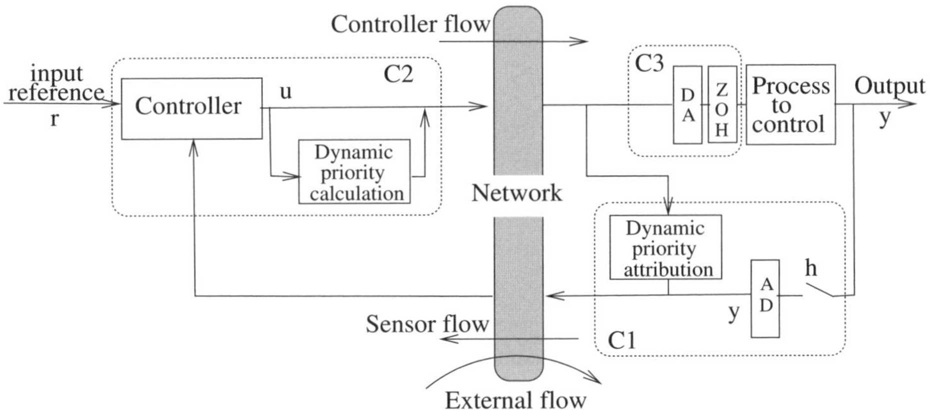

Figure 3.7. *Implementation of the dynamic priority mechanism*

We can have, for example, this scenario for the dynamic priority values at three successive re-evaluation instants [JUA 07a]: the highest value at the first re-evaluation instant, then an intermediary value at the second, and again the highest value at the third re-evaluation instant, etc. Such an oscillatory behavior shows that the control of a situation requiring a big value of the dynamic priority is inadequate in terms of the maintenance of this big value, since after leaving this value for an intermediary one, at the second re-evaluation instant, we come back to this big value at the third re-evaluation instant. The observation of this phenomenon suggests increasing the duration of the dynamic priority with a big value in order to improve transient behavior.

The (hp+sts) scheme is then the following. In contrast to the scheme hp, where the dynamic priority is re-evaluated in the controller site after each reception of an f_{sc} frame, the instant of the re-evaluation is no longer so closely related to the sampling instants. Here, the duration of the time interval between two successive re-evaluations depends on the value of the dynamic priority at the beginning of the time interval. This duration must be relevant, in particular, from the point of view of the transfer function of the process control application and, more precisely, of its transient behavior (defined before its implementation through the network). We considered the following algorithm:

– if the dynamic priority has a value between the highest priority (P_{max}) and half the highest priority ($\frac{1}{2}P_{\mathrm{max}}$), we keep this value for four sampling intervals, and we re-evaluate the dynamic priority afterwards; this duration is equal to the rise time t_r (we have chosen $h = \frac{tr}{4}$) which represents a good characteristic of a transient behavior).

– if the dynamic priority has a value inferior to half the highest priority, we re-evaluate it after each sampling instant, as in the previous algorithm.

Note that the implementation of the dynamic priority is like the one represented in Figure 3.7 except that now we have a comparison with the priority $\frac{1}{2}P_{\mathrm{max}}$ and the new re-evalution strategy in the controller site.

3.2.2.3. *(hp+dts) scheme*

A criticism of the (hp+sts) scheme is the static aspect of the time strategy for re-evaluating the dynamic priority. The goal of this new scheme is to have a behavior which is flexible enough to adapt to different transient situations.

3.2.2.3.1. Main ideas

Fist, we define [JUA 08], which we call the reference profile of the dynamic priorities. This reference profile expresses the dynamic priority values, which must be used at the successive sampling instants of a transient situation (i.e. after an input change or a disturbance or after successive input changes and/or disturbances) from the beginning of such a situation till the establishment of the permanent behavior. This expression is made in function of a time domain which is a virtual view of the sampling process during a transient behavior.

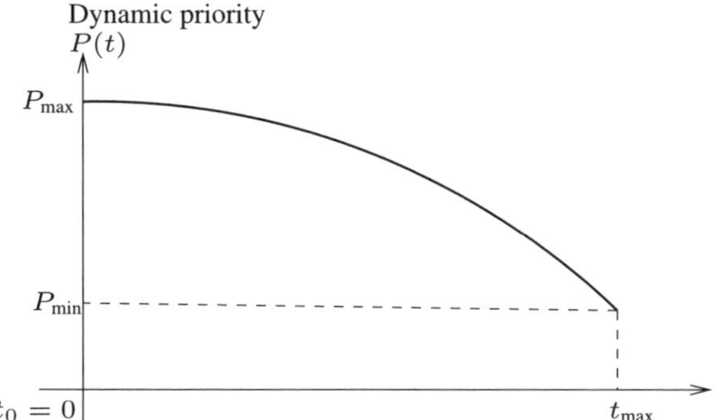

Figure 3.8. *Reference profile*

The reference profile that we are considering (Figure 3.8) consists of a decreasing continuously function $P(t)$:

$$P(t) = P_{\max} - (P_{\max} - P_{\min})(\frac{t}{t_{\max}})^2 \qquad 0 \leq t \leq t_{\max}.$$

The values of $P(t)$ decreases from a priority at time 0 (maximum dynamic priority at the beginning of the hardest transient situation) to a priority $P_{\min}$ (this priority is used in three situations: at the end of a transient behavior, during a permanent behavior and at the configuration of the system) at time $t_{\max}$. This time value must be compatible with the dynamic of the process control application; here we consider that $t_{\max}$ is the response time at 5% of the process control application (it is an arbitrary choice). When we are in permanent behavior (point $P_{\min}$, $t_{\max}$), as soon as we have a transient situation (input change or disturbance), we have a movement on the left of the curve $P(t)$ (if it is a very significant transition situation, we go to (point $P_{\max}$, 0)).

The dynamic priority decreases slowly from the value $P_{\max}$ at the beginning of a transient behavior (in order to be as reactive as possible). Note that different functions of $P(t)$ can be studied.

The time domain ($0 \leq t \leq t_{\max}$) does not express the ordered sampling instants, but it allows for situating each virtual sampling instant t_k with respect to the previous virtual sampling instant t_{k-1}, and then to deduce the dynamic priority $P(t_k)$.

Initially, just after the appearance of an input change or a disturbance (movement to the left on the curve i.e. dynamic priority increase), we could think that we then only have movements showing a decrease in dynamic priority, but this is not a correct interpretation. What we have results from the influence of the network (variable loads) and also the possibility of successive fast input changes or disturbances in the application which lengthen the transient behavior. Thus, since the evolution of the dynamic

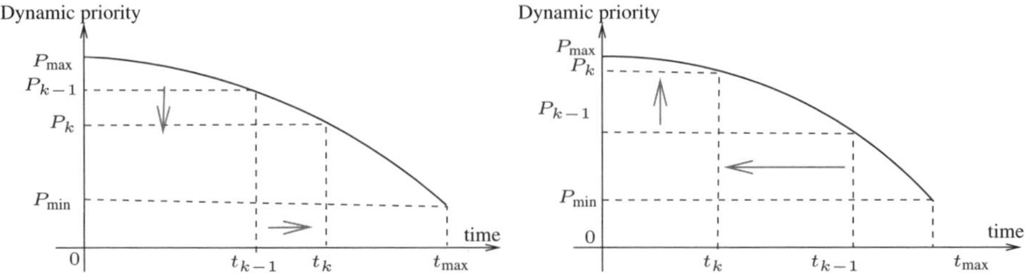

Table 3.2. *Increase and decrease of the priority*

priorities cannot be continually decreasing i.e. being at a virtual sampling instant, we can, by considering the reference profile curve, move back to a dynamic priority value higher than the present value. The evolution of the dynamic priorities between two successive virtual sampling instants can then be as is shown in Figure 3.2.

So, in order to take into account this behavior when computing the virtual sampling instants, we have to add a component called on-line temporal supervision, in addition to the sampling period h. This on-line temporal supervision is based on a function of the control signal $(g(u))$ which will correct the positioning of the virtual sampling instant.

We use (Figure 3.9) the function $g(u)$ here with $g(u) \in [0, t_{\max}]$:

$$g(u) = \begin{cases} t_{\max} \sqrt{\dfrac{|u|}{\frac{2}{3}|u|_{\max}}}, & 0 \leq |u| \leq \frac{2}{3}|u|_{\max} \\ t_{\max}, & |u| > \frac{2}{3}|u|_{\max}. \end{cases}$$

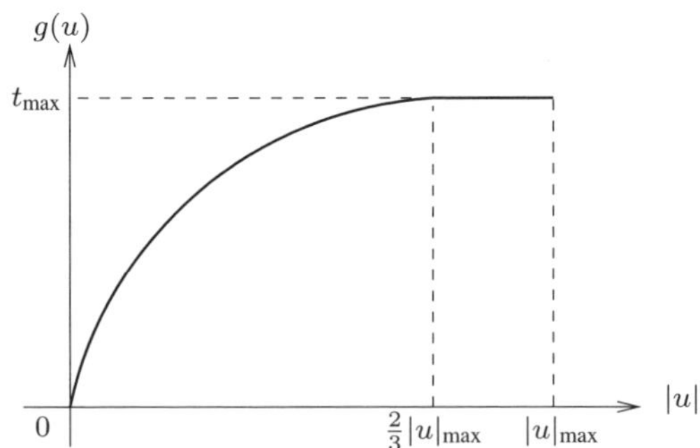

Figure 3.9. *The considered function $g(u)$*

3.2.2.3.2. Algorithm for computing the dynamic priority at any sampling instant t_k

($t_k \in [0, t_{\max}]$) The operations of the algorithm consist in positioning, when the controller receives the f_{sc} frame, at an instant in the interval of time $[0, t_{\max}]$ of the definition of the reference profile and in detecting the value of the dynamic priority to use from this instant. The value of the instant depends on the value of $g(u)$ at the reception of the f_{sc} frames.

Initially (configuration of the system), the reference profile is at (point ($P_{\min}, t_{\max}$)) i.e. $t_k = t_{\max}$. Then, upon the reception of an f_{sc} frame, the controller:

1) computes $g(u)$;

2) computes $x = t_k - \alpha g(u)$, where x is an intermediate variable, α is a coefficient, defined by $\alpha = \frac{t_k}{t_{\max}}$ ($0 \leq \alpha \leq 1$) which balances the influence of $g(u)$ by increasing this influence even more because the dynamic priority is low (when the dynamic priority is low, a large value of $g(u)$ must induce greater feedback; it is not as necessary when the priority is already high)

- if $x \leq 0$ then $x = 0$; we go to the time 0 on the reference profile (priority $P_{\max}$),

- if $0 \leq x \leq t_{\max}$, we go to the time x in the reference profile and get the priority $P(x)$;

3) re-initializes the virtual time for the next sampling $t_k = x + h$ (if $t_k > t_{\max}$ then $t_k = t_{\max}$). This value will be used for computing the dynamic priority on the reception of the next f_{sc} frame.

3.2.3. *Study of the three schemes based on hybrid priorities*

3.2.3.1. *Study conditions*

We consider the process control application which was presented in section 3.2.1.1. The input is a position step which starts at time 0, and we study the transient behavior until it reaches permanent behavior.

The QoS parameters, which need to be taken into consideration, are the mean delay $\bar{D}$ of the control loop and its standard deviation σ. The QoC parameter is the response time at 5% (noted res_t) which is obtained directly from the tool TrueTime.

In order to evaluate the QoS parameters, we use the message exchange temporal diagrams which are also provided by TrueTime, and the value of res_t.

From the message exchange temporal diagrams, we can get the delay in the control loop (delay of the message of the flow f_{sc} + delay of the message of the flow f_{ca} + $D_{\mathrm{sc}} + D_{\mathrm{ca}}$) for each sampling period (call D_i this delay for the sampling period i).

URF (%)	Multiple of $\frac{1}{h}$	T_{ex} (ms)
99.2	9	1.1111
89.6	8	1.25
80	7	1.4286
70.4	6	1.6667
60.8	5	2.0
51.2	4	2.5
41.6	3	3.3333
32	2	5.0
22.4	1	10.0

Table 3.3. *Different* URF*s*

Counting the number n of sampling periods in the response time res_t, we deduce the value of $\bar{D}$ and σ by the formulas: $\overline{D} = \frac{\sum_{i=1}^{n}(D_i)}{n}$ and $\sigma = \sqrt{\frac{\sum_{i=1}^{n}(D_i - \overline{D})^2}{n}}$.

In order to make a quantitative analysis, we cause a variation in the network load (URF) by varying the period T_{ex} of the external flow: we consider an external flow, the frequency of which (noted $\frac{1}{T_{\text{ex}}}$) is a multiple of the sampling frequency ($\frac{1}{h}$). The different URFs being considered are given in Table 3.3.

The following important points must still be emphasized:

– the flows f_{sc} (which are generated at the sampling times) and f_{ex} are synchronous (starting at the same time) and as we consider the cases where the frequency of f_{ex} is a multiple of the sampling frequency, then their medium access attempts coincide at every sampling time;

– up to the value 70.4% of the URF (value of 1.6667 ms for Tex), we can see that during Tex, one frame of each flow can access the medium: 0.96 ms $+ 0.64$ ms $=$ 1.6 ms < 1.6667 ms (the third flow can begin to be transferred and then cannot be interrupted). This remark is very important for the analysis which is done in section 3.2.3.3;

– a last point must be still noted: at the beginning of a transient behavior, as the control signal is at a maximum, the dynamic priority of the flows of the process control application is P_{max}. This point also is important for the analysis in sections 3.2.3.2, 3.2.3.3, and 3.2.3.4.

3.2.3.2. *hp scheme*

As concerns the process control application, we give $\bar{D}$ and σ in Table 3.4 and res_t in Table 3.5. The values depend on the network load URF (which depends on the frequency f_{ex}), and on the priority threshold Pr_th (which depends on the importance we give to f_{ex}).

	Pr_th					
URF	$0.9P_{\max}$		$0.5P_{\max}$		$0.25P_{\max}$	
(%)	$\bar{D}$	σ	$\bar{D}$	σ	$\bar{D}$	σ
99.2	5.333	1.680	3.743	2.262	1.804	1.380
89.6	3.264	1.286	2.240	1.228	1.629	0.846
80.0	2.48	0.887	1.978	0.828	1.5418	0.592
70.4	1.891	0.462	1.716	0.478	1.472	0.384
51.2	1.891	0.462	1.716	0.478	1.472	0.384
22.4	1.891	0.462	1.716	0.478	1.472	0.384

Table 3.4. *hp scheme: $\bar{D}$ and σ (ms)*

Concerning the values of $\bar{D}$, we observe the following main points:

– for each value of Pr_th:

- for URF ≤ 70.4 %, we note that we have the same values of $\bar{D}$ and σ whatever the value of URF is. This is a consequence of the fact that (cf. remark in the study condition) the two frames of f_{sc} and f_{ca}, during each sampling period, can be sent during the period of f_{ex}, which is not the case with URF > 70.4 % where $\bar{D}$ and σ increase with the value of URF (see in table 3.4 ,URF $= 80$ %, 89.6 %, 99.2 %).

- We explain the difference (URF ≤ 70.4 % and URF > 70.4 %) by means of two exchange temporal diagrams provided by TrueTime (Figures 3.10 and 3.11 for the case of $Pr\_th = 0.9P_{\max}$). In Figure 3.10, we see that the frames f_{sc} or f_{ca} can be delayed, during a sampling period, at the very most for the duration of one frame of f_{ex} (0.96 ms). In Figure 3.11, we see that the two frames of f_{sc} and f_{ca} can be delayed, and the delays for the frame of f_{ca} can be more than the duration of one frame of f_{ex}.

- Note then, when URF $> 70.4\%$ and for increasing values of URF, $\bar{D}$ increases because the network load increases (then more chances to delay the frames of f_{sc} and f_{ca}).

– For increasing values of Pr_th, $\bar{D}$ also increases because the dynamic priorities of the frames of f_{sc} and f_{ca} have fewer chances of being higher (except at the beginning of a transient behavior) than the threshold.

URF	Pr_th		
(%)	$0.9P_{\max}$	$0.5P_{\max}$	$0.25P_{\max}$
99.2	359	228	105
89.6	148	110	103
80.0	111	108	101
70.4	107	105	99
51.2	107	105	99
22.4	107	105	99

Table 3.5. *hp scheme: res_t (ms)*

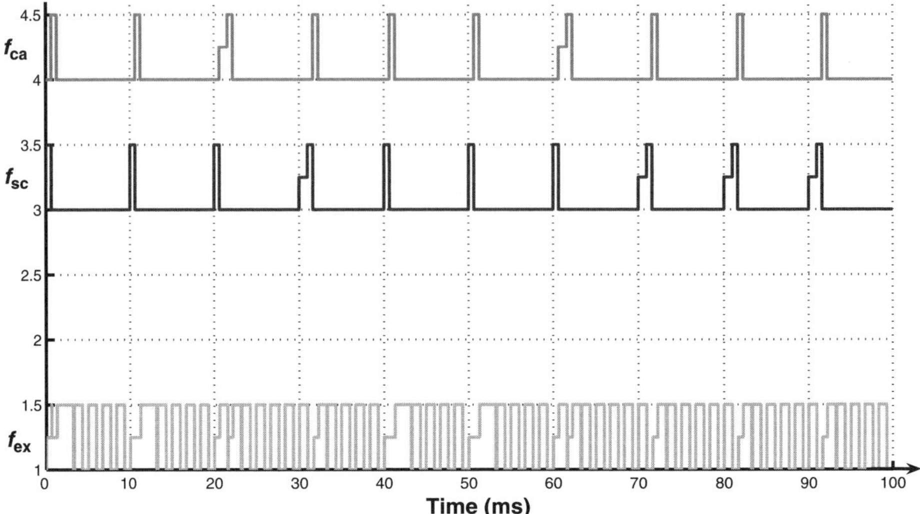

Figure 3.10. *hp scheme,* $URF = 70.4\%$, $Pr\_th = 0.9P_{\max}$

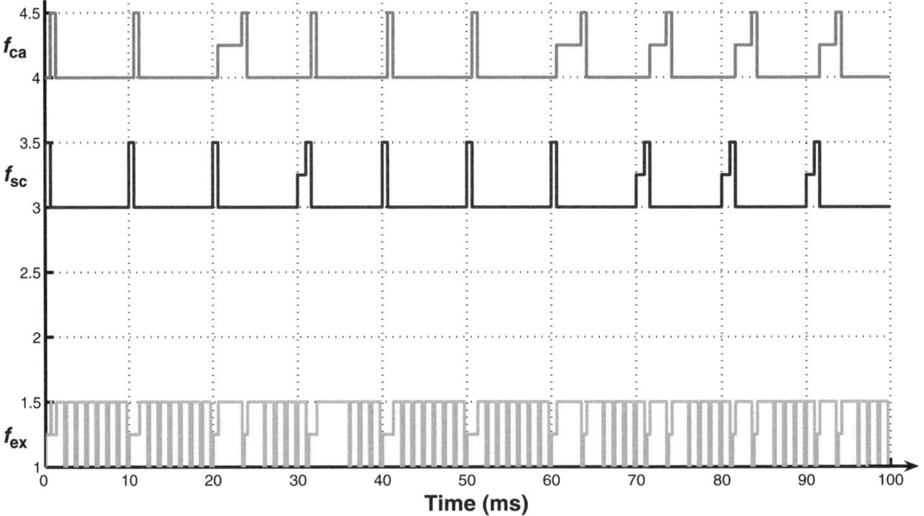

Figure 3.11. *hp scheme,* $URF = 89.6\%$, $Pr\_th = 0.9P_{\max}$

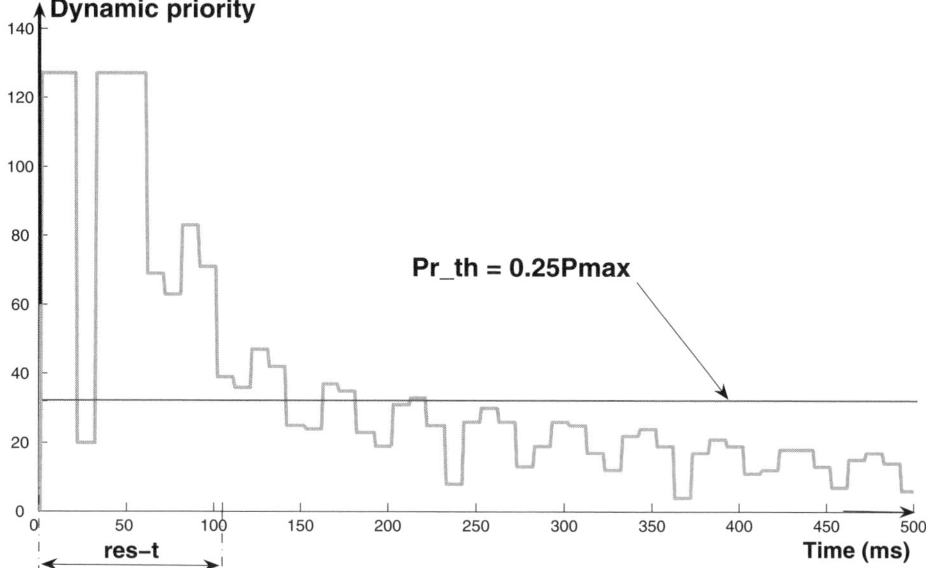

Figure 3.12. *hp scheme,* $URF = 99.2\%$, $Pr\_th = 0.25 P_{\max}$

– Concerning the values of σ, we have the following comments: for each value of URF, the variation of σ, when Pr_th increases, presents a maximum (which occurs for a value of Pr_th around $Pr\_th = 0.5 P_{\max}$). The explanation is given by means of Figures 3.12–3.14 (which represent the dynamic priority variation for $Pr\_th = 0.25 P_{\max}$, $Pr\_th = 0.5 P_{\max}$, and $Pr\_th = 0.9 P_{\max}$). These figures allow us to evaluate the number of times where, during the res_t, the frames of f_{ca} have a higher or lower priority than the threshold (a higher priority means a lower delay; a lower priority means a bigger delay). Then we can see that we have for $Pr\_th = 0.5 P_{\max}$, the maximum value of σ (the number of times where the dynamic priorities are higher than the threshold $\approx$ the number of times where the dynamic priorities are lower than the threshold). For $Pr\_th = 0.25 P_{\max}$ ($Pr\_th = 0.9 P_{\max}$), the number of times where the dynamic priorities are higher (lower) than the threshold is much greater than the number of times where the dynamic priorities are lower (higher) than the threshold. Thus, we have values of σ smaller than with $Pr\_th = 0.5 P_{\max}$ (in the case of $Pr\_th = 0.25 P_{\max}$ with a small value of $\bar{D}$; in the case of $Pr\_th = 0.9 P_{\max}$ with a higher value of $\bar{D}$).

Obviously, for each value of Pr_th, σ increases with URF (the reason is still the increase of the network load).

Important remark: for $Pr\_th \leq 0.15 P_{\max}$ i.e. low threshold (we have not represented the results for reasons of limited space), we have the minimal value for $\bar{D}$

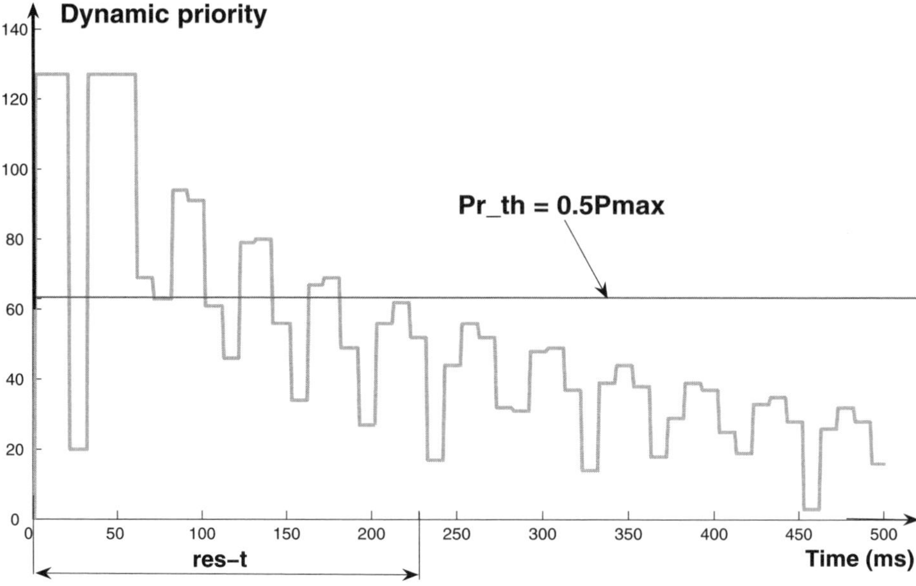

Figure 3.13. *hp scheme,* $URF = 99.2\%$, $Pr\_th = 0.5P_{\max}$

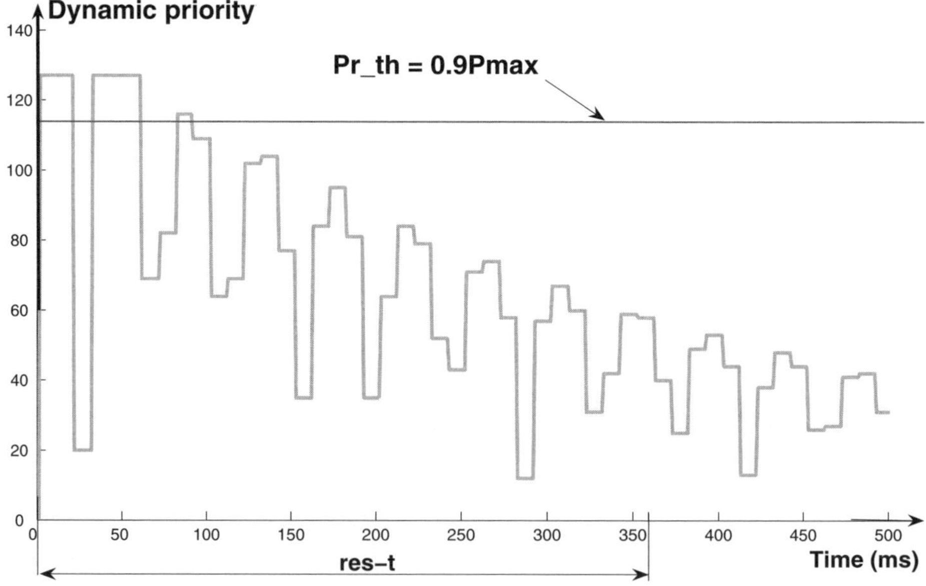

Figure 3.14. *hp scheme,* $URF = 99.2\%$, $Pr\_th = 0.9P_{\max}$

URF	Pr_th					
	$0.9P_{\max}$		$0.5P_{\max}$		$0.25P_{\max}$	
(%)	$\bar{D}$	σ	$\bar{D}$	σ	$\bar{D}$	σ
99.2	2.589	2.138	2.589	2.138	1.28	0.0
89.6	1.856	1.152	1.856	1.152	1.28	0.0
80.0	1.664	0.768	1.664	0.768	1.28	0.0
70.4	1.28	0.0	1.28	0.0	1.28	0.0
51.2	1.28	0.0	1.28	0.0	1.28	0.0
22.4	1.28	0.0	1.28	0.0	1.28	0.0

Table 3.6. *(hp+sts) scheme: $\bar{D}$ and σ (ms)*

(1.28 ms i.e. a frame of f_{sc} (0.64 ms) and then a frame of f_{ca} (0.64 ms) always use the medium before the frames of f_{ex} because the dynamic priority is always higher than Pr_th during the response time). Then, of course, $\sigma = 0$.

3.2.3.3. *(hp+sts) scheme*

For the hp scheme, we give $\bar{D}$ and σ in Table 3.6 and res_t in Table 3.7. The values are obviously a function of URF and Pr_th.

We can see important differences with the hp scheme:

– for URF $\leq$ 70.4 %, $\bar{D}$ is now always constant, whatever the Pr_th is (this is for two reasons: the first reason is because of the consequence of the property (URF $\leq$ 70.4 %) indicated in section 3.2.3.1; the second is the fact that now, at the beginning of the transient behavior, the dynamic priority is used by the flows f_{sc} and f_{ca} for a duration, at least, equal to $4h$). Obviously, as $\bar{D}$ is constant, $\sigma = 0$.

– For $Pr\_th = 0.25P_{\max}$, we have $\bar{D}$ which is constant for all URF values (this means that, on all the network load conditions, the dynamic priority is higher than the threshold). The explanation is given by the exchange temporal diagram on Figure 3.15.

– Analysis of a row of Table 3.6 (in the case where $Pr\_th > 0.25P_{\max}$): we have the same values of $\bar{D}$ and σ whatever the value of Pr_th. The explanation is given by the exchange temporal diagrams of Figures 3.16 and 3.18 where we consider URF $= 99.2$ %. These diagrams are identical.

URF	Pr_th		
(%)	$0.9P_{\max}$	$0.5P_{\max}$	$0.25P_{\max}$
99.2	103	103	46
89.6	100	100	46
80.0	98	98	46
70.4	46	46	46
51.2	46	46	46
22.4	46	46	46

Table 3.7. *(hp+sts) scheme: res_t (ms)*

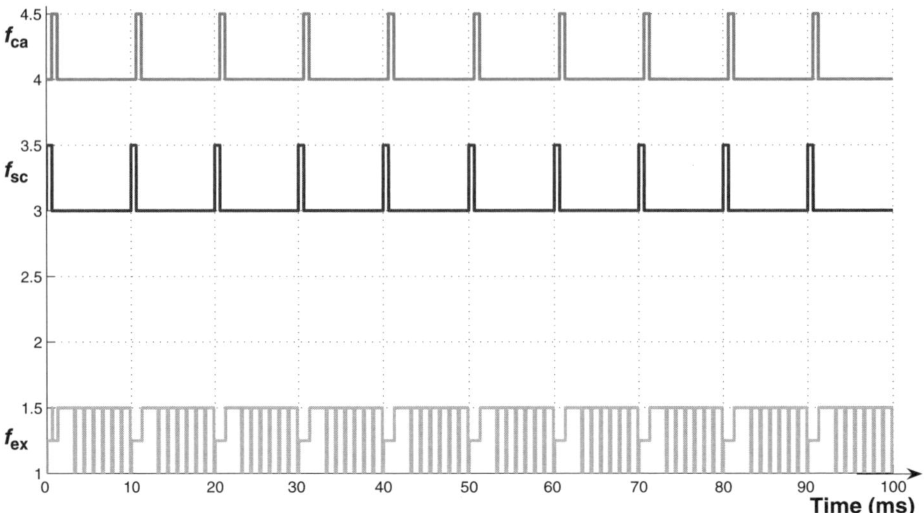

Figure 3.15. *(hp+sts) scheme,* $URF = 99.2\%$, $Pr\_th = 0.25 P_{\max}$

– Analysis of a column of Table 3.6 (in the case where URF > 70.4 %): we note an increase of $\bar{D}$ and σ with URF (the explanation is given by Figures 3.17 and 3.18); the delay of the frame f_{ca} (sampling periods 8 and 9) in Figure 3.18 is higher than in Figure 3.17).

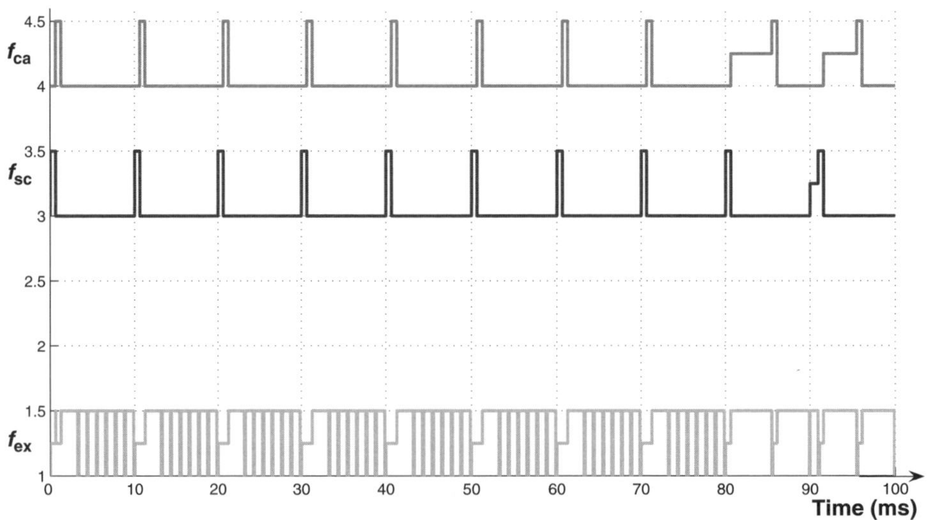

Figure 3.16. *(hp+sts) scheme,* $URF = 99.2\%$, $Pr\_th = 0.5 P_{\max}$

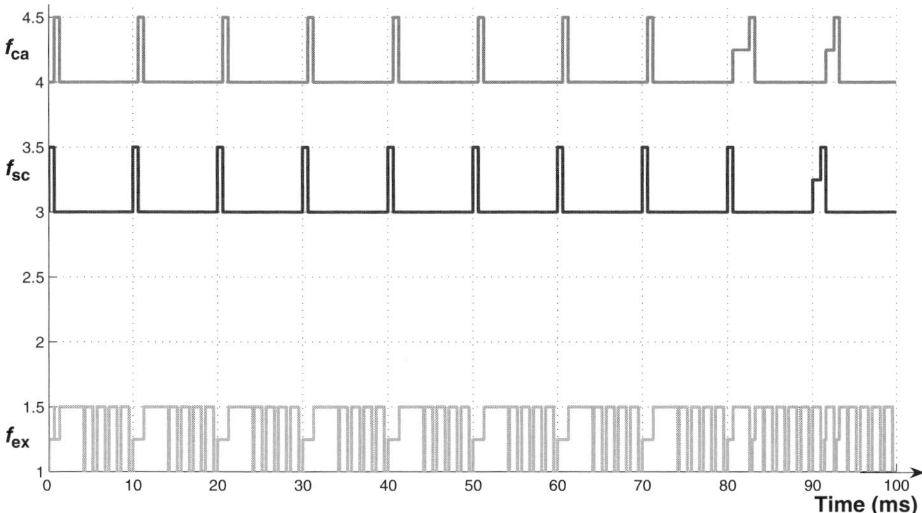

Figure 3.17. *(hp+sts) scheme,* $URF = 80\%$, $Pr\_th = 0.9P_{\max}$

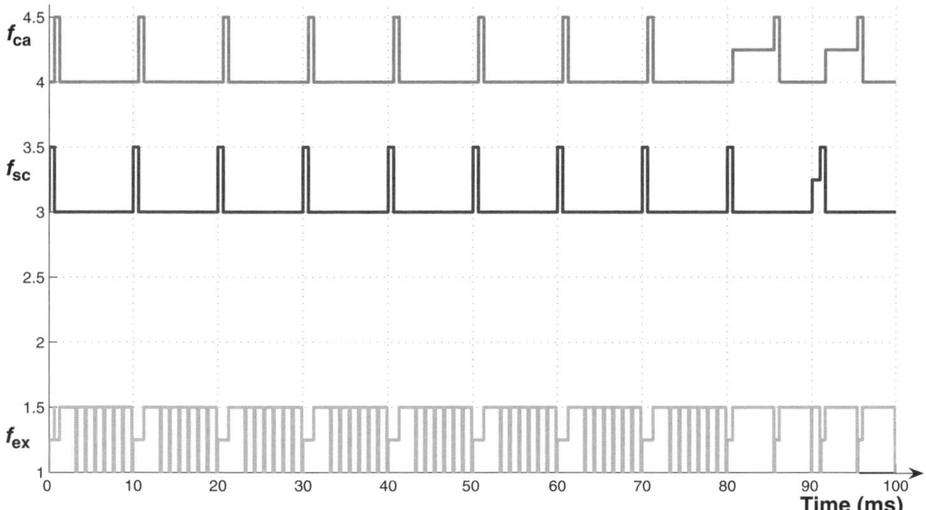

Figure 3.18. *(hp+sts) scheme,* $URF = 99.2\%$, $Pr\_th = 0.9P_{\max}$

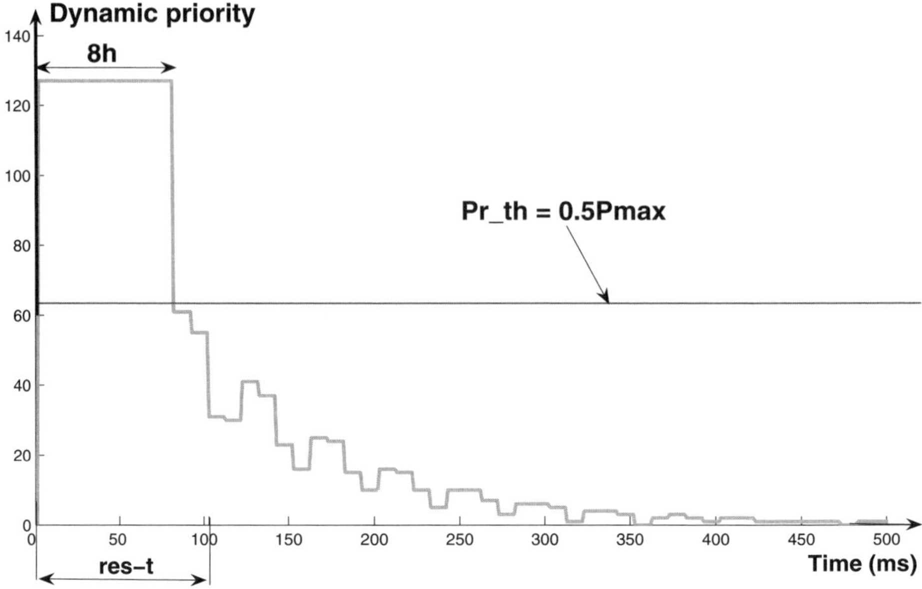

Figure 3.19. *(hp+sts) scheme,* $URF = 99.2\%$, $Pr\_th = 0.5P_{\mathrm{max}}$

With respect to the hp scheme, all the improvements (which give best response time for the process control application) result from the fact that the dynamic priority P_{max} is used for a longer time. In Figure 3.19 (compare with Figure 3.13), we have an example of the evolution of the dynamic priority (we have P_{max} during $8h$).

3.2.3.4. *(hp+dts) scheme*

We give, as for the previous schemes, $\bar{D}$ and σ in Table 3.8 and res_t in Table 3.9.

We can see now that we always have the minimum constant value $\bar{D}$ (duration of the f_{sc} frame (0.64 ms) + duration of the f_{ca} frame (0.64 ms)), then $\sigma = 0$, and the best response time (46 ms). This is a consequence of the fact that the dynamic priority is continuously controlled (by the control signal u) and that it is higher than the threshold for a time longer than the res_t (see Figure 3.20).

3.2.4. *QoC visualization*

We represent, in Figures 3.21–3.23, the time response to an input step for the three schemes in the following conditions: $URF = 99.2\%$ and $Pr\_th = 0.9P_{\mathrm{max}}$. The

URF	Pr_th					
	$0.9P_{\max}$		$0.5P_{\max}$		$0.25P_{\max}$	
(%)	$\bar{D}$	σ	$\bar{D}$	σ	$\bar{D}$	σ
99.2	1.28	0.0	1.28	0.0	1.28	0.0
89.6	1.28	0.0	1.28	0.0	1.28	0.0
80.0	1.28	0.0	1.28	0.0	1.28	0.0
70.4	1.28	0.0	1.28	0.0	1.28	0.0
51.2	1.28	0.0	1.28	0.0	1.28	0.0
22.4	1.28	0.0	1.28	0.0	1.28	0.0

Table 3.8. *(hp+dts) scheme:* $\bar{D}$ *and* σ *(ms)*

oscillatory transient behavior clearly shows the performances of the three schemes in terms of overshoot and damping. The conditions of a big network load and a high threshold still show the interest of the two schemes with a time strategy (hp+sts, hp+dts) to get good performances. The dynamic aspect of the time strategy in the scheme hp+dts shows in the end that it is the best scheme.

3.2.5. *Comment*

We have considered three hybrid priority schemes and we have demonstrated the particular interest of a scheme, call (hp+dts), with a double aspect: dynamic priority based on a temporal supervision function of the control signal of the process control application. We have also evaluated, on the one hand, the QoS in terms of the mean delay and its standard deviation, and, on the other hand, the QoC in terms of the response time at 5%, and the relation between QoS and QoC (overshoot, damping). Concerning the results which have been obtained, we want to emphasize that, even with a big extended load (consider Figures 3.21–3.23, with URF = 99% whereas the load of the process control application is only 12.8%), the process control application gets very good results (especially with the last two schemes).

URF	Pr_th		
(%)	$0.9P_{\max}$	$0.5P_{\max}$	$0.25P_{\max}$
99.2	46	46	46
89.6	46	46	46
80.0	46	46	46
70.4	46	46	46
51.2	46	46	46
22.4	46	46	46

Table 3.9. *(hp+dts) scheme:* res_t *(ms)*

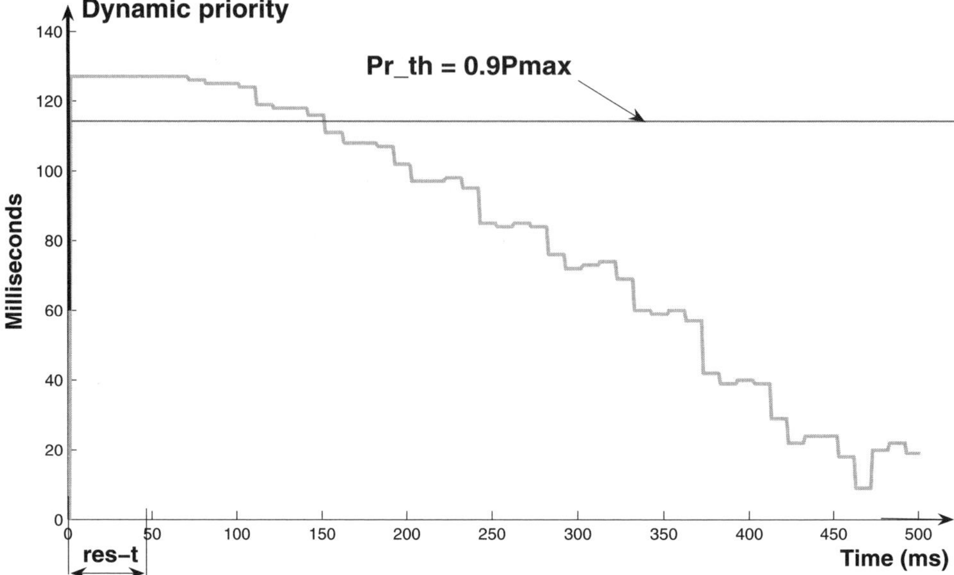

Figure 3.20. *(hp+dts) scheme,* $URF = 99.2\%$, $Pr\_th = 0.9P_{\max}$

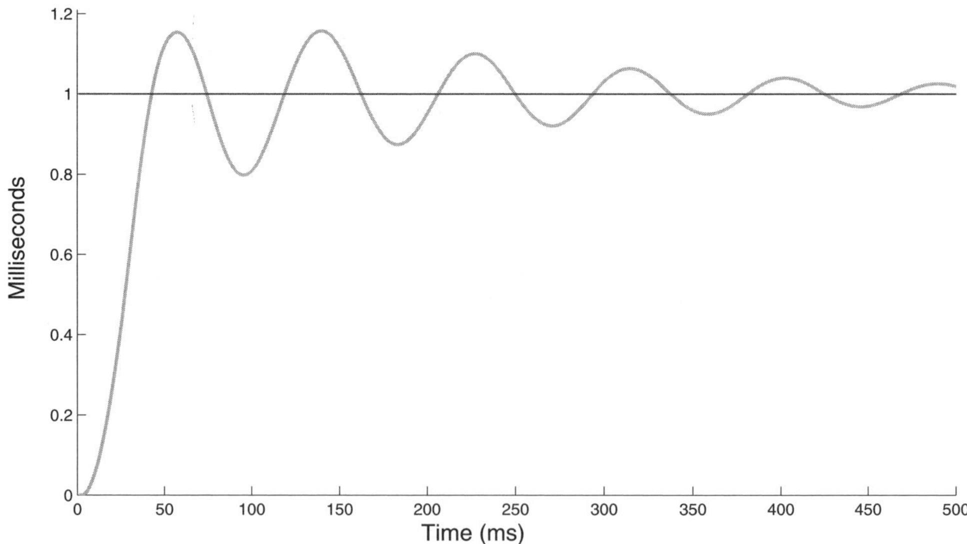

Figure 3.21. *hp scheme: response time to an input step*

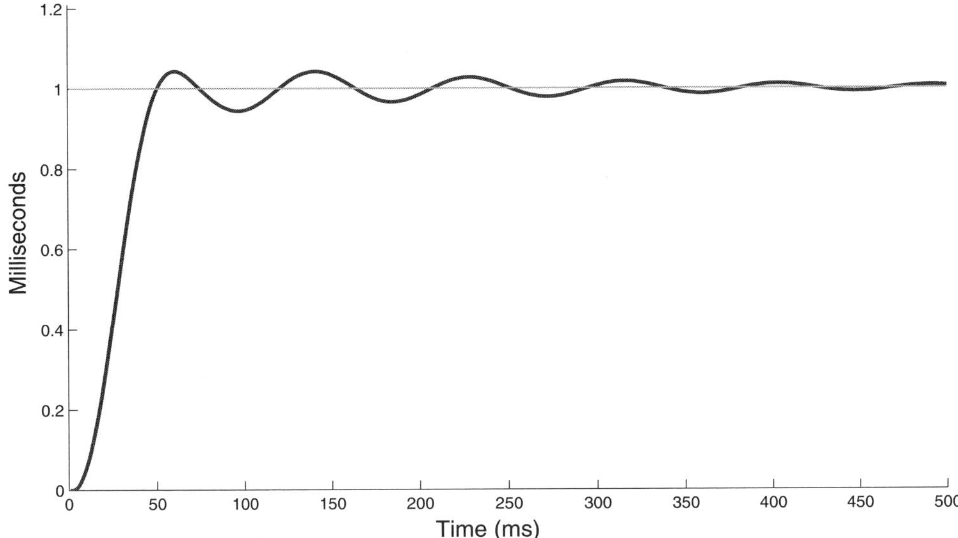

Figure 3.22. *hp+sts scheme: response time to an input step*

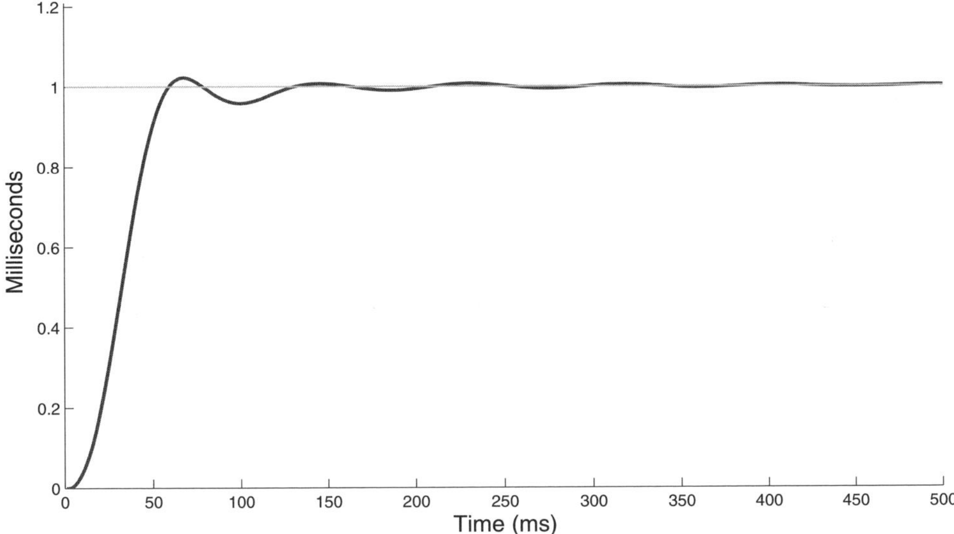

Figure 3.23. *hp+dts scheme: response time to an input step*

3.3. Bandwidth allocation control for switched Ethernet networks

Compared to the CAN bus protocol studied in section 3.2, the native Ethernet does not implement any priority mechanism. Non-standardized solutions have been proposed: adapting the inter-frame gap (smaller for high priority frames), modifying the *Binary Exponential Backoff* algorithm (the waiting time is not randomly calculated, but in relation with the priority), or using a variable length of the preamble (smaller for high priorities). Another approach consists of using a time division multiple access method over the native CSMA/CD protocol: pre-allocated time slots are defined for the transmission of time-critical data.

Nevertheless, the evolution of Ethernet to segmented architectures and the definition of the *Virtual Local Area Networks* (VLAN) have led to the birth of a new standards set (802.1D/p, 802.1Q) in which new encapsulation fields are added to the classical frame [IEE 03]. One of these fields is specified in order to support eight priority levels associated with eight types of applications (voice, video, network management, best effort, etc.). The number of classes of service may be different to the number of priority levels, and also different for each port. That is why the standard also recommends a mapping between classes, priority, and ports queues.

The next point is the scheduling policy used to forward the frames at the output port regarding their dedicated priorities. [IEE 03, section 8.6.6] defines two items:

– for a given supported value of traffic class, frames are selected from the corresponding queue for transmission only if all queues corresponding to numerically higher values of traffic class supported by the port are empty at the time of selection;

– for a given queue, the order of which frames are selected shall maintain the incoming ordering.

It means that the scheduling policy defined is the *Strict Priority (SP)* algorithm, and the policy must be FIFO for a given queue. But the standard enables to implement other algorithms. The main drawback of the SP algorithm is that it can lead to the impossibility for the lowest priority queues to be served. It corresponds to famine situations for the non-real-time applications. To resolve it, CoS switches implement a supplementary policy: the *Weighted Fair Queuing (WFQ)*. In the fair queuing algorithms, the service offered to the high priority queues is moderated as follows. A weight is associated with each queue. Then the scheduler gives to each queue (from the highest priority to the lowest) a bandwidth determined by its associated weight.

The WFQ, initially proposed in [DEM 89], is also known as the packetized generalized processor sharing (PGPS). It is based on the conceptual algorithm called generalized processor sharing (GPS) [PAR 93]. However, practical implementations of WFQ in today's switch products are based on its simplified version named WRR.

In a round robin policy, packets are pushed in queues according to their priority level. Then the server pools the different queues according to a cyclic sequence (using a pre-computed order defined by the queues priorities) in an attempt to serve one packet for each non-empty queue. Even if this algorithm respects the fairness quality, no flexibility is integrated. Moreover, the fairness can be damaged with variable packet lengths. To improve the lack of flexibility of a simple round robin policy, the WRR [DEM 89; KAT 91] associates a weight w_i with each flow i. Now, the WRR server will attempt to serve a flow i with a rate of $\frac{w_i}{\sum_j w_j}$ before looking for the following queue. Comparing to PGPS, delays could be more important since if the system is heavily loaded and a frame just misses its slot, it will have to wait for its next slot i.e. a cycle.

In the following, a WFQ policy *based on a per-priority queuing and a WRR scheduling* is studied. This implementation is typical of switch products, like the *Cisco Catalyst 2950*. The WRR assigns a priority i to each flow. It serves all the flows in a cyclic way, from the queue with the highest priority to the lowest. The number of frames that will be forwarded by the server for a queue i is bounded by the number ω_i. When the queue is empty, the scheduling protocol immediately processes the next queue.

In the context of NCS, the CoS is interesting since it enables to change the best effort service and the FIFO scheduling by a differentiated service and advanced scheduling. This approach is illustrated in Figure 3.24. For instance, the WRR policy [DEM 89] manages the network performances by adjusting the number of frames forwarded for each flow according to the frames priority. To illustrate the interest of CoS, a modification of the TrueTime kernel was proposed in [DIO 08] in order to allow the simulation of the WRR over switched Ethernet networks. It was then used to differentiate the service offered to the frames on the embedded network of the quadrotor.

The performances are by means of hard deadline computation algorithm which is treated in the following section.

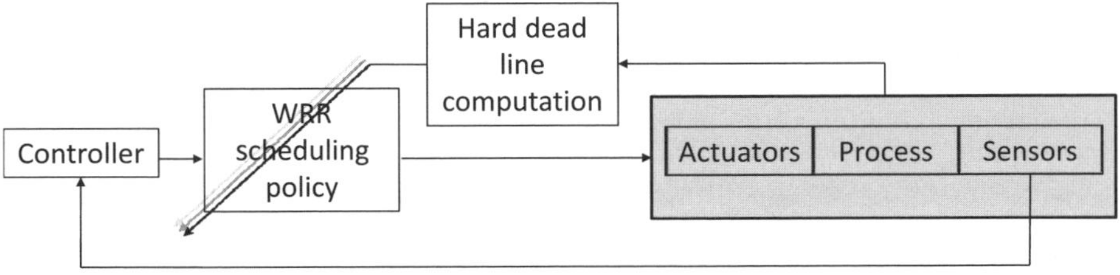

Figure 3.24. *Controller PN model*

3.3.1. *NCS performance analysis*

In this section, we analyze network-induced delay effects on system stability for linear, time-invariant control systems. The state space equations are modified according to induced delay, NT_s, where N varies from 1 to D. Then DT_s is the hard deadline which represents the critical value of induced delays beyond which the stability of the overall system is not guaranteed. The pole positions of the augmented state equation are tested to derive the necessary conditions for asymptotic system stability. [KAN 92]define the hard deadline as follows. Let X_A and U_A be the allowed state space and the admissible input space, respectively. Suppose the state, x, is evolved from time k_o in the presence of a computation-time delay N according to

$$x(k) = \Phi \left\{ k, k_0, x(k_0), u(k - N) \right\}, \tag{3.1}$$

where Φ is the state transition map and u is the control signal. Then, the hard deadline is given by

$$D(x(k_0)) = \sup_{u(k-N)\in U_A} \left\{ N : \Phi(k, k_0, x(k_0), u(k - N)) \in X_A. \right. \tag{3.2}$$

Due to the delay in the transmission, it is assumed that the control input is updated at time mTs. The augmented state space equation becomes

$$x((m + 1)N) = A^N x(mM) + \sum_{j=N-i}^{N-1} A^j Bu(mN)$$
$$+ \sum_{j=0}^{N-i-1} A^j Bu(mN + N - j - 1) \tag{3.3}$$

The hard deadline is derived from equation 3.3 by iteratively testing the current pole location of the closed-loop-augmented system. In this case, the hard deadline is expressed in the number of sampling period.

3.3.2. *NCS modeling*

3.3.2.1. *Introduction*

PN is able to formally represent the behavior of any kind of system. It has the capability to express many mechanisms usually used in distributed environments such as parallelism, synchronization, concurrency, and resource sharing [DAV 04; JEN 92; JUA 04]. Moreover, there are many references in the literature where PN is employed to model network protocols in order to validate protocol specifications, as in [BIL 82] and [LAI 89]. The objective of this research is to show that PN can also be used to model NCS in order to assess its behavior in a common formal language.

NCS are complex to model since they integrate different components such as controllers, actuators, sensors, the plant, and the network. In order to simplify the NCS model, it is necessary to split it into several sub-models. This means that the PN used to model NCS has to be based on the concept of Hierarchy defined in Hierarchy PN (HPN). In this case, a transition of the PN model can represent a sub-PN model. This particular transition is named substitution transition. The advantage of using HPN is also to build models in modular way. For example, the change of protocols in the NCS model consists only in replacing the sub-model describing the current protocol by the new one. Another difficulty during the NCS modeling is to be able to differentiate the messages according to its time constraints. The solution is to use the Colored HPN (CHPN). In this case, the color allows message tagging in regard to its time specifications. Finally, the time properties are obviously defined in NCS modeling. Thus, the Timed CHPN (TCHPN) is chosen to model NCS. This is the formalism defined by [JEN 92] and implemented in CPNTools developed by the Aarhus University [JEN 07].

3.3.2.2. *Network modeling*

Network modeling is usually achieved in two steps: traffic modeling and network device modeling (router, switch, etc.). Traffic modeling has to define the frame format according to the kind of protocol used. This modeling can be simplified by specifying only the format size and the information that is useful in the NCS context. The following list of colors is used to represent a frame:

```
Colset inp = with I1|I2;       /* Source Address */
Colset outp = with O1|O2;      /* Destination Address */
Colset prio = int with L...H;  /* Frame Priority: Low, Medium, High
                                  (Val H=2,Val M =1,Val L =0) */
Colset data = INT;             /* Data exchanged between
                                  controller and Sensor-Actuator */
```

There are many ways to model a network device (ref). Firstly, they all depend on memory allocation management but this memory can be located at the input, at the output or in the middle of the device. The last approach also called shared memory architecture is the one most often implemented by network constructors, and is the one retained in this paper. Secondly, the communication device can manage differentiated services. It requires defining at each output because there are as many buffers as specified differentiated services. This services integrate the used mechanisms to switch to the messages in the output buffer according to both their destination and their priority, and implement the scheduler. Thirdly, the wires have to be considered. The full-duplex mode is used to model a link. Figure 3.25 shows the general functions of a communication nodes, and Figure 3.26 shows its corresponding TCHPN model.

The switching modeling describes two functions in one:

– basic switching, which ensures frame transfer between the input port and the output port of the network device;

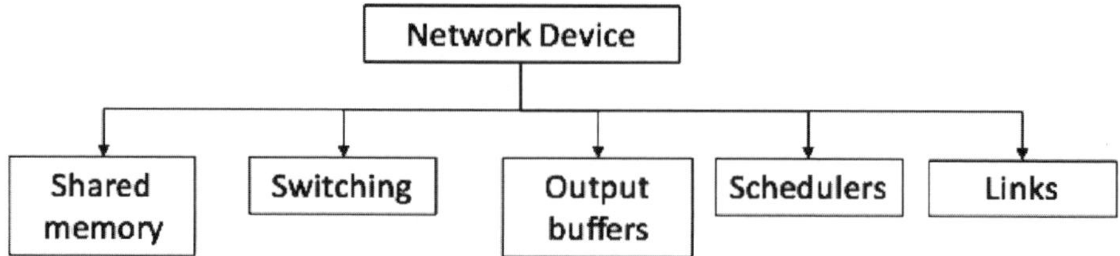

Figure 3.25. *General structure of the communication system*

– and the switching between the different buffers associated with one output port, which is done through analyzing the priority level of the frame. This function is called the classification step.

Figure 3.27 shows a simplified TCHPN model of a network device including two output ports able to differentiate three service levels. This means that three buffers are defined per output port. The frames waiting inside the shared memory and extracted in the FIFO queue model are firstly sent to the appropriate output port and then they are sent to the buffer corresponding to the frame priority (high, medium, and low).

Finally, managing the frames stored in the output buffers depends on the selected scheduling policy. The constructors of network devices mainly implement two kinds of schedulers: strict priority policy and the WRR policy. The general rule applied to manage the output buffers in the strict priority mode is that the frames stored in a buffer are processed only if all the buffers with higher classes do not contain a frame. The advantage of the strict priority policy is its simplicity to model (Figure 3.28) and

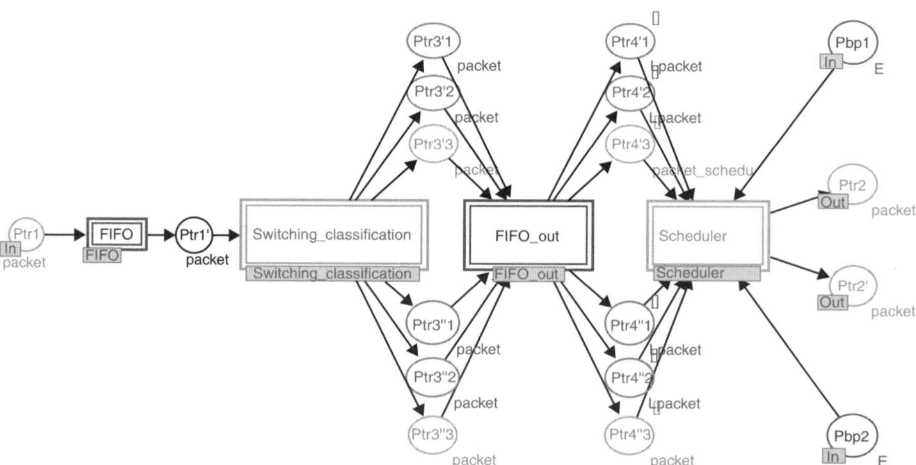

Figure 3.26. *TCHPN model of a communication node*

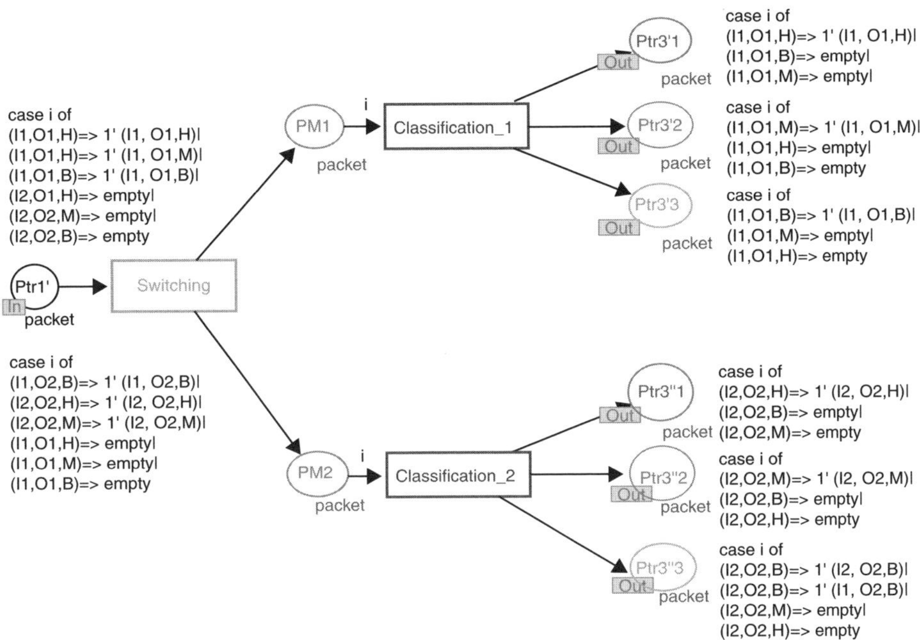

Figure 3.27. *CoS mechanism TCHPN model*

then to code in network devices. The problem is the generation of famine situation, because if the high class buffer continually receives frames, the frames waiting in the lower class buffers are not processed.

On the other hand, the WRR scheduler is more complex to model (Figure 3.29), but avoids the famine problem. The WRR scheduler cyclically the output buffers, and the number of frames processed is relative to the weight associated with each buffer.

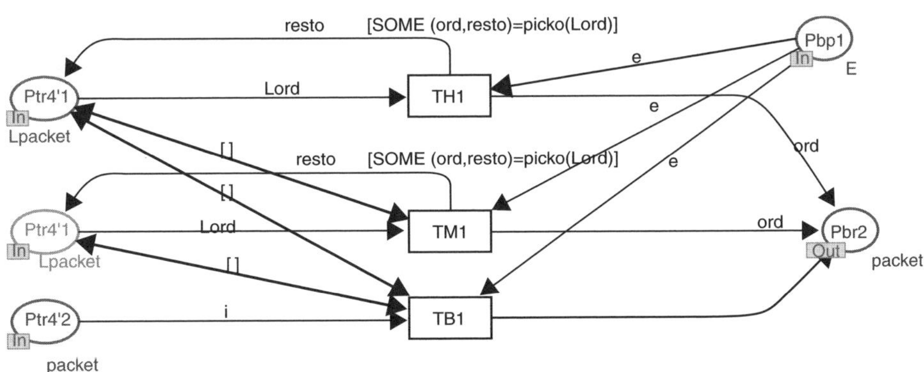

Figure 3.28. *Strict priority TCHPN model*

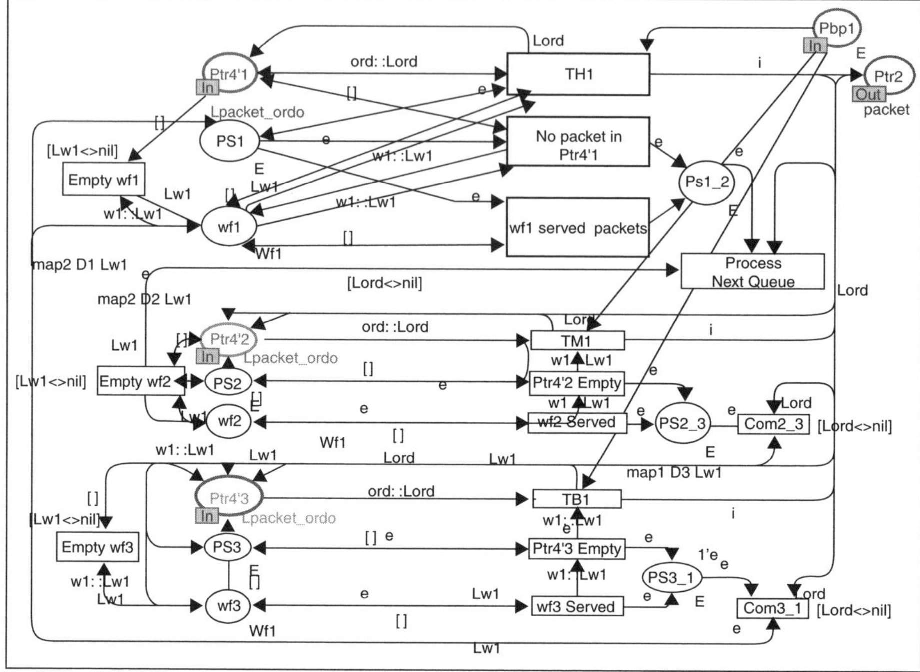

Figure 3.29. *WRR TCHPN model*

The weight of the high class buffer should be larger than the others in order to offer more bandwidth for the high priority frames. If a buffer is empty, the WRR scheduler automatically works on the following buffer, and so on.

The implementation of schedulers in the network devices is crucial in the context of NCS, since it allows for differentiating the services offered by the network according to the temporal constraints of the application. But, the difficulty is to analyze the relationships between the network tuning and controller specifications. The main interest of proposing an integrated approach for modeling NCS is to be able to study and to adjust the network parameters by observing their impacts on the plant's behavior.

3.3.2.3. *System modeling*

The NCS structure discussed in this section is composed of the plant, sensors, controllers and actuators which are spatially distributed and closed over the network. It is supposed that a sensor is time-driven with an identical sampling period h. By event-triggered controller or actuator, it is meant that the calculation of the new control or actuator signal is started as soon as information concerning the new control arrives,

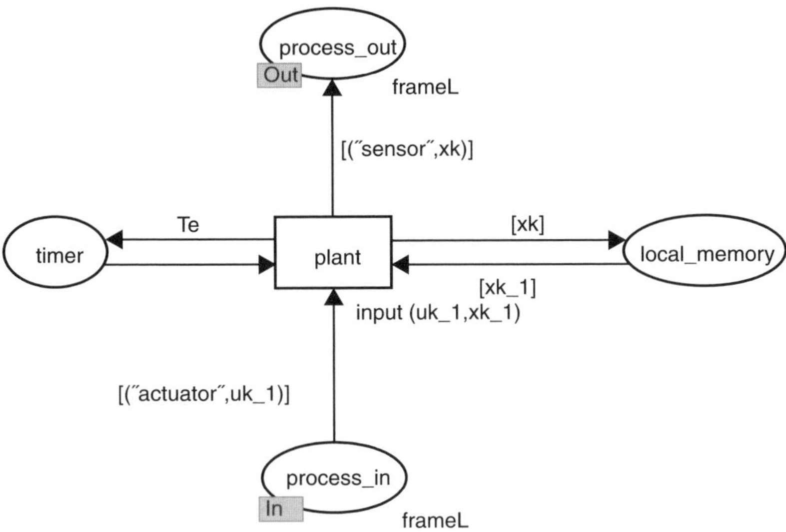

Figure 3.30. *Process PN model*

similar to [HAL 88]. Suppose that an LTI dynamical discrete time model is described as follows:

$$\begin{cases} x_{k+1} = \Phi\, x_k + \Gamma u_k \\ y_k = PX_k \end{cases} \tag{3.4}$$

where $x_k \in \Re^n$ is the state vector, $y_k\,(k) \in \Re^m$ the output vector and $u_k \in \Re^p$ the input vector. Φ; Γ; C are all real constant matrices and matrix of appropriate dimensions.

The controller is defined by the following equation

$$u\,(k) = -Kx\,(k)$$

where K is the controller gain matrix to be designed. Each of the components involved in the NCS are assigned to a specific task which is structured in code segments. The model used to define the state space equation is represented in Figure 3.30. The functions `process_` in and `process_out` corresponds to the actuator input and sensor output respectively. The current state value is computed at each sample time and stored in local memory when the code associated with the transition `plant` is executed.

In this chapter, only the controller is detailed. The other models (actuators, sensors, etc.) are explained in [BRA 07].

3.3.2.4. *Controller modeling*

The model of the controller is shown at Figure 3.31. The node `net_output2` receives the value `xk` from the sensor through the network. This value allows the transition `controller` to be sent on, then the associated segment code is executed as represented by the following commands.

```
input (cons , uk_1 , xk );
output (uk );
action
let
val  convert_string_to_real  =Option . valOf  oReal . fromString ;
val  cons1=convert_string_to_real (cons );
val  uk1_1=convert_string_to_real (uk_1 );
val  xk1=convert_string_to_real (xk );
val  k1_1=convert_string_to_real (K1 );
val  uk1=  k1_1 *( cons1 −xk1 );
val  uk=Real . toString (uk1 );
in
uuk :=[ uk ];
uk
end ;
```

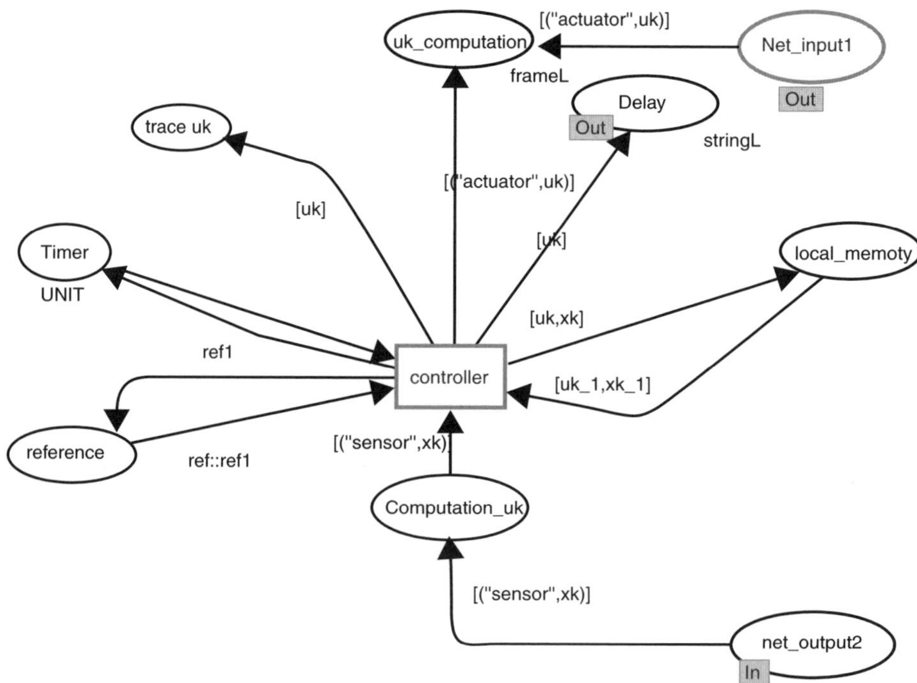

Figure 3.31. *Controller PN model*

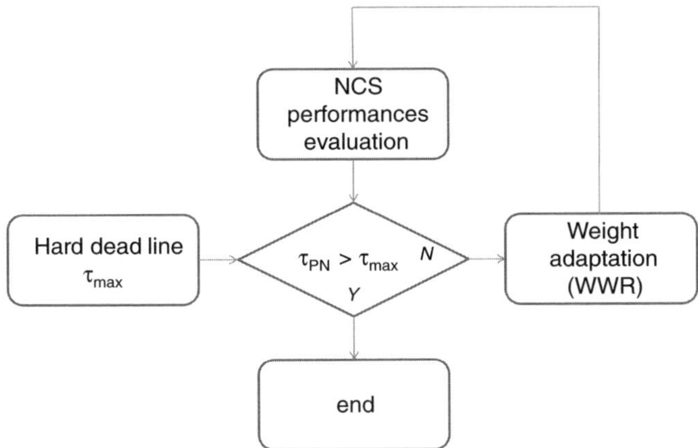

Figure 3.32. *General scheme of network adaptation mechanism*

where parameter k1_1 is the control feedback gain and ref1 is the reference value. At this stage, the control or sensor signals have to be embedded in order to be transmitted through the network.

3.3.3. *Network adaptation mechanism*

Figure 3.32 describes the global procedure for NCS analysis based on PN modeling. The maximum acceptable delay, $\tau_{\max}$, is determined by means of the *hard deadline* algorithm. Then, $\tau_{\max}$ is compared to delay τ_{PN} which is provided by the PN model. If necessary, the weights are re-assigned off-line.

3.3.4. *Example*

3.3.4.1. *Maximum delay computation*

Consider the discrete-time system:

$$x\left(k+1\right) = 0.98x\left(k\right) + 1.6u\left(k\right)$$
$$u(k) = -Kx(k)$$

Where K the state feedback matrix is determined in such a way so as to minimize the following cost function

$$J(u) = \sum_{n=1}^{\infty} \left[x(n)^T Q x(n) + u(n) R^T u(n) + 2x(n)^T N x(n) \right]$$

Where $Q \in \Re^{n \times n}$ and $R \in \Re^{l \times l}$ are positive semidefinite and positive definite, respectively. K is obtained by solving the associated discrete Riccati equation. For this example $Q = 2$ and $R = 4$. Then $K = 0.402$. The pole of the closed loop system is given by

$$x((m+1)N) = A^N x(mM) - \sum_{j=N-i}^{N-1} A^j BK x(mN)$$

$$- \sum_{j=0}^{N-i-1} A^j BK x(mN + N - j - 1)$$

is computed for different values of N. The system becomes unstable for $N = 3$. The sampling period is fixed at $1\ ms$. Then the maximum acceptable delay is $\tau_{\max} = 3\ ms$.

3.3.4.2. *Results*

The objective of this section is to show the interest of Petri Nets approach to simulate NCS. The system being considered is the same as the one before. Ethernet uses 10BT links. In these simulations, two kinds of traffic are considered. Real-time traffic between the controller and the sensors/actuators is periodically sent at Te and the size of the message correspond enough to the minimal Ethernet frame size to be able to transport the output and input information. The second type of traffic is called background traffic and is used to load the network. It is not time-constrained. This traffic allows for simulating the context of a shared network where the network can be used by other applications. The real time frames are tagged with a high priority field, and the background frames are tagged with a low priority field. The Ethernet switch implements the WRR scheduler and it is tuned to offer 10% of the total bandwidth to the real-time frames and 90% to the background traffic. This WRR configuration is called (A).

The first scenario generates background frames using 30% of the network bandwidth. Figure 3.33 a shows that this load does not disturbed the process system performances since the real-time traffic delays observed in Figure 3.33b are low (less than $0.5\ ms$).

The second scenario increases the network load at 70% inducing progressively (due to the time to fill the buffers) a degradation on the real time frame delays (Figure 3.34b). The observed delays are up to $3\ ms$ and generate instability on the process control.

The NCS modeling offers a double view in the same environment (Petri Nets), allowing designers to better understand the interactions between the process control and the network. In this case, the delays induced by the network have to be mitigated to ensure the stability of the system. Firstly, the hard-deadline method is used to estimate the delay threshold acceptable for the process controller. The result obtained that is the delay has to be lower than $3\ ms$. Secondly, the network parameters have to be

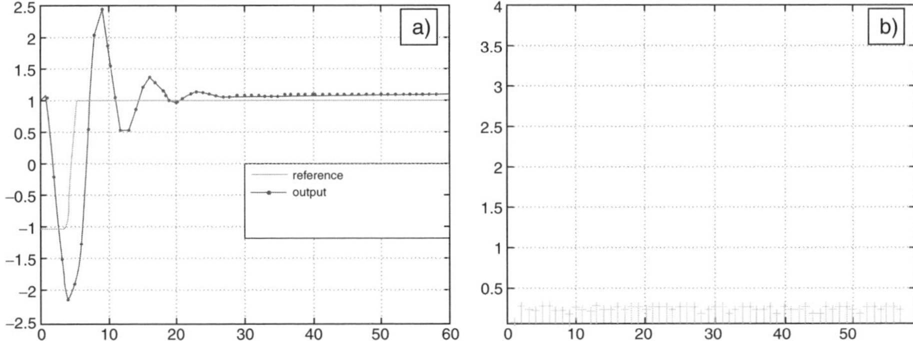

Figure 3.33. *WRR configuration (A) with an overload of 30%: (a) system output, (b) real-time frame delay in ms*

tuned. Moreover, the weights of the WRR scheduler have to be changed in order to offer more bandwidth to real-time traffic. Secondly, the Petri net model is then run in an iterative way to find a "good" configuration. The solution called the WRR configuration (B) is to provide 99% of the bandwidth for the real-time traffic and only 1% for the background traffic. Finally, a simple algorithm (implementing in the Ethernet device) allows to dynamically switch between the (A) and (B) configurations according to the real-time frame delays. Figure 3.35 shows the results when this algorithm is applied. When a real-time frame delay is greater than $3\ ms$ (Figure 3.35b), the configuration (B) is selected in order to reduce this delay and to maintain the stability of the system (Figure 3.35a).

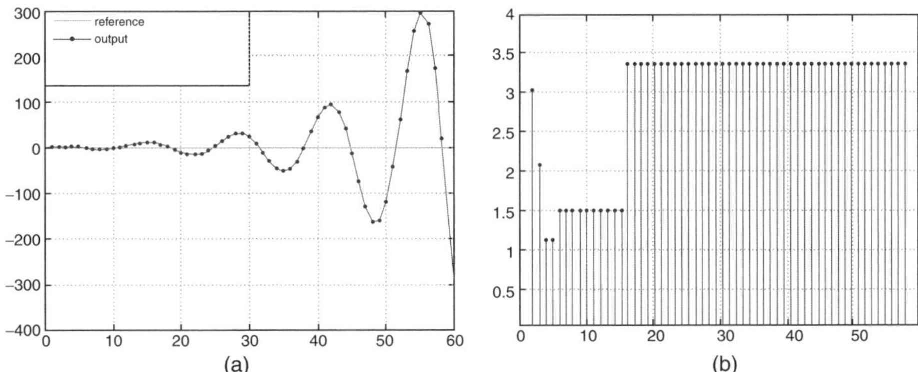

Figure 3.34. *WRR configuration (A) with an overload of 70%: (a) system output, (b) real-time frame delay in ms*

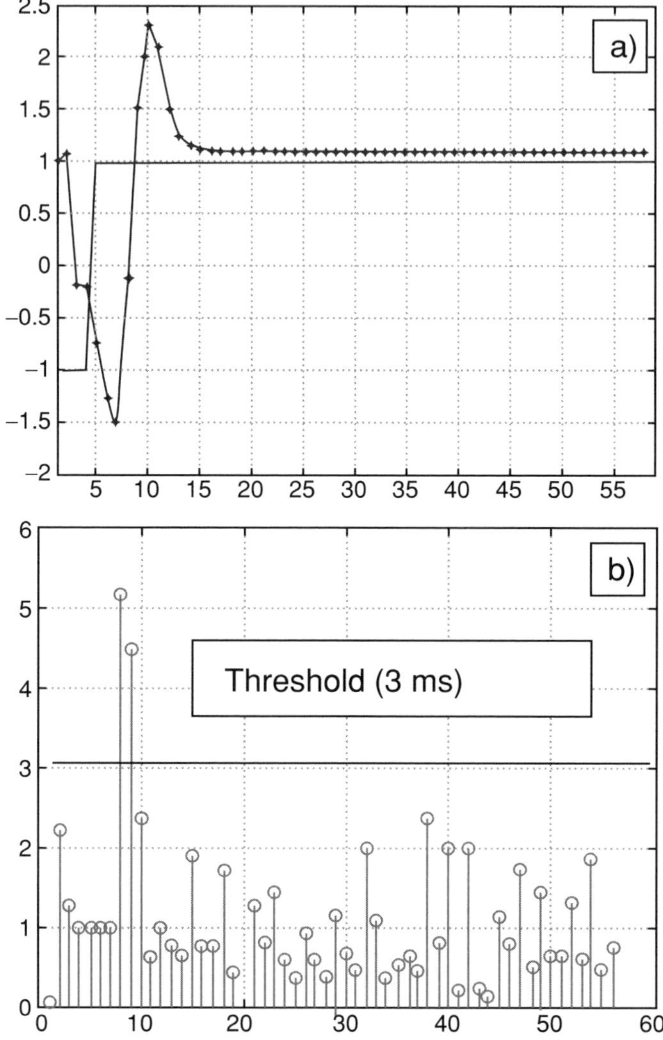

Figure 3.35. *Dynamic network reconfiguration - (a): system output*
(b): real-time frame delay in ms

3.4. Conclusion

Two approaches for network resource adaptation have been presented in this chapter. The first part shows the interest of a hybrid priority strategy for message scheduling on a network. Two applications with different needs in terms of transmission urgency in their messages flows (one with variable transmission urgency for its messages; another with constant needs) are distributed through the network. An important characteristic in an NCS context is the capacity to implement process control applications with good performances, whatever the network load is. We have precisely shown that message scheduling strategies, based on hybrid priority schemes, allow for implementing a distributed process control application, even if the network load is heavy. NCS requires a global modeling of all its components in order to be able to precisely evaluate its performances. The second part of the chapter shows that the PN formal language can express the behavior of communications, controllers and plants. It allows for easily tuning the network parameters by directly analyzing their effects on the system to be controlled. Moreover, PN can be used to test new QoS adaptive algorithms embedded inside Ethernet switches by dynamically tracking the current QoC level. For both approaches, the network is adapted according to the performances of the applications, which are expressed in terms of stability.

3.5. Bibliography

[ÅST 97] ÅSTRÖM K., WITTENMARK B., *Computer-Controlled Systems*, Information and System Sciences Series, Prentice Hall, 3rd edition, 1997.

[BIL 82] BILLINGTON J., Specification of the Transport service using Numerical Petri Nets, *Second International workshop specification, Testing and Verification, IFIP*, West Lafayette, USA, October 1982.

[BOS 91] BOSCH, *CAN specification 2.0 (A)*, www.semiconductors.bosch.de/pdf/can2-spec.pdf, 1991.

[BRA 07] BRAHIMI B., Integrated approach based on high level Petri nets to simulate and evaluate the networked control systems, PhD thesis, Henri Poincaré University Nancy I, France, 2007.

[CER 03] CERVIN A., HENRIKSSON D., LINCOLN B., EKER J., ÅRZÉN K.-E., How Does Control Timing Affect Performance? Analysis and Simulation of Timing Using Jitterbug and TrueTime, *IEEE Control Systems Magazine*, vol. 23, num. 3, p. 16–30, June 2003.

[CIA 02] CIA, *CAN, CANopen, DeviceNet*, URL www.CAN-CiA.de, 2002.

[DAV 04] DAVID R., ALLA H., *Discrete, Continuous, and Hybrid Petri Nets*, Springer-Verlag, Berlin, 2004.

[DEM 89] DEMERS A., KESHAV S., SHENKER S., Analysis and simulation of a fair queueing algorithm, *ACM SIGCOMM Computer Communication Review*, vol. 19, num. 4, p. 1–12, September 1989.

[DIO 08] DIOURI I., BERBRA C., GEORGES J.-P., GENTIL S., RONDEAU E., Evaluation of a Switched Ethernet network for the control of a quadrotor, *16th Mediterranean Conference on Control and Automation (MED'08)*, Ajaccio, France, p. 1112–1117, June 2008.

[HAL 88] HALEVI Y., RAY A., Integrated communication and control systems: Part I- Analysis, *ASME Journal of Dynamic Systems, Measurement and Control*, vol. 110, num. 4, p. 367–373, 1988.

[IEE 02] IEEE COMPUTER SOCIETY, IEEE standard for information technology - Telecommunications and information exchange between systems - Local and metropolitan area networks - Specific requirements - Part 3: Carrier sense multiple access with collision detection (CSMA/CD) access method and physical layer specifications, IEEE standard 802.3, Edition 2002, 2002.

[IEE 03] IEEE COMPUTER SOCIETY, IEEE Standards for local and metropolitan area networks - Virtual bridged local area networks, IEEE standard 802.1Q, Edition 2003, 2003.

[JEN 92] JENSEN K. (ed.), *Application and Theory of Petri Nets*, vol. 616 of *Lecture Notes in Computer Science*, Springer, 1992.

[JEN 07] JENSEN K., KRISTENSEN L., WELLS L., Coloured Petri Nets and CPN Tools for modelling and validation of concurrent systems, *International Journal on Software Tools for Technology Transfer (STTT)*, vol. 9, num. 3, p. 213–254, June 2007.

[JUA 04] JUANOLE G., DIAZ M., VERNADAT F., Réseaux de Petri étendus et méthodologie pour l'analyse de performances, Report num. LAAS 03480, Laboratoire d'Analyse et d'Architecture des Systèmes, Toulouse, France, 2004.

[JUA 05] JUANOLE G., MOUNEY G., CALMETTES C., PECA M., Fundamental Considerations for Implementing Control Systems on a CAN Network, *FET2005, 6th IFAC International conference on Fielbus Systems and their Applications*, Puebla, Mexico, November 2005.

[JUA 07a] JUANOLE G., MOUNEY G., Networked Control Systems: Definition an Analysis of a Hybrid Priority Scheme for the Message Scheduling, *Proceedings of the* 13th *IEEE conference on Embedded and Real-Time Computing Systems and Applications (RTCSA2007)*, Daegu, Korea, August 2007.

[JUA 07b] JUANOLE G., MOUNEY G., Using an hybrid traffic scheduling in networked control systems, *Proceedings European Control Conference 2007*, Kos, Greece, July 2007.

[JUA 08] JUANOLE G., MOUNEY G., CALMETTES C., On different priority schemes for the message scheduling in Networked Control Systems: Definition an Analysis of a Hybrid Priority Scheme for the Message Scheduling, *Proceedings of the* 16th *Mediterranean Conference on Control and Automation, MED'08*, Ajaccio, France, June 2008.

[JUR 58] JURY E. I., *Sampled-Data Control Systems*, Wiley, New York, 1958.

[KAN 92] KANG S., HAGBAE K., Derivation and Application of Hard Deadlines for Real-Time Control Systems, *IEEE Transaction On Systems, Man and Cybernetics*, vol. 22, num. 6, p. 1403–1413, 1992.

[KAT 91] KATEVENIS M., SIDIROPOULOS C., COURCOUBETIS C., Weighted round-robin cell multiplexing in a general purpose ATM switch chip, *IEEE Journal on Selected Areas in Communications*, vol. 9, num. 8, p. 1265–1279, October 1991.

[LAI 89] LAI R., DILLON T.-S., PARKER K.-R., Application of numerical Petri nets to specify ISO FTAM Protocol, *Singapore International Conference on Networks*, Singapore, July 1989.

[MAR 04] MARTI P., YEPEZ J., VELASCO M., VILLA R., FUERTES J., Managing quality-of-control in network-based control systems by controller and message scheduling co-design, *Industrial Electronics, IEEE Transactions on*, vol. 51, num. 6, p. 1159–1167, December 2004.

[OHL 07] OHLIN M., HENRIKSSON D., CERVIN A., TrueTime 1.5 – Reference Manual, Lund Institute of Technology, Sweden, January 2007.

[PAR 93] PAREKH A., GALLAGER R., A generalized processor sharing approach to flow control in integrated services networks: The single node case, *IEEE/ACM Transactions on Networking*, vol. 1, num. 2, p. 344–357, June 1993.

[WAL 01] WALSH G., YE H., Scheduling of networked control systems, *Control Systems Magazine, IEEE*, vol. 21, num. 1, p. 57–65, February 2001.

[ZAM 08] ZAMPIERI S., Trends in Networked Control Systems, *17th IFAC World Congress*, Seoul, Korea, p. 2886–2894, July 2008.

[ZUB 97] ZUBERI K., SHIN K., Scheduling messages on controller area network for real-time CIM applications, *IEEE Transactions On Robotics And Automation*, vol. 13, num. 2, p. 310–314, 1997.

[ZUB 00] ZUBERI K., SHIN K., Design and Implementation of Efficient Message Scheduling for Controller Area Network, *IEEE Transactions on Computers*, vol. 49, num. 2, p. 182–188, 2000.

Chapter 4

Plant-state-based Feedback Scheduling

4.1. Overview

A constant challenge in embedded systems development is represented by computational resource limitations. In fact, economic constraints impose the desired functionalities to be performed with the lowest cost. These limitations call for a more efficient use of the available resources. In this context, integrated control and scheduling methodologies have been proposed in order to allow a more flexible and efficient utilization of the computational resources [ÅRZ 00].

The problem of optimal sampling period selection, subject to schedulability constraints, was first introduced in [SET 96]. Considering a bubble control system benchmark, the relationship between the control cost (corresponding to a step response) and the sampling periods were approximated using convex exponential functions. Using the Karush–Kuhn–Tucker (KKT) first-order optimality conditions, the analytic expressions of the optimal off-line sampling periods were established. The problem of the joint optimization of control and off-line scheduling has been studied in [REH 04; LIN 02; BEN 06c].

The idea of feedback scheduling was introduced in [EKE 00; LU 02]. First approaches in feedback scheduling considered feedback from resource utilization (for example task execution times) in order to optimize the control performance [EKE 00; CER 02], or to minimize a deadline miss ratio in soft real-time systems [LU 02]. Naturally, the on-line adjustment of sampling periods calls for optimal sampling periods assignment. The approaches in [EKE 00; CER 02] used a similar method to [SET 96]

Chapter written by Mongi BEN GAID, David ROBERT, Olivier SENAME and Daniel SIMON.

in order to find analytic expressions of the optimal sampling periods, under cost approximation assumptions (linear or quadratic approximation of the cost as a function of the sampling period). The experimental evaluation of the feedback scheduling concept was undertaken in [SIM 05]. The issue of guaranteeing the stability and performance of the controlled systems, when their sampling periods are varied on-line (by a feedback scheduler, for example), was addressed in [ROB 07a], using the H_∞ approach for linear parameter varying systems.

Later, it was pointed out that the optimal sampling frequencies are also dependent on the controlled system actual state [MAR 02; MAR 04], and not only on off-line considerations. The problem of the optimal integrated control and scheduling was formalized (using a hybrid system approach) and solved in [BEN 06b]. Heuristics for integrated control and non-pre-emptive scheduling were proposed, in particular the OPP [BEN 06b] and RPP [BEN 06c] algorithms as well as the relaxed dynamic programming-based scheduling strategy [CER 06]. A common point to these heuristics is that the scheduling decisions (which task to execute or message to send) are determined on-line through the comparison of a finite number of quadratic functions of the extend state (actual state extended by previous controls). These quadratic cost functions are pre-computed off-line based on the intrinsic characteristics of the controlled systems. Another common point is that concurrency was modeled in a finely grained way. Related approaches were proposed in [DAČ 07], and where scheduling decisions are based on the discrepancies between current and the most recently transmitted values of nodes' signals. These latter results may be applied to the problem of dynamic scheduling of CAN networks.

Other approaches relied on the notion of periodic tasks and used task periods as variable scheduling decision modification. [SHI 99] considers adaptive scheduling for a set of controllers. With each controller is associated a cost function to model the quality of control (QoC) as a function of its period. In response to processor failures or to variations in the computing duties assigned to processors, heuristics assign control tasks to processors and adapt the control periods to optimize the global QoC.

In [HEN 05], the problem of the optimal sampling period selection of a set of LQG controllers, based on plant states knowledge, was studied. It has been shown that the optimal solution to this problem is too complicated. The optimal LQG cost, as a function of the sampling period, was depicted for some selected numerical examples. Explicit formulas, relating the optimal sampling periods to the plant state, were derived in the case of the minimum variance control of first-order plants. The issue of the choice of the feedback scheduler period was also studied. The same setting was considered in [CAS 06]. The on-line sampling period assignment was based on a look-up table, which was constructed off-line, for predefined values of the sampling periods. A heuristic procedure, allowing the construction of this look-up table, was also proposed.

Other approaches of state-based resource allocation were proposed, as in [TAB 07] and [LEM 07]. Although these approaches do not aim to optimize a global cost function, their objective is to allocate the computational resources in order to achieve other control objectives such as the asymptotic stability [TAB 07] or a specified l_2 attenuation level [LEM 07].

4.2. Adaptive scheduling and varying sampling robust control

A variable sampling rate appears to be a decisive actuator in scheduling and CPU load control. Although it is quite conservative, the LPV/H_∞-based design developed in section 2.4 guarantees plant stability and performance level, whatever the speed of variation of the control period inside its predefined range. Hence, the control task periods of such controllers can be adapted on-line by an external loop (the feedback scheduler) on the basis of resource allocation and global quality of service (QoS), with no further problems concerning the process control stability. Hence a quite simple scheduling controller can be used, e.g. like a simple re-scaling as proposed in [CER 03], or an elastic scheduler as in [BUT 00].

Indeed, besides the flexibility and robustness provided by an adaptive scheduling, a full benefit would come by taking into account directly the controlled process state in the scheduling loop. It has been shown in [EKE 00] that even for simple cases the full theoretical solution based on optimal control was too complex to be implemented in real time.

However, it is possible to sketch effective solutions suited for specific case studies, as depicted in Figure 4.1 taken from [ROB 07b].

A computing resource is shared between several process controllers. The computing power distribution between the process controllers is on-line adapted by a feedback scheduler. However, conversely, with the robot controller in section 1.4.2.3, the load allocation ratio between the control components is no longer constant and defined at

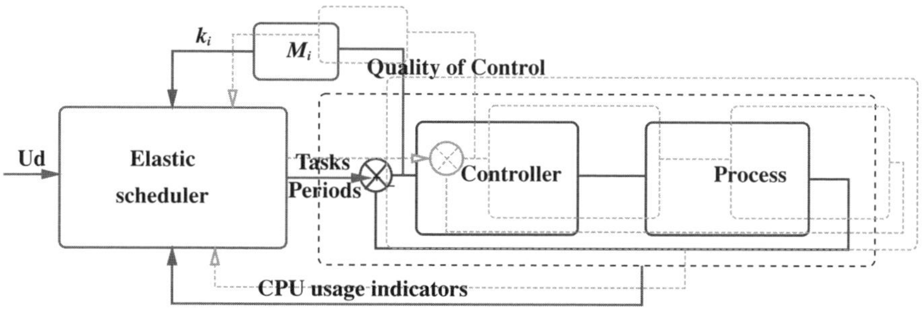

Figure 4.1. *Integrated control and scheduling loops*

design time. It is made dependent on the measure of the QoC to give advantage to the controller with higher control error.

4.2.1. *Extended elastic tasks controller*

The approach relies on a modified elastic scheduler algorithm [BUT 00], whose original objective is a distribution of the CPU utilization between n tasks, acting on their periods under the following constraints:

– the overall CPU load is smaller than the reference U_d: $\sum_{i=1}^{n} U_i \leq U_d$,

– the period of a control task is bounded: $h_{i_{min}} \leq h_i \leq h_{i_{max}}$,

– the CPU load distribution is balanced thanks to weights k_i.

The on-line scheduling adaptation reacts to changes in the overall load reference U_d, to variations in a task's execution time or to variations in a weight k_i. The scheduling is updated thanks to an iterative algorithm described in [BUT 00], where the CPU utilization U_d is shared in proportion of the weights k_i, accounting for the period bounds: for example Figure 4.2 depicts the temporal behavior of three tasks with initial weights $k_i = 1$, $i = 1,\ldots,3$. The horizontal axis represents the CPU use shared between the three tasks, along several load and weights configurations. In a first step, decreasing U_d induces a decreasing in the individual loads in equal proportion. In the second step, increasing the weight k_3 from 1 to 2 increases the CPU time allocated to task 3 while slackening equally the CPU power allocation for tasks 1 and 3. Increasing again k_3 from 2 to 3 cannot compress again task 1 CPU allocation which already reached its lower bound, therefore only task2'CPU allocation is reduced.

Indeed, the tasks set's behavior mimics the one of a springs chain with overall length U_d, where every spring has a stiffness k_i and a length U_i bounded between $U_{i_{min}}$ and $U_{i_{max}}$: recalling that the computing load U_i and the period h_i of a control

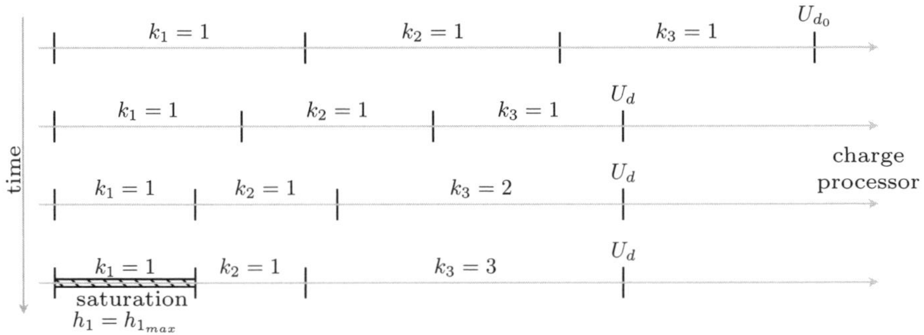

Figure 4.2. *Example of elastic tasks scheduling*

task are linked with $U_i = c_i/h_i$, where c_i is the execution time of an instance of task i), allows for computing the actual task periods.

To adapt the scheduling algorithm and task periods w.r.t. the actual performance of the process controllers, it is now necessary to enhance the elastic tasks algorithm to make the weights k_i depend on the measured QoC, as in the structure in Figure 4.1. The main problem consists of measuring an adequate image of the control performance and finding a function to link it with k_i.

In this first approach the QoC is measured via the mean square tracking error. It is evaluated on a time window equal to the feedback scheduler period. To keep the scheduling cost low this period is chosen larger than the process control periods.

The k_i weights are functions of the QoS measurements and are handled by the M_i components in Figure 4.1; in the simpler case they can be only static gains. The choice of the M_i gains must provide identical weights k_i for controllers with similar performance. However, the QoC measures are normalized to well balance the CPU allocation between process controllers with different dynamics.

The adaptation of the k_i weights as functions of the control performance is a feedback, whose stability and dynamics must be investigated. The relationship between k_i weights and the corresponding QoC would be analyzed, which appear to be very complex as it involves the behaviors of both the elastic scheduler and the process controllers. The scheduler is a nonlinear system where the task period are functions of the CPU load reference, the k_i weights, the period bounds and the varying execution time of the control tasks. The relationship between the control intervals and the corresponding control performance is also difficult to quantify as it depends on the process itself, on the controller and on exogenous signals. Note that even for a constant control period the quadratic tracking errors vary with exogenous signals and disturbances.

In consequence, it still out of the scope of the present analysis to rigorously chose the feedback scheduler gains. However, these gains may be empirically chosen as in the following example, studied in simulation using again TrueTime to take into account the coupling between continuous process control and real-time scheduling.

4.2.2. *Case study*

The case study consists of the control of two pendulums sharing a common computing resource, one is the "T" pendulum already used in the example of section 2.4.4, the other one is a straight (stable) pendulum. Each pendulum is controlled by a LPV/H_∞ controller designed as in section 2.4, so that the stability of the position control loops is guaranteed whatever the variations of the control intervals, provided that they stay inside the bounds use for the synthesis. The allowed control intervals are chosen according to the desired closed-loop bandwidth and capabilities of the process, they are [1,3] ms for the T pendulum and [4,12] ms for the stable one. As in the case detailed in section 2.4.4 Taylor's expansion is truncated at the order 2, leading to reduced polytopes with three vertex and to a simple convex combination of three state-feedback elementary controllers at run time.

The pendulum controllers are implemented as real-time tasks running under control of a pre-emptive and fixed priority RTOS. The control performance of each pendulum is measured by the mean quadratic tracking error over a feedback scheduler's period (5 s). The pendulums are driven by a sinusoidal reference. At $t = 12$ s, a disturbing task with intermediate priority appears. Noise is injected on the stable pendulum measure at time $t = 20$ s. Simulation results are plotted in Figure 4.3 for the scheduling parameters and in Figure 4.4 for the plant outputs, quadratic tracking error and computing resource share between the controllers. This latter plot is equal to 1 when all the CPU is allocated to the T pendulum and 0 when it is assigned only to the stable pendulum.

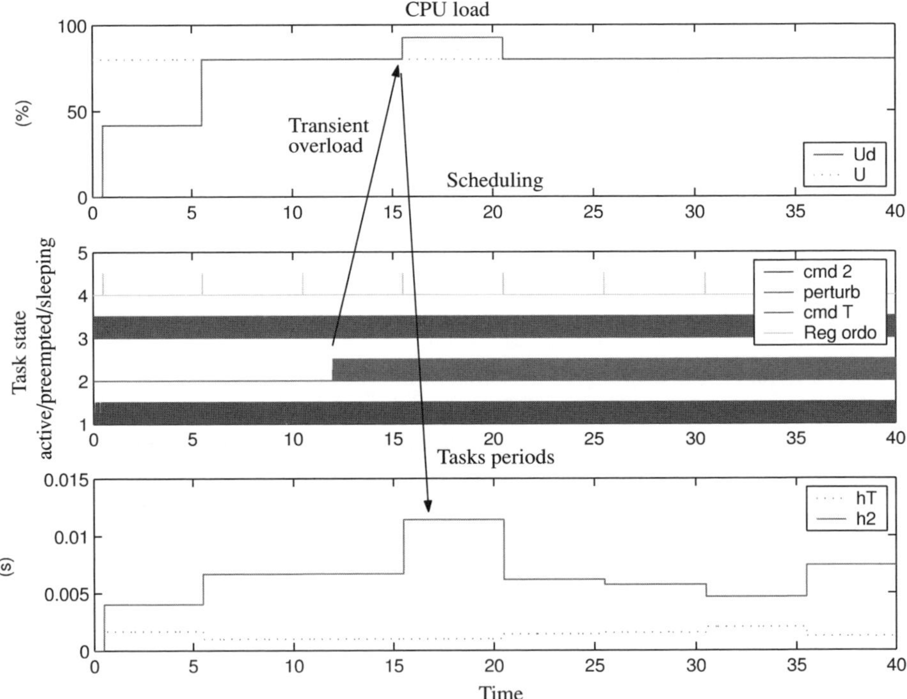

Figure 4.3. *Scheduling behavior*

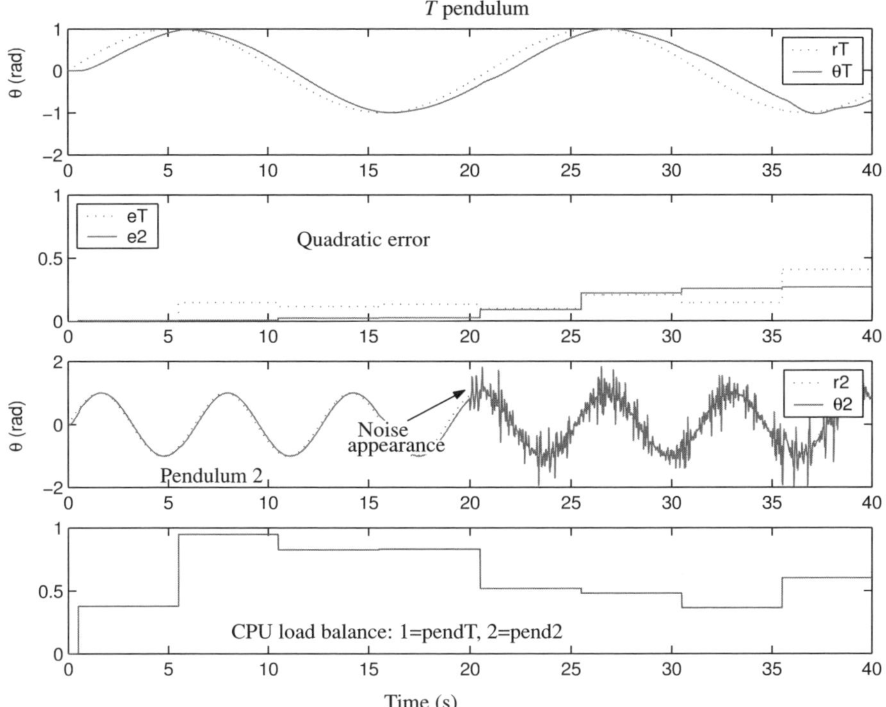

Figure 4.4. *Pendulums behavior*

The appearance of the disturbing task at $t = 12$ s induces a CPU overload which is rapidly canceled by the feedback scheduler, in one scheduling interval, as there is no filtering in this elastic scheduler. At this time, the CPU share is mainly allocated to the T pendulum so that it is weakly disturbed by the added task, while the second pendulum is subject to a larger control interval increase. The noise added to the second pendulum at $t = 20$ s increases the corresponding tracking error, therefore inducing a re-allocation of CPU power in favor of the second pendulum. This is made at the cost of increasing the control period of the first pendulum which in turns increases its performance index, thus claiming for additional computing power (at $t = 36$).

The approach is simple to implement and, even if only tested in simulation up to now, has shown significant performance improvements compared with more simple (i.e. control quality unaware) resource allocation. However, the behavior and performance of the overall control system depend on numerous parameters, such as the period of the feedback scheduler, normalization factors between the plants, optional filters, etc.

A rational setting of these parameters needs a better understanding of the coupling between the control performance and the scheduling parameters. As said before, analyzing the stability of this feedback scheduling loop, gathering the complex dynamics of the plants and of the control system, remains to be done. It requires an adequate modeling of the relationships between the control quality and the scheduling parameters and seems to be out of reach in a general case. However, some restrictive assumptions on the plant model (e.g. linearity) and on the control algorithms (e.g. LQ control) may lead for tractable solutions as shown in the following section.

4.3. MPC-based integrated control and scheduling

Model predictive control (MPC) has received increased industrial acceptance during recent years, mainly because of its ability to handle constraints explicitly and the natural way in which it can be applied to multi-variable processes [GAR 89]. MPC is based on iterative, finite horizon optimization of a plant model. At time t the current plant state is sampled, and a cost minimizing control strategy is computed (via a numerical minimization algorithm) for a receding horizon in the future: $[t, t + T]$. Only the first step of the control strategy is implemented, then the plant state is sampled again, and the calculations are repeated starting from the new current state, yielding a new control and new predicted state path.

The computational requirements of MPC, where typically a quadratic optimization problem is solved on-line in every sample, have previously prohibited its application in areas where fast sampling is required. Therefore, MPC has traditionally only been applied to slow processes, mainly in the chemical industry. However, the advent of faster computers and the development of more efficient optimization algorithms, e.g. [CAN 01], has led to applications of MPC to processes governed by faster dynamics. However, much still remains to be done to develop efficient real-time implementations of MPC.

The execution of an MPC controller is based on two main parameters: the sampling period and the receding horizon, for which optimization is computed. From a temporal point of view, MPC controllers are characterized by large execution times, but also by large variations of these durations from sample to sample. Hence, the large variations in execution time for MPC tasks make a real-time design based on worst-case bounds very conservative and give an unnecessary long sampling period. As usual, the robustness of closed-loop control can be exploited, so that more flexible implementation schemes are expected to provide better use of the execution resources and make MPC applicable to a larger scope than in the current case.

Feedback scheduling has been, for example, applied to MPC in [HEN 02; HEN 06]. The method uses feedback information from the optimization algorithm to find when to terminate the current iterative optimization. The goal is to find the best trade-off between performance increase due to numerous iterations and the degradation due to very long computations and induced latencies.

Joint control and scheduling may combine both control laws, e.g. based on the MPC concept, working together with an existing scheduling policy to manage the network QoS. Dynamic feedback scheduling policies combined with predictive control have been proposed in [ZHA 08] to cope with network induced delays in the control loops, where the scheduling policy, e.g. Rate Monotonic and dynamic feedback policies, allows us to constrain the delay upper bound. [MIL 08] presents a model predictive controller with an implementation scheme based on a queuing-selecting method and an estimator to compensate for data losses when data packets are dropped. The associated feedback scheduler is designed to minimize the traffic over the network due to the measurement flow needed by the controller.

Control and scheduling co-design, combining the MPC approach within the framework of resource-constrained systems, has been successfully developed in [BEN 06a; BEN 06b; BEN 09]. A summary of this work is given in the following sections.

4.3.1. *Resource constrained systems*

In this section, an abstract view of a distributed embedded control system operating under communication constraints is presented. This abstract view is described by the class of *computer-controlled systems*, which was introduced by Hristu in [HRI 99]. This class allows us to model, in a finely grained and abstract way, the impact of the resource limitations on the behavior of the controlled system. In the following, we will rather use the term of *resource-constrained systems* to refer to this class of systems. In [BEN 06a], it has been shown that a resource-constrained systems may be modeled in the *mixed logical dynamical* (MLD) framework, which represents a modeling framework for hybrid systems, and which was introduced by Bemporad and Morari in [BEM 99]. A summary of these results is given in the following. Consider the continuous-time LTI plant described by

$$\dot{x}_c(t) = A_c x_c(t) + B_c u_c(t) \qquad (4.1)$$

$$y_c(t) = C_c x_c(t), \qquad (4.2)$$

where $x_c(t) \in \mathbb{R}^n$, $u_c(t) \in \mathbb{R}^m$, and $y_c(t) \in \mathbb{R}^p$ represent, respectively, the state, the command input, and the output. The plant is controlled by a discrete-time controller, with sampling period T_s. The plant (4.2) and the controller are connected through a limited bandwidth communication bus. At each sampling instant $t = kT_s$ ($k \in \mathbb{N}$), the bus can carry at most b_r measures and b_w control commands, with $b_r \leq p$ and $b_w \leq m$. The input to the plant is preceded by a zero-order holder, which maintains the last received control commands constant until new control values are received. Let

$u(k)$ be the input of the zero-order holder at instant kT_s, then its output is given by

$$u_c(t) = u(k) \quad \text{if } kT_s \leq t < (k+1)T_s. \tag{4.3}$$

Let $x(k) = x_c(kT_s)$ and $y(k) = y_c(kT_s)$ be respectively the sampled values of the state and the output. A discrete-time representation of the plant (4.2) at the sampling period T_s is given by

$$\begin{aligned} x(k+1) &= Ax(k) + Bu(k) \tag{4.4} \\ y(k) &= Cx(k), \tag{4.5} \end{aligned}$$

where $A = e^{A_c T_s}$, $B = \int_0^{T_s} e^{A_c \tau} B_c \, d\tau$ and $C = C_c$.

It is assumed, throughout this section, that the pairs (A, B) and (A, C) are, respectively, reachable and observable. These assumptions are systematically satisfied if the pair (A_c, B_c) is reachable, the pair (A_c, C_c) is observable, and the sampling period T_s *non-pathological*. A sampling period is said pathological if it causes the loss, for the sampled-data model, of the reachability and observability properties, which were verified by the continuous model before its discretization. In [KAL 63], Kalman *et al.* have proved that the set of pathological sampling period is countable, and uniquely depends on the eigenvalues of the state matrix A_c. Consequently, in order to avoid the loss of reachability and observability, which may be caused by the sampling, it is sufficient to choose T_s outside this set.

Communication constraints may be formally described by introducing two vectors of Booleans $\sigma(k) \in \{0,1\}^{b_r}$ and $\delta(k) \in \{0,1\}^{b_w}$, defined for each sampling instant k.

DEFINITION.– *The vector $\sigma(k)$ defined by*

$$\begin{cases} \sigma_i(k) = 1 & \textit{if } y_i(k) \textit{ is read by the controller at instant } k, \\ \sigma_i(k) = 0 & \textit{otherwise} \end{cases}$$

is called sensors-to-controller scheduling vector at instant k.

DEFINITION.– *The vector $\delta(k)$ defined by*

$$\begin{cases} \delta_i(k) = 1 & \textit{is } u_i(k) \textit{ updated at instant } k, \\ \delta_i(k) = 0 & \textit{otherwise} \end{cases}$$

is called controller-to-actuators scheduling vector at instant k.

The vector $\sigma(k)$ indicates the measures that the controller may read at instant k. In a similar way, the $\delta(k)$ indicates the control inputs to the plant that the controller may update at instant k. The introduction of the scheduling vectors allows us to model in a simple way the communication constraints. The limitations that affect the transmission of the measures to the controller may be described by the following inequality:

$$\sum_{i=1}^{p} \sigma_i(k) \leq b_r. \tag{4.6}$$

In a similar way, the limitations concerning the sending of the control commands to the actuators may be modeled by

$$\sum_{i=1}^{m} \delta_i(k) \leq b_w. \tag{4.7}$$

The last received control inputs (through the communication bus) are kept constant. Consequently, if a control input is not updated at the kth sampling period, then it is maintained constant. This assertion may be modeled by the logic formula

$$\delta_i(k) = 0 \implies u_i(k) = u_i(k-1). \tag{4.8}$$

The plant, the analog-to-digital and digital-to-analog converters, the communication bus, and the controller are schematically depicted in Figure 4.5. In this figure, $\eta(k) \in \mathbb{R}^{b_r}$ represents the vector of partial measurements that the controller receives (through the communication bus) at the sampling period k. In a similar way, vector $v(k) \in \mathbb{R}^{b_w}$ represents the vector of partial control commands that the controller may send to the actuators (through the limited bandwidth communication bus) at the sampling period k. Blocks D/A and A/D, respectively, represent the digital-to-analog and analog-to-digital converters. The controller may also assign the values of the sensors-to-controller scheduling vector ($\sigma(k)$) as well as the controller-to-actuators scheduling vector ($\delta(k)$).

Knowing $v(k)$ and relation (4.8), $u(k)$ is given by

$$\begin{cases} u_i(k) = v_j(k) & \text{if } \delta_i(k) = 1 \text{ and } \sum_{l=1}^{i} \delta_l(k) = j, \\ u_i(k) = u_i(k-1) & \text{otherwise.} \end{cases} \tag{4.9}$$

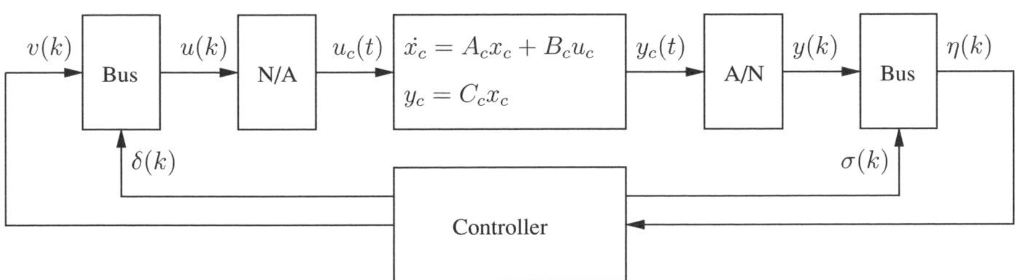

Figure 4.5. *Schematic representation of a resource-constrained system*

In the same way, the input $\eta(k)$ to the controller is defined by

$$
\begin{cases}
\eta_i(k) = y_j(k) & \text{if } \sigma_j(k) = 1 \text{ and } \sum_{l=1}^{j} \sigma_l(k) = i, \\
\eta_i(k) = 0 & \text{otherwise.}
\end{cases}
\tag{4.10}
$$

Equations (4.5), (4.6), (4.7), (4.9), and (4.10) describe a model where dynamics and plant performance are tightly coupled with the assignment of communication resources. In the particular case where $b_r = p$, $b_w = m$, $\sigma(k) = 1_{p,1}$ and $\delta(k) = 1_{m,1}$, for all $k \in \mathbb{N}$, this model coincides with the classical model of a sampled-data system. The presence of the communication bus moves the classical frontier between "the plant" and "the controller". In fact, for sampled-data systems, this frontier lies at the digital-to-analog and analog-to-digital converters. In the considered model, this frontier moves to the communication bus interface. The *resource-constrained system* is defined as the entity constituted by the sampled-data model of the plant and the communication bus. The formal definition of a resource-constrained system is given thereafter.

DEFINITION.– *A resource-constrained system is a mixed logical dynamical system having three inputs: the command input $v(k)$, the scheduling vector of the sensors-to-controller link $\sigma(k)$ and the scheduling vector of the controller-to-actuators link $\delta(k)$. It has one output denoted $\eta(k)$. Its mathematical model is defined by:*

– a recurrent equation (4.5) describing the sampled dynamics of the plant,

– inequality constraints (4.6) and (4.7) expressing the limitations of the communication medium,

– logic formulas (4.9) describing the mapping of the computed controller outputs $v(k)$ to plant inputs $u(k)$, knowing the scheduling decisions $\delta(k)$,

– logic formulas (4.10) describing the mapping of the sampled plant outputs $y(k)$ to the controller's inputs $\eta(k)$, knowing the scheduling decisions $\sigma(k)$.

The particularity of a resource-constrained system, compared to a sampled-data system, is that at each sampling period, it is important to determine:

– the measures that should be acquired (it is only possible to acquire at most b_r measures, defined by the scheduling function $\sigma(k)$),

– the control commands that should be applied (it is only possible to apply at most b_w control commands, defined by the scheduling function $\delta(k)$),

– the value of the applied control commands.

4.3.2. *Optimal integrated control and scheduling of resource constrained systems*

This section considers a resource-constrained system $\mathcal{S}$ where the full state vector $x(k)$ is available to the controller at each sampling period. It is shown how the use of the model predictive control approach may be seen as an algorithmic solution allowing computing on-line, at the same time, the optimal values of the control signals and the communication scheduling of resource-constrained systems.

Using MPC, an optimal control problem is solved on-line at each sampling period T_s. It aims at finding the optimal control values sequence

$$\hat{u}^{N-1^*} = (\hat{u}^*(0), ..., \hat{u}^*(N-1))$$

and the optimal communication sequence $\hat{\delta}^{N-1^*} = (\hat{\delta}^*(0), ..., \hat{\delta}^*(N-1))$, which are solutions of the following optimization problem:

$$\begin{cases} \min_{\hat{u}^{N-1}, \hat{\delta}^{N-1}} \sum_{h=0}^{N-1} \begin{bmatrix} \hat{x}(h) \\ \hat{u}(h) \end{bmatrix}^T Q \begin{bmatrix} \hat{x}(h) \\ \hat{u}(h) \end{bmatrix} + \hat{x}^T(N) Q_0 \hat{x}(N) \\ \text{subject to} \\ \hat{x}(0) = x(k) \\ \hat{x}(h+1) = A\hat{x}(h) + B\hat{u}(h), \, h \in \{0, \ldots, N-1\} \\ \sum_{i=1}^{m} \hat{\delta}_i(h) = b, \, h \in \{0, \ldots, N-1\} \\ \hat{\delta}_i(0) = 0 \Longrightarrow \hat{u}_i(0) = u_i(k-1) \\ \hat{\delta}_i(h) = 0 \Longrightarrow \hat{u}_i(h) = \hat{u}_i(h-1), \, h \in \{1, \ldots, N-1\}. \end{cases} \qquad (4.11)$$

The solution of this problem is based on the prediction of the future evolution of the system over a horizon of N sampling periods. This predicted evolution is calculated according to the model of the plant, knowing the current state $x(k)$ of the system. The variables $\hat{x}(h)$, $h \in \{0, \ldots, N\}$ represent the predicted values of system states $x(k+h)$. The sequences $(\hat{u}(0), \ldots, \hat{u}(N-1))$ (virtual control sequence) and $(\hat{\delta}(0), \ldots, \hat{\delta}(N-1))$ (virtual communication sequence) are called virtual sequences, because they are based on the predicted evolution of the system. The resolution of this problem aims at finding the optimal virtual control sequence $(\hat{u}^*(0), \ldots, \hat{u}^*(N-1))$ and the optimal virtual communication sequence $(\hat{\delta}^*(0), \ldots, \hat{\delta}^*(N-1))$ that minimize a quadratic cost function over a finite horizon of N sampling periods. Assuming that the optimal virtual sequences exist, the actual control commands are obtained by setting

$$v(k) = M_{\hat{\delta}^*}(0)\hat{u}^*(0) \qquad (4.12)$$

and

$$\delta(k) = \hat{\delta}^*(0) \qquad (4.13)$$

and disregarding the remaining elements

$$(\hat{u}^*(1), \ldots, \hat{u}^*(N-1)) \text{ and } (\hat{\delta}^*(1), \ldots, \hat{\delta}^*(N-1)).$$

At the next sampling period (step $k+1$), the whole optimization procedure is repeated, based on $x(k+1)$.

The optimality of the model predictive controlled may be proved if an infinite horizon cost function $J(\tilde{x}, v, \delta, 0, +\infty) = J(x, u, 0, +\infty)$ is used and if the prediction horizon N is chosen infinite. At each time step k, and for any extended state $\tilde{x}(k)$, the model predictive controller over an infinite horizon computes the optimal solutions $v^*(k)$ and $\delta^*(k)$ that minimizes the cost function $J(\tilde{x}, v, \delta, k, +\infty)$, subject to the communication constraints. Its optimality directly results from the Bellman optimality principle, which states:

DEFINITION.– *An optimal policy has the property that whatever the initial state and the initial decision are, the remaining decisions must constitute an optimal policy with respect to the state resulting from the first decision.*

In many practical situations, it is sufficient to choose the prediction horizon N bigger enough than the response time of the system to get a performance that is close to the optimality. This is possible when the virtual sequences of the optimal control commands, which are computed at each sampling period, converge exponentially to zero as the horizon increases. The obtained finite horizon solution will then approximate the optimal infinite horizon solution.

However, the on-line solving of the optimization algorithm, which is required by the MPC approach, is very costly. For that reason, an on-line scheduling algorithm, called OPP was proposed in [BEN 06a; BEN 06b]. While being based on a pre-computed optimal off-line schedule, OPP makes it possible to allocate on-line the communication resources, based on the state of the controlled dynamical systems. It was shown that under mild conditions, OPP ensures the asymptotic stability of the controlled systems and enables in all the situations the improvement of the control performance compared to the basic static scheduling. Furthermore, under these conditions, the determination of OPP control and scheduling amounts to comparing a limited number of quadratic functions of the state.

4.4. A convex optimization approach to feedback scheduling

4.4.1. *Problem formulation*

Consider a collection of N continuous-time LTI systems $\{\mathcal{S}_i\}_{1 \leq i \leq N}$. Each system $\mathcal{S}_i$ is described by the state space representation

$$\dot{x}_i(t) = A_i x_i(t) + B_i u_i(t), \tag{4.14}$$

where $x_i \in \mathbb{R}^{n_i}$ and $u_i \in \mathbb{R}^{m_i}$. An infinite horizon continuous-time cost functional J_i, defined by

$$J_i(x_i, u_i) = \int_0^\infty \left(x_i^T(t) Q_i x_i(t) + u_i^T(t) R_i u_i(t) \right) dt, \tag{4.15}$$

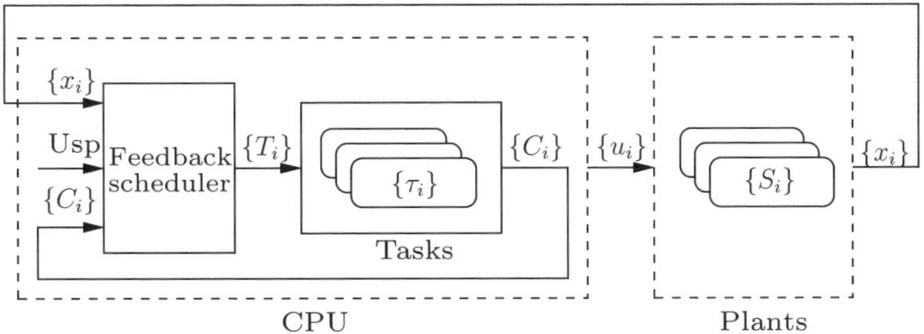

Figure 4.6. *Integrated plant-state and execution-time feedback scheduling*

is associated with $\mathcal{S}_i$, and represents the design specifications of its ideal controller. It is assumed that Q_i and R_i are positive definite matrices of appropriate dimensions and that the pair (A_i, B_i) is reachable. Each system $\mathcal{S}_i$ is controlled by a control task τ_i, characterized by a period h_i and an execution-time C_i. These two parameters may be time varying. The N control tasks $\{\tau_i\}_{1 \leq i \leq N}$ are executed on the same processor. A global cost functional $J(x_1, \ldots, x_N, u_1, \ldots, u_N)$, defined by

$$J(x_1, \ldots, x_N, u_1, \ldots, u_N) = \sum_{i=1}^{N} \omega_i J_i(x_i, u_i), \tag{4.16}$$

is associated with the entire system, allowing the evaluation of its global performance. Constants $\{\omega_i\}_{1 \leq i \leq N}$ are weighting factors, representing the relative importance of each control loop.

The main objective of this paper is to design a feedback scheduler, allowing task periods to be assigned $\{h_i\}_{1 \leq i \leq N}$ that optimize the global control performance (defined by J), subject to processor utilization constraints (defined by U_{sp}), and based on both task execution time $\{C_i\}_{1 \leq i \leq N}$ and plant state measurements $\{x_i\}_{1 \leq i \leq N}$, as shown in Figure 4.6.

Remark. U_{sp} represents the desired processor utilization of tasks whose periods are controlled by the feedback scheduler. It may be chosen by the designer in order to cope with the presence of other tasks, whose processor utilization is not controlled by the feedback scheduler. In practice, even in the situations where all the tasks are controlled by the feedback scheduler, choosing U_{sp} less than schedulable utilization bound allows a "utilization margin" to be obtained and to avoid overruns that may result from the variations of task execution times.

4.4.2. *Cost function definition and approximation*

4.4.2.1. *Cost function definition*

For a given fixed sampling period h_i of system $\mathcal{S}_i$, assume that there exists an optimal sampled-data controller u^*_{i,h_i}, defined by the state-feedback control gain $K^*_{h_i}$, and which minimizes the cost functional (4.15), subject to the plant model (4.14) and to zero-order hold constraints

$$u_{i,h_i}(t) = u_{i,h_i}(kh_i) \quad \text{for } kh_i \leq t < (k+1)h_i. \tag{4.17}$$

The expression of $K^*_{h_i}$ may be found in control textbooks, for example [ÅST 97]. The computation of $K^*_{h_i}$ requires the resolution of an algebraic Riccati equation (ARE).

Let $t_i(k)$ be the kth instant where the control input u_i is updated and

$$J_i(t_i(k), x_i, u^*_{i,h_i}) = \int_{t_i(k)}^{\infty} \left(x_i^T(t)Q_i x_i(t) + u^{*T}_{i,h_i}(t)R_i u^*_{i,h_i}(t) \right) dt. \tag{4.18}$$

An interesting property in optimal LQ sampled-data control is that the cost functional $J_i(t_i(k), x_i, u^*_{i,h_i})$ may be characterized by a unique positive definite matrix $S_i(h_i)$ of size $n_i \times n_i$, which is the solution of the ARE. This property considerably simplifies the computation of the cost function (4.18), when the optimal sampled-data control u^*_{i,h_i} is used. In fact, instead of simulating the evolution of the sampled-data system (4.14), (4.17) and using equation (4.18) for cost computation, it suffices to use the formula

$$J_i(t_i(k), x_i, u^*_{i,h_i}) = x_i(t_i(k))^T S_i(h_i) x_i(t_i(k)).$$

In the following, the solution of the ARE associated with the problem of finding the optimal continuous time controller $u^*_{i,c}$ is denoted by S^c, which minimizes the cost functional (4.15), subject to plant dynamics (4.14). The QoC measure, associated with each system, will be the difference between the optimal sampled-data cost $J_i(x_i, u^*_{i,h_i})$ and the optimal continuous-time cost $J_i(x_i, u^*_{i,c})$:

$$\begin{aligned} J_i^*(x_i(0), h_i) &= J_i\left(x_i, u^*_{i,h_i}\right) - J_i\left(x_i, u^*_{i,c}\right) \\ &= x_i(0)^T \left(S_i(h_i) - S_i^c\right) x_i(0), \end{aligned}$$

and similarly

$$\begin{aligned} J_i^*(x_i(t_i(k)), h_i) &= J_i\left(t_i(k), x_i, u^*_{i,h_i}\right) - J_i\left(t_i(k), x_i, u^*_{i,c}\right) \\ &= x_i(t_i(k))^T \left(S_i(h_i) - S_i^c\right) x_i(t_i(k)). \end{aligned}$$

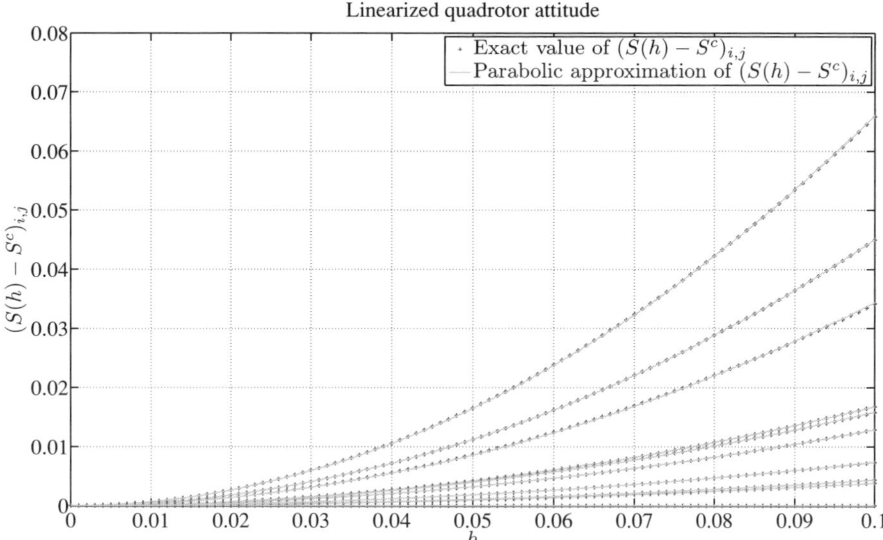

Figure 4.7. *X4 quadrotor: cost coefficients vs. sampling period*

4.4.2.2. *Introductory example: quadrotor attitude control*

Consider the linearized model of the attitude of the quadrotor, which was described in section 2.3.7 of Chapter 2.

The blue "+" marks in Figure 4.7 represent the values of the different coefficients of matrix $(S(h) - S^c)$, as a function of the sampling period. It is easy to see that the coefficients of $(S(h) - S^c)$ may be approximated as a parabolic function of h. The green curve in Figure 4.7 represents the mean square best-fitting parabola of $(S(h) - S^c)$. Using a basic linear least-squares method, $S(h) - S^c$ was approximated as

$$S(h) - S^c \approx \Theta h^2,$$

where

$$\Theta = \begin{bmatrix} 4.5028 & 0 & 0 & 1.2894 & 0 & 0 \\ 0 & 6.6017 & -0.0000 & 0 & 1.6797 & -0.0000 \\ 0 & -0.0000 & 3.4407 & 0 & -0.0000 & 1.5872 \\ 1.2894 & 0 & 0 & 0.3815 & 0 & 0 \\ 0 & 1.6797 & -0.0000 & 0 & 0.4336 & -0.0000 \\ 0 & -0.0000 & 1.5872 & 0 & -0.0000 & 0.7338 \end{bmatrix}.$$

This parabolic evolution of the cost was observed in two other benchmarks: a linearized model of an unstable pendulum (for $0 \leq h \leq 100$ ms) and a 14-order car active suspension system [BEN 06b] (for $0 \leq h \leq 15$ ms), as illustrated in [BEN 08].

These examples illustrate that in many situations, it is possible to approximate the relationship between the solutions of the Riccati equation and the sampling period, using polynomial interpolations, over a defined range of sampling periods. In the remainder of this paper, it is assumed that for $0 \leq h_i \leq h_i^{\max}$:

$$S_i(h_i) - S_i^c \approx \Theta_i h_i^2. \tag{4.19}$$

Note that although only this approximation is considered in this paper, the obtained results may be easily generalized to other polynomial approximations.

It is worth remarking that

– the choice of $h_i^{\max}$ depends on the quality and the validity of approximating the true values of the Riccati matrix coefficient using parabolic functions;

– based on Riccati equation solution approximations, analytic expressions of the control gains as a function of the sampling period may easily be deduced;

– the relationship between the cost function and the sampling frequencies may become more complicated, and even non-convex, when the frequencies are decreased to near the Nyquist rate, as illustrated in [EKE 00].

4.4.3. *Optimal sampling period selection*

4.4.3.1. *Problem formulation*

Let $f_i = \frac{1}{h_i}$ be the sampling frequency (corresponding to the sampling period h_i). Assume that C_i is constant (the following subsection shows how the time variations of C_i may be handled). The optimal sampling period frequency selection problem may be formulated as follows:

$$\left\{ \begin{array}{l} \min\limits_{f_1,\ldots,f_N} \sum\limits_{i=1}^{N} \omega_i \dfrac{x_i^T \Theta_i x_i}{f_i^2} \\ \text{subject to:} \\ \sum\limits_{i=1}^{N} C_i f_i \leq U_{sp} \\ f_i \geq f_i^{\min} \text{ for } i = 1,\ldots,N, \end{array} \right. \tag{4.20}$$

where $f_i^{\min} = \frac{1}{h_i^{\max}}$.

Remark. In optimization problem (4.20), the values of the sampling frequencies are implicitly upper-bounded by the processor utilization constraint. Furthermore, it is also straightforward to add upper-bound constraints on the sampling frequencies (i.e. by adding constraints $f_i \leq f_i^{\max}$ or equivalently $h_i^{\min} \leq h_i$). However, these additional constraints may lead to a slight increasing complexity of the problem resolution.

4.4.3.2. *Problem solving*

Problem (4.20) has a convex objective function and affine inequality constraints. Consequently, if the feasibility region is non-empty, its optimal solution will exist, and may be computed analytically using the Karush–Kuhn–Tucker (KKT) conditions [BOY 04]. Analysis of the different conditions of KKT conditions leads to the following algorithm (4.1) for the computation of the optimal sampling frequencies:

Algorithm 4.1: Optimal sampling frequencies computation

Compute $\beta_i = \omega_i x_i^T \Theta_i x_i$;

Compute $\Gamma_i = \dfrac{2\beta_i}{C_i f_i^{\min 3}}$;

Sort Γ_i in the increasing order (i.e. find the permutation φ so that $\Gamma_{\varphi(1)} \le \Gamma_{\varphi(2)} \le \cdots \le \Gamma_{\varphi(N)}$)

Determine the largest integer p so that

$$\sum_{i=1}^{p} C_{\varphi(i)} f_{\varphi(i)}^{\min} + \sum_{i=p+1}^{N} C_{\varphi(i)} \Big(\frac{\beta_{\varphi(i)} C_{\varphi(p)}}{\beta_{\varphi(p)} C_{\varphi(i)}}\Big)^{\frac{1}{3}} f_{\varphi(p)}^{\min} \ge U_{sp} \ ;$$

if $1 \le i \le p$ **then**

$\quad\big|\quad f_{\varphi(i)}^{*} = f_{\varphi(i)}^{\min}$;

endif

if $p + 1 \le i \le N$ **then**

$$\quad\big|\quad f_{\varphi(i)}^{*} = \Big(\frac{\beta_{\varphi(i)}}{C_{\varphi(i)}}\Big)^{\frac{1}{3}} \frac{U_{sp} - \sum_{j=1}^{p} C_{\varphi(j)} f_{\varphi(j)}^{\min}}{\sum_{j=p+1}^{N} \beta_{\varphi(j)}^{\frac{1}{3}} C_{\varphi(j)}^{\frac{2}{3}}} \cdot \ ;$$

endif

Algorithm 4.1 results from application of KKT conditions (see [BEN 08] for a complete proof).

4.4.3.3. *Feedback-scheduling algorithm deployment*

The feedback scheduler is executed as a periodic task, with period h_{fbs}. The choice of this period is a trade-off between the complexity of the feedback scheduler and the performance improvements it brings, as illustrated in [HEN 05].

Task execution times may be estimated on-line and smoothed using a first-order filter

$$\hat{C}_i(kh_{\mathrm{fbs}}) = \lambda \hat{C}_i((k-1)h_{\mathrm{fbs}}) + (1 - \lambda)C_i(kh_{\mathrm{fbs}}), \tag{4.21}$$

where λ is a forgetting factor, $\hat{C}_i(kh_{\mathrm{fbs}})$ and $C_i(kh_{\mathrm{fbs}})$ are, respectively, the estimated and the measured execution times at instant kh_{fbs}.

In practice, using algorithm 4.1, the optimal sampling frequencies of a plant tend to be reduced to zero as the plant approaches the equilibrium. This has the drawback of reducing the disturbance rejection abilities of that plant. Another drawback is that when all the plants approach equilibrium, coefficients β_i tend to approach zero. The optimal sampling frequencies assignment may result in an undetermined form 0/0. Fortunately, all these issues may be solved if a constant term representing "a prediction of the cost of future disturbances" is added to the cost functions 4.15. This amounts to replacing

$$\beta_i = \omega_i x_i^T \Theta_i x_i,$$

by

$$\beta_i = \omega_i x_i^T \Theta_i x_i + \bar{\beta}_i,$$

where $\bar{\beta}_i$ is a constant coefficient, which have to be chosen off-line, according to the future disturbances that a given plant may be subjected to. These coefficients may be chosen by trial and error, until the best behavior is obtained. A small value of $\bar{\beta}_i$ increases the sensitivity of the optimal sampling period h_i^* with respect to state values. A larger value reduces this state sensitivity.

Remark. The expression of $\bar{\beta}_i$ may be explicitly computed if a linear quadratic Gaussian formulation and a finite optimization horizon are adopted in the optimal sampling frequency assignment problem (instead of the deterministic infinite horizon formulation that was adopted in this section).

4.4.4. *Application to the attitude control of a quadrotor*

In this section, the proposed feedback scheduling approach is applied to the attitude control of the quadrotor. As illustrated in Chapter 2, the roll, pitch and yaw control loops of the attitude controller are independent. In fact, the linearized model of the quadrotor (2.35) is made of three independent second-order sub-systems. For this reason, each loop may be implemented by an independent control task. The attitude controller will contain three control tasks τ_ϕ, τ_θ, and τ_ψ (respectively the roll, pitch and yaw control tasks). Task execution times are equal to $C_\phi = C_\theta = C_\psi = C = 10$ ms. Task respective periods will be, respectively, denoted by h_ϕ, h_θ, and h_ψ. The desired utilization of these three task is $U_{sp} = 80\%$. Task periods h_ϕ, h_θ, and h_ψ that may be assigned by the feedback-scheduling algorithm verify

$$0 \leq h_\phi \leq h_\phi^{\max}, 0 \leq h_\theta \leq h_\theta^{\max} \text{ and }, 0 \leq h_\psi \leq h_\psi^{\max},$$

where $h_\phi^{\max} = h_\theta^{\max} = h_\psi^{\max} = 100$ ms. Constants $\bar{\beta}_\phi$, $\bar{\beta}_\theta$, and $\bar{\beta}_\psi$ are equal to 10^{-8}. The period of the feedback scheduler is $h_{\text{fbs}} = 100$ ms.

In these simulations, the attitude of the quadrotor has to follow the specified set points depicted in Figure 4.8. Roll and pitch angle set points from instant $t = 0$ s to instant $t = 20$ s are sine signals with respective amplitudes 17° and 5° and periods 10 s and 20 s. From instant $t = 20$ s, roll and pitch set points are zero. Yaw angle set

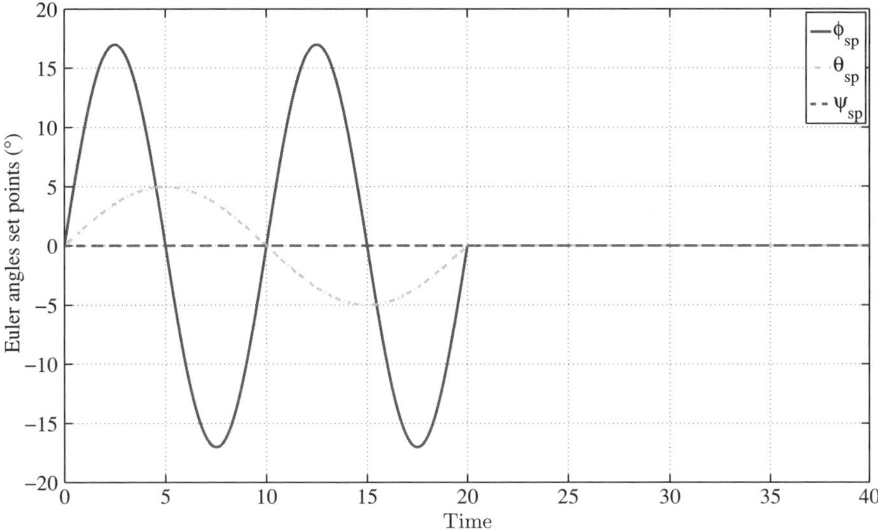

Figure 4.8. *Set points of the attitude controller*

point is zero. From instant $t = 20$ s, a yaw torque disturbance, consisting of a band limited white noise, with period 10^{-3} s and noise power 2×10^{-4}, is applied to the quadrotor. Note that in this simulation example, tasks execution times were assumed to be constant. Although the control design was based on the quadrotor linearized model, the simulations were applied on the nonlinear models (2.33) and (2.34).

Quadrotor Euler angles as well as their associated tasks sampling periods, assigned by the feedback scheduler, are depicted in Figure 4.9. This figure illustrates how the feedback-scheduling algorithm reduces the sampling period of the control task that has the most important needs of computing resources in order to improve the global control performance. From instants $t = 0$ s to $t = 20$ s, the sampling period of the yaw control task is set to around the maximal allowed value (100 ms). Roll and pitch control task period reductions are correlated with the rate of the change of their corresponding roll and pitch angles. In fact, roll control task period is minimal when the roll angle crosses zero. The pitch control task period is augmented when the pitch angle reaches its minimum or maximum values. The same observation holds for the roll angle. Since the roll set point has the greatest amplitude and rate of variation, the feedback scheduler assigned the most important parts of the computational resources to the roll control task.

From instant $t = 20$ s to instant $t = 30$ s, the specified set points for the three Euler angles are zero. Since $\bar{\beta}_\phi = \bar{\beta}_\theta = \bar{\beta}_\psi$, and no torque disturbance is applied to the quadrotor, the assigned sampling periods converge smoothly to $\frac{3C}{U_{sp}} = 37.5$ ms.

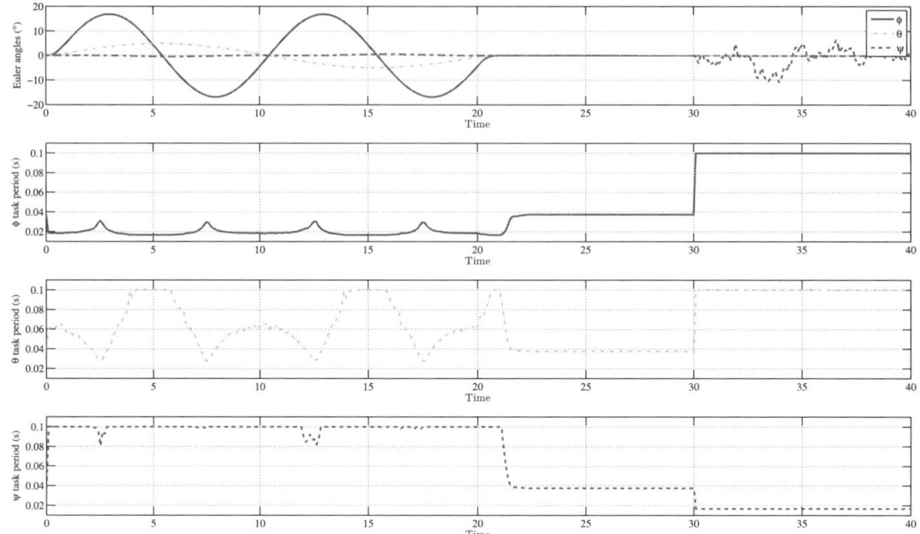

Figure 4.9. *Feedback scheduling of the attitude controller*

At instant $t = 30$ s, when the yaw torque disturbance starts to be applied, the feedback reduces smoothly the period of the yaw control task to the minimum value (16.6 ms), and augments the periods of roll and pitch tasks to the maximum allowed values (100 ms), in order to achieve a better reduction of the yaw torque disturbance.

4.5. Control and real-time scheduling co-design via a LPV approach

The feasibility of a feedback scheduler to manage the real-time parameters of a robot arm controller has been shown in section 1.4.2.3. In this preliminary example, the nonlinear nature of the process forbids to find simple loss functions to relate the controller's tasks sampling periods and the tracking performance. A very rough model has been extracted from simulations, providing a static CPU relative allocation between three compensation tasks. These loss functions are globally evaluated for a whole trajectory. The resulting feedback scheduler efficiently controls the overall CPU load, but only from the measured tasks CPU loads. However, even for a given trajectory, the contribution of each task to the tracking performance of the controller likely varies along the trajectory. For example, the disturbances due to Coriolis and centrifugal forces increase with the velocity, and the CPU allocated to compute the corresponding compensation action should be increased at high speed to better cancel this disturbance.

The following sections summarize a first approach to take into account the non-linear plant's state to dynamically adapt the scheduling parameters of the robot arm controller; this approach is exposed in more details in [SEN 08].

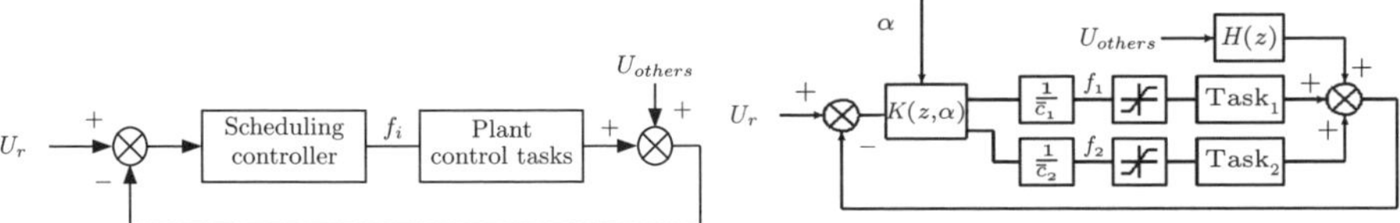

Figure 4.10. *Feedback-scheduling block diagram; control scheme for CPU resources*

4.5.1. *A LPV feedback scheduler sensible to the plant's closed-loop performances*

Feedback scheduling is a dynamic approach allowing a better usage of the computing resources, in particular when the workload changes (e.g. due to the activation of an admitted new task). The CPU activity is controlled according to the resource availability by adjusting scheduling parameters (e.g. the control intervals) of the plant control tasks, as recalled in section 1.4. However, as the goal of the controllers is the achievement of a requested performance level, the use of computing resources should be also linked to the dynamic behavior of the plant(s) to be controlled. The main result given in this section consists in deriving a new feedback-scheduling controller which depends on the plant trajectory, in view of an "optimal" resource sharing. It is designed in the LPV/H_∞ framework for polytopic systems.

Following previous results in [SIM 05] and section 1.4.2.3, the feedback scheduler, as illustrated in Figure 4.10, is a dynamic system between the control task frequencies and the processor utilization. As far as the adaptation of the control tasks is concerned, the load of the other tasks is seen as an output disturbance.

The CPU utilization is assumed to be measured or estimated, and the scheduling is here limited to periodic (or more exactly recurrent) tasks. In this case, the processor load induced by a task is defined by $U = \frac{c}{h}$, where c and h are the execution time and period of the task. Hence, as in [CER 02], the processor load induced by a task is estimated for each period h_s of the scheduling controller as

$$\hat{U}_{kh_s} = \lambda\,\hat{U}_{(k-1)h_s} + (1-\lambda)\,\frac{\overline{c}_{kh_s}}{h_{(k-1)h_s}}, \tag{4.22}$$

where h is the sampling frequency currently assigned to the plant control task (i.e. at each sampling instant kh_s), and $\overline{c}$ is the mean of its measured job execution time. λ is a forgetting factor used to smooth the measure (here $\lambda = 0.3$).

For an n-multi-task control system, we should note that, as in [SIM 03], if the execution times are constant, then the relationship $U = \sum_{i=1}^{n} C_i f_i$, where $f_i = 1/h_i$ is the frequency of the task, is a linear function (which is not the case if expressed as a function of the task periods). Therefore, using (4.22), the estimated CPU load is given as

$$\hat{U}(kh_S) = \frac{(1-\lambda)}{z-\lambda} \sum_{i=1}^{n} \bar{c}_i(kh_S) f_i(kh_S). \tag{4.23}$$

However, in practice, the execution time of the control tasks may vary according to the run-time environment, e.g. actual processor speed. As proposed in [SIM 05], a "normalized" linear model of the task i (i.e. independent of the execution time), G'_i, is used for the scheduling controller synthesis, where $\bar{c}$ is omitted and will be compensated by on-line gain-scheduling $(1/\bar{c})$ as shown below:

$$G'_i(z) = \frac{\hat{U}(z)}{f_i(z)} = \frac{1-\lambda}{z-\lambda}, \quad i = 1, \ldots, n. \tag{4.24}$$

Also, as explained above, the use of computing resources is chosen to depend on the plant trajectory. Hence, the control scheme of computing resource control is illustrated in Figure 4.10 for a two control-tasks system for simplicity.

In Figure 4.10, the interval of frequencies is limited by the "saturation" block, α represents a set of real parameters $\{\alpha_1, \alpha_2, \ldots, \alpha_n\}$ dedicated to the set of control tasks $\{U_1, U_2, \ldots, U_n\}$. These parameters will be used to make the resource sharing vary according to the plant trajectory. In the two control-tasks system, where $U = U_1 + U_2$, it is required that

$$U_1 = \alpha U \tag{4.25}$$

$$U_2 = (1-\alpha)U, \tag{4.26}$$

with α being a varying parameter. This makes the control scheme flexible enough to distribute on-line the use of computing resources to the different control tasks. The choice of the value of the time-varying parameters $\{\alpha_1, \alpha_2, \ldots, \alpha_n\}$ can be done in many different ways, e.g. from the on-line computation of optimal cost functions or from a dependency on the control effort. It will be illustrated in details in section 4.5.2 for the robot-arm control example.

Here, the design of the controller $K(\alpha)$ is done using the H_∞ control approach for LPV systems. The H_∞ control scheme to synthesize the controller $K(\alpha)$ is given in Figure 4.11.

In Figure 4.11, G' is the model of the scheduler, the output of which is the vector of all task loads. To get the sum of all the task loads as in (4.24), $C' = [1 \ldots 1]$ is used. The H transfer function represents the sensor dynamic behavior which measures the load of the other tasks; it may be a simple first-order filter. The template W_e specifies the performances on the load-tracking error. It is chosen in the continuous-time domain as

$$W_e(s) = \frac{s/M_s + \omega_b}{s + \omega_s \epsilon}, \tag{4.27}$$

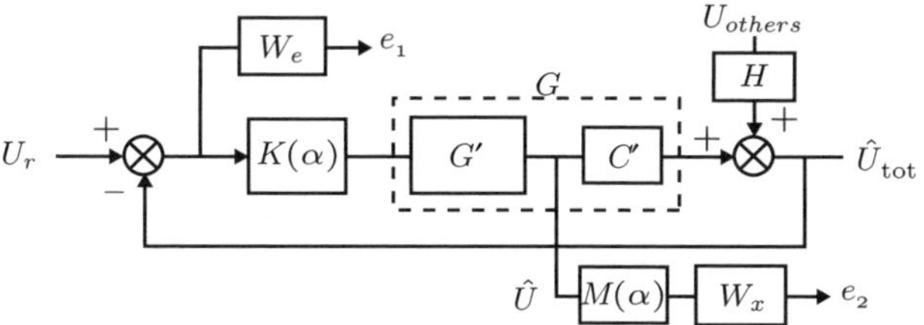

Figure 4.11. *A LPV / H_∞ controller for CPU resources*

with $M_s = 2$, $\omega_s = 10$ rad s^{-1}, $\epsilon = 0.01$ to obtain a closed-loop settling time of 300 ms, a static error less than 1 % and a good robustness margin.

The resource distribution is done through the $M(\alpha)$ matrix defined below. Note that for an n-multi-tasks system

$$U = U_1 + U_2 + \cdots + U_n \tag{4.28}$$

$$U = (\alpha_1 + \alpha_2 + \cdots + \alpha_n)U, \tag{4.29}$$

where $\alpha_1 + \alpha_2 + \cdots + \alpha_n = 1$. Then

$$U_1 = \alpha_1 U \tag{4.30}$$

$$U_2 = \alpha_2 U = \frac{\alpha_2}{\alpha_1} U_1 \tag{4.31}$$

$$U_3 = \alpha_3 U = \frac{\alpha_3}{\alpha_2} U_2 \tag{4.32}$$

$$\vdots \qquad \vdots \tag{4.33}$$

$$U_n = \alpha_n U = \frac{\alpha_n}{\alpha_{n-1}} U_{n-1}. \tag{4.34}$$

Then to ensure the on-line distribution of the computing resources M is chosen as

$$M = \begin{bmatrix} -\alpha_2 & \alpha_1 & 0 & \cdots & \cdots & 0 \\ 0 & \alpha_3 & \alpha_2 & 0 & \cdots & 0 \\ \vdots & \cdots & \cdots & \cdots & \cdots & 0 \\ \cdots & \cdots & \cdots & \cdots & -\alpha_n & \alpha_{n-1} \end{bmatrix} \tag{4.35}$$

$$= \alpha_1 M_1 + \alpha_2 M_2 + \cdots + \alpha_n M_n. \tag{4.36}$$

Using [APK 95] the LPV controller $K(\alpha)$ is obtained through the solution of the H_∞ control problem for polytopic systems, and consists in solving 2 LMIs. Then the design of $K(\alpha)$ can be done directly either in the discrete-time domain or in the continuous-time one and then discretized. For this example, $K(\alpha)$ has been synthesized in the continuous-time domain using the H_∞ control approach for polytopic systems, as described in details in [SEN 08].

By solving the H_∞ problem for the LPV system using the Yalmip interface and Sedumi solver [STU 99; LOF 04], one obtains $\gamma_{\text{opt}} = 1.8885$, and a controller of order 7.

4.5.2. *Application to a robot-arm control*

The seven degrees of freedom Mitsubishi PA10 robot arm already used in section 1.4.2.3 is again considered. The problem under consideration is tracking a desired trajectory for the position of the end effector. Using the Lagrange formalism, the following model can be obtained:

$$\Gamma = M(q)\ddot{q} + \text{Gra}(q) + C(q, \dot{q}), \tag{4.37}$$

where q stands for the positions of the joints, M is the inertia matrix, Gra is the gravity forces vector and C gathers Coriolis, centrifugal and friction forces.

The structure of the ideal linearizing controller includes a compensation of the gravity, Coriolis/centrifugal effect, and inertia variations as well as a proportional-derivative (PD) controller for the tracking and stabilization problem, of the form

$$\Gamma = \text{Gra}(q) + C(q, \dot{q}) + K_p(q_d - q) + K_d(\dot{q}_d - \dot{q}), \tag{4.38}$$

leading to the linear closed-loop system $M(q)\ddot{q} = K_p(q_d - q) + K_d(\dot{q}_d - \dot{q})$, where q_d and $\dot{q}_d$ stand for the reference trajectory positions and velocities.

As in 1.4.2.3 the controller is split into five tasks, i.e. a specific task is considered for the PD control, the trajectory generation and for the gravity, inertia, and Coriolis compensations, which are implemented as a multi-rate controller. In this feedback scheduling scheme, only the periods of the compensation tasks are adapted, as they are time consuming compared with the PD task while being less critical for the stability.

4.5.2.1. *Performance evaluation of the control tasks in view of optimal resource distribution*

In order to associate the use of computing resources with the robot trajectory, the contribution of each of the three control tasks to the closed-loop system performances has been evaluated as a function of its execution period.

The methodology is the following. Assuming a nominal sampling period for each task of 1 ms, the period of each compensation control task is changed, and new simulations are performed during which the following cost is computed:

$$J = \int_{t_i}^{t_f} (P_{h_i, h_g, h_c}(t) - P_{\text{ref}}(t))^2 - \int_{t_i}^{t_f} (P_c(t) - P_{\text{ref}}(t))^2, \tag{4.39}$$

where P_{ref} is the desired position in the operational space of the end tip, computed from q_d using the geometric model. P_c is the position obtained when all the control tasks act with the minimal sampling period of 1 ms. Finally, P_{h_i,h_g,h_c} is the position obtained when the sampling period of one of the compensation tasks is increased from 1 to 30 ms.

Simulations are performed for a particular robot trajectory, defined by the reference vector $(q_d, \dot{q}_d)$ for all the robot joints. Here q_d goes from $\pi/2$ to $-\pi/2$. Figure 4.12 shows the evolution of the cost function J for the three compensation control tasks.

4.5.2.1.1. Discussion

It is difficult to infer the relations between the compensation task execution period and the trajectory tracking performance, anyway a natural interpretation can be proposed. First, the gravity compensation effect is very sensitive to the increase of the sampling period at the end of the trajectory, as the cost increases in the second part of the trajectory (first part of the graph as the trajectory goes from $\pi/2$ to $-\pi/2$). It is desired to ensure the availability of CPU resources for this task in a linear way with the trajectory position. The situation is almost opposite for the inertia effect. Finally, even if some variations can be observed, a constant use of CPU resources of the Coriolis compensation task, all along the trajectory is still needed.

Then the distribution of the control task periods is chosen as

$$U_I = \alpha_I U, \ \ U_G = \alpha_G U, \ \ U_C = \alpha_C U, \tag{4.40}$$

where $\alpha_C = 0.25$, $\alpha_I = 1 - \alpha_G$, and α_G is linked to the plan trajectory by

$$\alpha_G = \alpha_{\min} + (\alpha_{\text{Max}} - \alpha_{\min}) \times \frac{q_d - q_{\text{end}}}{q_{\text{ini}} - q_{\text{end}}}, \tag{4.41}$$

where $[\alpha_{\min}; \ \alpha_{\text{Max}}] = [0.1; \ \ 0.65]$, q_{ini} is the initial position, and q_{end} is the final trajectory position.

4.5.2.2. *Simulation with TrueTime*

TrueTime is a free toolbox for MATLAB/Simulink aimed to ease the simulation of the temporal behavior of a multi-tasking real-time system executing controller tasks [OHL 07]. In this application, the period of the feedback scheduler has been fixed to 30 ms to be larger than the robot control task periods, whose limits have been set from 1 ms to 30 ms.

In the experiment depicted in Figure 4.14, the desired CPU usage is initially set to 50% of the maximum usage. The upper plots show the tasks periods and CPU usage. The PD loop period is fixed at 1 ms and the trajectory generator at 5 ms.

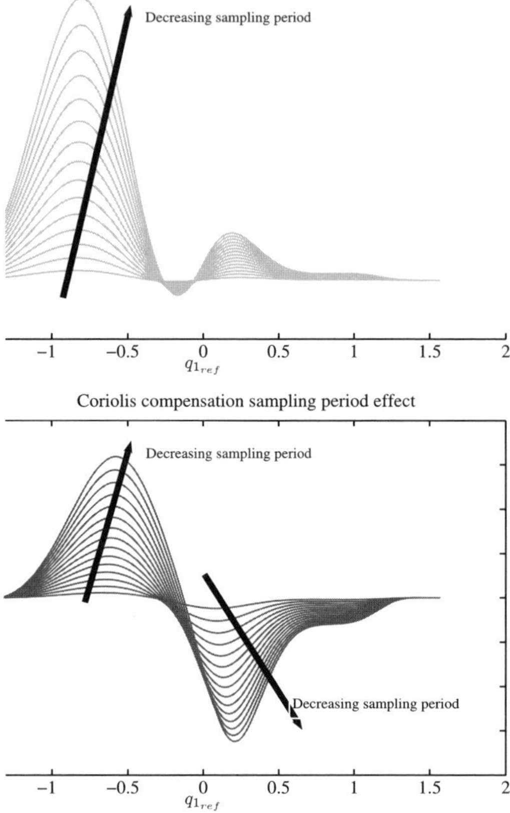

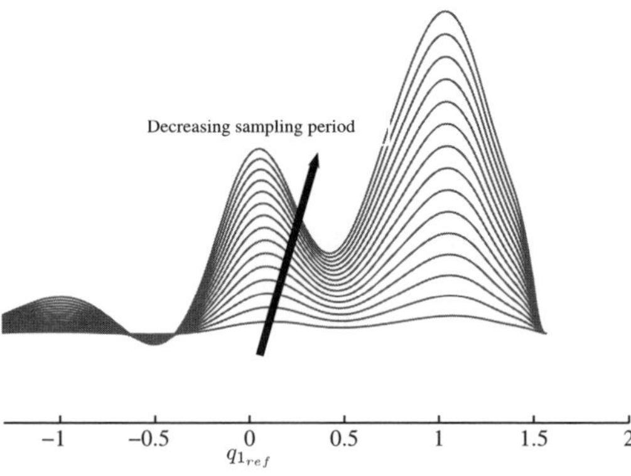

Figure 4.12. *Cost variation due to varying sampling for gravity, Coriolis, and inertia compensation task*

As seen in Figure 4.14(a), the load of the compensation tasks (gravity, Coriolis, and inertia) vary on-line as expected according to the parameter α_I (see Figure 4.15a). The corresponding evolution of the task periods is shown in Figure 4.14(b). Moreover, in Figure 4.15b, the adaptive LPV case (α varying) is compared with the constant case ($\alpha = 0.375$). It can be seen that the LPV case leads to a smaller cost function which emphasizes the real interest of the provided approach.

4.5.2.3. *Feasibility and possible extensions*

Note that, as explained in [SIM 05] and depicted in Figure 1.10, the scheduling feedback loop can be easily implemented on top of an off-the-shelf real-time operating system (e.g. Posix) in the form of an additional real-time periodic task, i.e. a control module whose function is specified and encoded by the control designer. The inputs are the measured execution times of the control tasks, and the set point is a desired global computing load. Outputs are the sampling intervals of the gravity, Coriolis, and inertia control tasks which are triggered by programmable timers provided by the operating system as illustrated in section 7.4.1.

Thanks to the use of a hierarchical control structure, the given results may also be integrated with existing methods for the design of varying sampled controllers, as in [TAN 02] or using the LPV approach of section 2.4, which makes this integrated approach somewhat generic.

4.6. Summary

In this chapter, a few co-design approaches to integrate both control and implementation constraints have been presented. Even when the controlled plants are linear, theoretic optimal solutions are too computing intensive to be real-time compliant. Therefore, tractable solutions are designed under restrictive assumptions, leading to operational solutions with limited generality. For example, section 4.2 gives a integrated scheme combining varying sampled controllers for linear systems and elastic tasks models, leading to an effective and implementable solution for which no stability proof has been provided up to now. Conversely, the MPC-based approach in section 4.3, using hybrid modeling, is able to cope with nonlinear plants, state, and actuators constraints, and limitations in CPU power and networking bandwidth working on a slotted timescale. As optimal solutions are too complex to be used in real time, only sub-optimal controllers can be implemented. Other examples assume, in section 4.4, the convexity of all the cost functions associating the controllers performance with their execution periods, or are based in section 4.5 on cost functions measured on a specific trajectory of a given nonlinear plant. Note also that next Chapter 5 describes a control/scheduling co-design approach assuming LQ controllers associated with (m, k)-firm scheduling policies.

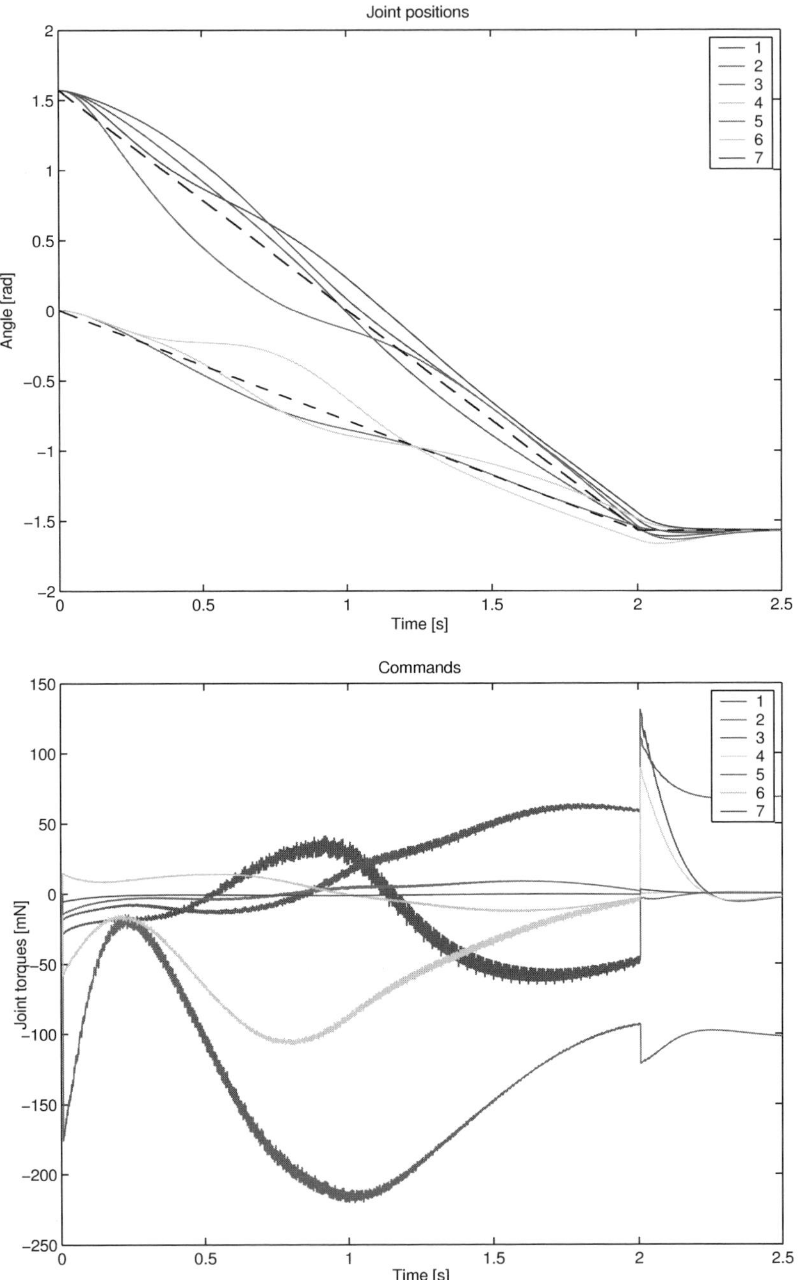

Figure 4.13. *Positions and control torques*

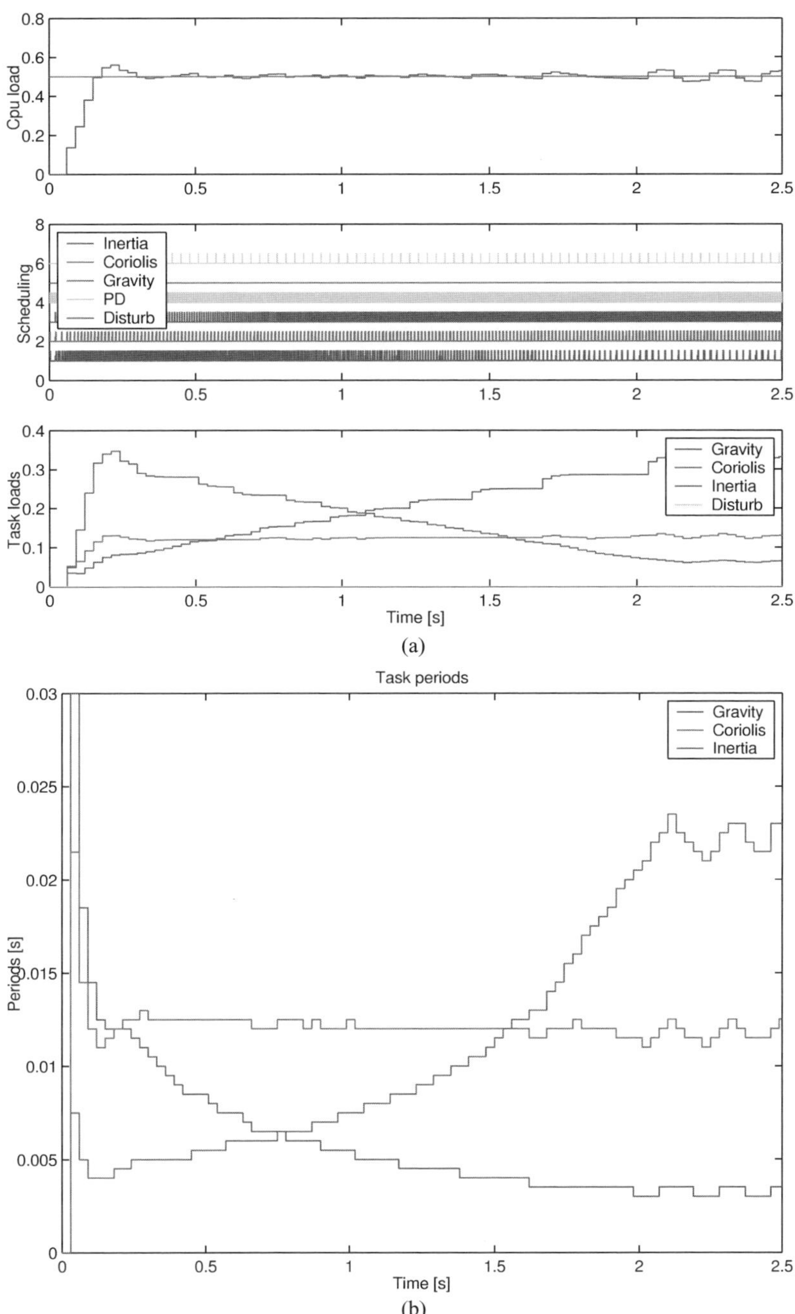

Figure 4.14. *TrueTime real-time parameters: (a) loads and (b) periods*

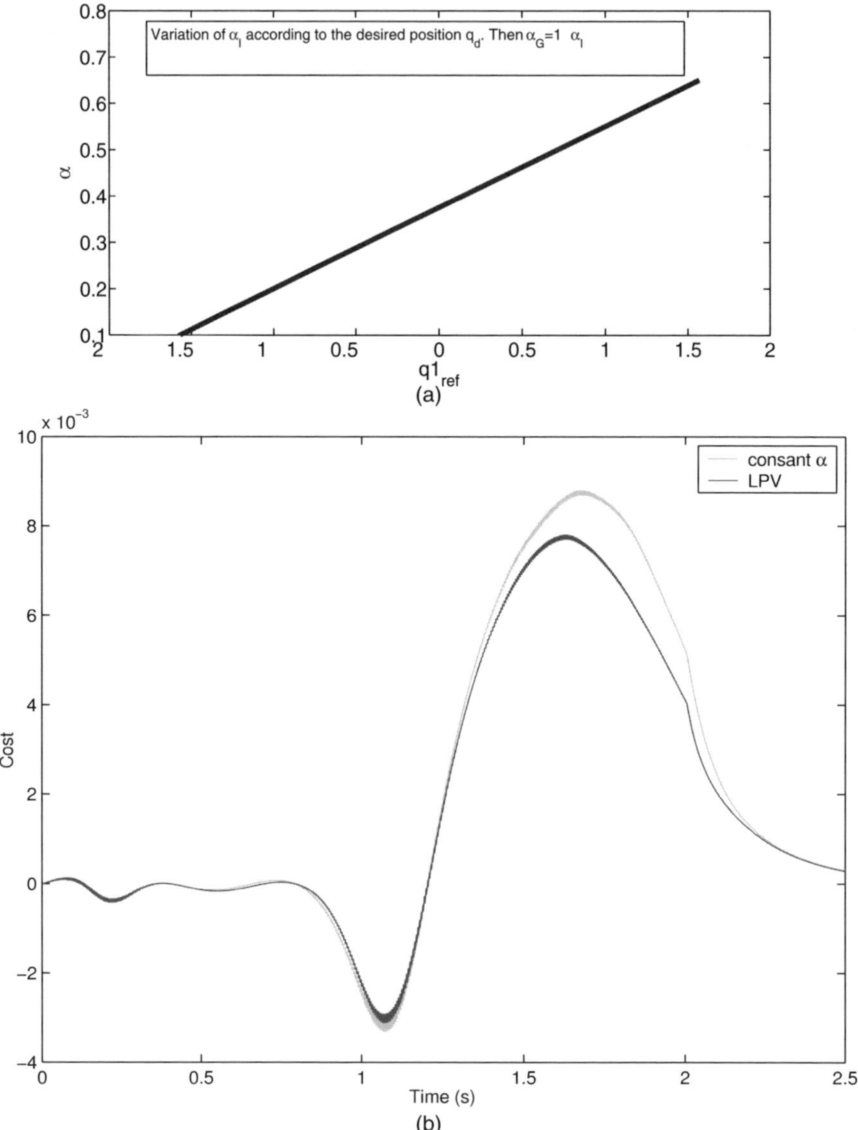

Figure 4.15. *(a) Variation of α_I and (b) total cost*

Indeed, control and scheduling co-design over networks handles complex and heterogenous systems, so that it is unlikely that a fully general, operational and unique theoretic framework emerge soon. Conversely, well chosen case studies are expected to bring effective solutions for some classes of systems and for some specific problems.

4.7. Bibliography

[APK 95] APKARIAN P. AND GAHINET P., A convex characterization of gain scheduled $\mathcal{H}_\infty$ controllers, *IEEE Transactions on Automatic Control*, vol. 40, p. 853–864, May 1995.

[ÅRZ 00] ÅRZÉN K.-E., CERVIN A., EKER J., AND SHA L., An introduction to control and scheduling co-design, *39th IEEE Conference on Decision and Control*, Sydney, Australia, December 2000.

[ÅST 97] ÅSTRÖM K. J., WITTENMARK B., *Computer-Controlled Systems*, Information and System Sciences Series, Prentice Hall, Englewood Cliffs, NJ, 3rd edition, 1997.

[BEM 99] BEMPORAD A., MORARI M., Control of systems integrating logic, dynamics, and constraints, *Automatica*, vol. 35, p. 407–427, 1999.

[BEN 06a] BEN GAID M., Optimal scheduling and control for distributed real-time systems, PhD thesis, University of d'Evry Val d'Essonne, France, 2006.

[BEN 06b] BEN GAID M.-M., ÇELA A., AND HAMAM Y., Optimal integrated control and scheduling of networked control systems with communication constraints: application to a car suspension system, *IEEE Transactions on Control Systems Technology*, vol. 14, p. 776–787, 2006.

[BEN 06c] BEN GAID M.-M., ÇELA A., HAMAM Y., AND IONETE C., Optimal scheduling of control tasks with state feedback resource allocation, *2006 American Control Conference ACC'06*, Minneapolis, USA, June 2006.

[BEN 08] BEN GAID M., SIMON D., AND SENAME O., A convex optimization approach to feedback scheduling, *16th IEEE Mediterranean Conference on Control and Automation MED'08*, Ajaccio, France, June 2008.

[BEN 09] BEN GAID M.-M., ÇELA A., AND HAMAM Y., Optimal real-time scheduling of control tasks with state feedback resource allocation, *IEEE Transaction on Control Systems Technology*, vol. 17, p. 309–326, March 2009.

[BOY 04] BOYD S. P., AND VANDENBERGHE L., *Convex Optimization*, Cambridge University Press, Cambridge, UK, 2004.

[BUT 00] BUTTAZZO G., AND ABENI L., Adaptive rate control through elastic scheduling, *39th Conference on Decision and Control*, Sydney, Australia, December 2000.

[CAN 01] CANNON M., KOUVARITAKIS B., AND ROSSITER J.-A., Efficient active set optimization in triple mode MPC, *IEEE Transactions on Automatic Control*, vol. 46, p. 1307–1312, August 2001.

[CAS 06] CASTAÑÉ R., MARTÍ P., VELASCO M., CERVIN A., AND HENRIKSSON D., Resource management for control tasks based on the transient dynamics of closed-loop systems, *18th Euromicro Conference on Real-Time Systems*, Dresden, Germany, July 2006.

[CER 02] CERVIN A., EKER J., BERNHARDSSON B., AND ÅRZÉN K.-E., Feedback-feedforward scheduling of control tasks, *Real-Time Systems*, vol. 23, p. 25–53, July 2002.

[CER 03] CERVIN A., Integrated control and real-time scheduling, PhD thesis, Department of Automatic Control, Lund Institute of Technology, Sweden, April 2003.

[CER 06] CERVIN A., AND ALRIKSSON P., Optimal on-line scheduling of multiple control tasks: a case study, *18th Euromicro Conference on Real-Time Systems*, Dresden, Germany, July 2006.

[DAČ 07] DAČIĆ D. B., AND NEŠIĆ D., Quadratic stabilization of linear networked control systems via simultaneous protocol and controller design, *Automatica*, vol. 43, p. 1145–1155, 2007.

[EKE 00] EKER J., HAGANDER P., AND ÅRZÉN K.-E., A feedback scheduler for real-time controller tasks, *Control Engineering Practice*, vol. 8, p. 1369–1378, 2000.

[GAR 89] GARCIA C.-E., PRETT D.-M., AND MORARI M., Model predictive control: theory and practice, *Automatica*, vol. 25, p. 335–348, 1989.

[HEN 02] HENRIKSSON D., CERVIN A., ÅKESSON J., AND ÅRZÉN K., On dynamic real time scheduling of model predictive controllers, *41st IEEE Conference on Decision and Control*, Las Vegas, USA, December 2002.

[HEN 05] HENRIKSSON D., AND CERVIN A., Optimal on-line sampling period assignment for real-time control tasks based on plant state information, *44th Conference on Decision and Control*, Sevilla, Spain, December 2005.

[HEN 06] HENRIKSSON D., Resource-constrained embedded control and computing systems, *PhD thesis*, Department of Automatic Control, Lund Institute of Technology, Sweden, January 2006.

[HRI 99] HRISTU D., Optimal control with limited communication, *PhD thesis*, Division of Engineering and Applied Sciences, Harvard University, June 1999.

[KAL 63] KALMAN R. E., HO B., AND NARENDRA K., Controllability of linear dynamical systems, *Contributions to Differential Equations*, vol. 1, p. 188–213, 1963.

[LEM 07] LEMMON M., CHANTEM T., HU X., AND ZYSKOWSKI M., On self-triggered full information H_∞ controllers, *Proceedings of Hybrid Systems: Computation and Control*, April 2007.

[LIN 02] LINCOLN B., AND BERNHARDSSON B., LQR optimization of linear system switching, *IEEE Transactions on Automatic Control*, vol. 47, p. 1701–1705, October 2002.

[LOF 04] LOFBERG J., AND YALMIP: A toolbox for modeling and optimization in Matlab, *Computer Aided Control System Design Conference*, Taipei, Taiwan, March 2004.

[LU 02] LU C., STANKOVIC J., TAO G., AND SON S., Feedback control real-time scheduling: framework, modeling and algorithms, *Special Issue of Real-Time Systems Journal on Control-Theoretic Approaches to Real-Time Computing*, vol. 23, p. 85–126, July 2002.

[MAR 02] MARTÍ P., FUERTES J., FOHLER G., AND RAMAMRITHAM K., Improving quality-of-control using flexible timing constraint: metric and scheduling issues, *23rd IEEE Real-Time Systems Symposium*, Austin, USA, December 2002.

[MAR 04] MARTÍ P., LIN C., BRANDT S., VELASCO M., AND FUERTES J., Optimal state feedback based resource allocation for resource-constrained control tasks, *25th IEEE Real-Time Systems Symposium*, Lisbon, Portugal, December 2004.

[MIL 08] MILLZAN P., JURADO I., VIVAS C., AND RUBIO F.-R., Algorithm for networked control systems with large data dropouts, *47th IEEE Conference on Decision and Control CDC'08*, Cancun, Mexico, December 2008.

[OHL 07] OHLIN M., HENRIKSSON D., AND CERVIN A., TrueTime 1.5—Reference Manual, January 2007.

[REH 04] REHBINDER H., AND SANFRIDSON M., Scheduling of a limited communication channel for optimal control, *Automatica*, vol. 40, p. 491–500, March 2004.

[ROB 07a] ROBERT D., SENAME O., AND SIMON D., A reduced polytopic LPV synthesis for a sampling varying controller: experimentation with a T inverted pendulum, *European Control Conference ECC'07*, Kos, Greece, July 2007.

[ROB 07b] ROBERT D., Contribution à l'interaction commande/ordonnancement, PhD thesis, INP Grenoble, France, January 2007.

[SEN 08] SENAME O., SIMON D., AND BEN GAID M., A LPV approach to control and real-time scheduling codesign: application to a robot-arm control, *Control and Decision Conference CDC'08*, Cancun, Mexico, December 2008.

[SET 96] SETO D., LEHOCZKY J. P., SHA L., AND SHIN K. G., On task schedulability in real-time control systems, *17th IEEE Real-Time Systems Symposium*, New York, USA, December 1996.

[SHI 99] SHIN K. G., AND MEISSNER C. L., Adaptation of control system performance by task reallocation and period modification, *Proceedings of 11th Euromicro Conference on Real-Time Systems*, York, UK, p. 29-36, June 1999.

[SIM 03] SIMON D., SENAME O., ROBERT D., AND TESTA O., Real-time and delay-dependent control co-design through feedback scheduling, *CERTS'03 Workshop on Co-design in Embedded Real-time Systems*, Porto, Portugal, July 2003.

[SIM 05] SIMON D., ROBERT D., AND SENAME O., Robust control/scheduling co-design: application to robot control, *11th IEEE Real-Time and Embedded Technology and Applications Symposium*, San Francisco, USA, March 2005.

[STU 99] STURM J. F., Using SeDuMi 1.02, a MATLAB toolbox for optimization over symmetric cones, *Optimization Methods and Software*, vol. 11/12, p. 625–653, 1999.

[TAB 07] TABUADA P., Event-triggered real-time scheduling of stabilizing control tasks, *IEEE Transactions on Automatic Control*, vol. 52, p. 1680–1685, 2007.

[TAN 02] TAN K., GRIGORIADIS K.-M., AND WU F., Output-feedback control of LPV sampled-data systems, *International Journal of Control*, vol. 75, p. 252–264, 2002.

[ZHA 08] ZHAO Y., LIU G., AND REES D., Integrated predictive control and scheduling co-design for networked control systems, *IET Control Theory Appl*, vol. 2, p. 7–15, 2008.

Chapter 5

Overload Management Through Selective Data Dropping

5.1. Introduction

During system and network overload periods, excessive delay or even data loss may occur. To maintain the quality of control of an NCS, the implementation system (including both computer and network) overload must be correctly handled. As we can see in the previous chapters, a common approach to dealing with this overload problem is to dynamically change the sampling period of the control loops. In this chapter, as an alternative to the explicit sampling period adjustment, we present an indirect sampling period adjustment approach which is based on selective sampling data dropping according to the (m, k)-firm model [HAM 94]. The interest of this alternative is its easy implementation despite having less adjustment quality, since only the multiples of the basic sampling period can be exploited. Upon overload detection, the basic idea is to selectively drop some samples according to the (m, k)-firm model to avoid long consecutive data drops. The consequence is that the shared network and processor will be less loaded. However, the control stability and performance must still be maintained to an acceptable level. This can be achieved by keeping either the total control tasks on a same processor or the messages sharing a same network bandwidth schedulable under the (m, k)-firm constraint.

In this chapter, we first give a sufficient condition for scheduling a set of control tasks under (m, k)-firm constraint (section 5.2). Then, through several examples, we

Chapter written by Flavia FELICIONI, Ning JIA, Françoise SIMONOT-LION and Ye-Qiong SONG.

present the methods for determining the value of k for a given control loop which will still maintain control loop stability (section 5.3), and the optimal values of m and the control gains which minimize a total cost function either in the presence of the control task configuration changes (section 5.4) or by further coping with the plant state changes (section 5.5). The problem of control loop stability with on-line parameter switching is also discussed in section 5.5.1.

To better illustrate this approach, let us first present the generic system architecture. Then, we identify the problem to solve and provide several notations.

5.1.1. *System architecture*

We consider a global control architecture that integrates a set of plants to control, each one being controlled by a discrete controller. We are interested in the deployment of the control application onto limited resources; for example, all the numerical controllers share the same processors or all the plant states sampled by sensors are transmitted to the controller through the same communication architecture. Moreover, we suppose that according to the global state of the plant, a supervisor chooses the current working mode of the global system. In particular, it can stop the control of a plant, start the control of a plant or modify the control strategy of a plant (control law, sampling period, etc.). The consequence of the transition between working modes is the modification of the set of active tasks/messages that can bring about

– some of them are stopped,

– new tasks are activated or new messages are transmitted,

– the characteristics of tasks may be modified, for example, changing their Worst Case Execution Time (WCET); the characteristics of messages can be modified, for example, their size or the sampling period can be transformed by using a new control strategy.

This means that the scheduling of the messages on a network or the scheduling of the tasks on a processor have to be redefined each time the supervisor modifies the global control mode. The schedulability analysis of a set of tasks or messages subject to hard real-time constraints (all the instances have to meet their deadline) can lead to over provisioning of the resources and this oversizing can be worse in the case of an architecture that implements several working modes as already mentioned before. Moreover, the relationship between the performance of the control and the scheduling of the activities is not well known quantitatively; therefore, the identification of the scheduling parameters relies generally on experiments and/or simulations [SIM 05] that are not exhaustive and so, cannot be generalized. So, a feasible scheduling provided by applying the relaxation of timed constraints as proposed in the classical solution does not lead systematically to the optimal performance of the control application.

Therefore, we propose a new scheduling architecture for handling such configurations as well as an adaptive technique that makes adjustments on-line for, on the one hand, the (m, k)-firm constraints of activities (data transmission and/or task implementing the control law) and, on the other hand, the parameters of the control laws. By doing so, the global performance of the application is fixed at an optimized level and the schedulability under (m, k)-firm constraints is guaranteed.

In the case study that will illustrate the adaptive approach, we consider plants that correspond to harmonic oscillators (as, for example, cart systems, a pendulum or an inverted pendulum). Furthermore, the solution is detailed for processor sharing and therefore, we will deal with the scheduling of tasks instead of messages.

The scheduling architecture is illustrated in Figure 5.1. The system to control is composed of a set of plants (Plant 1, Plant 2, ..., Plant n) that are assumed to be independent. Each plant is controlled by a controller (Controller 1, Controller 2, ..., Controller n). In this example, the controllers are deployed as a set of n tasks (τ_1, τ_2, .., τ_n) on the same processor. A task handler is dedicated to the identification of the optimal configuration of (m_i, k_i)-firm constraints for each task τ_i running on the processor. The system is observed by a supervisor, assumed implemented on another processor. The role of the supervisor is to observe the state of the plants and to decide at each instant which plants have to be controlled and which control laws have to be applied. Each time the supervisor modifies the configuration of the system, it sends the task handler a list of active controllers as well as their new characteristics, in

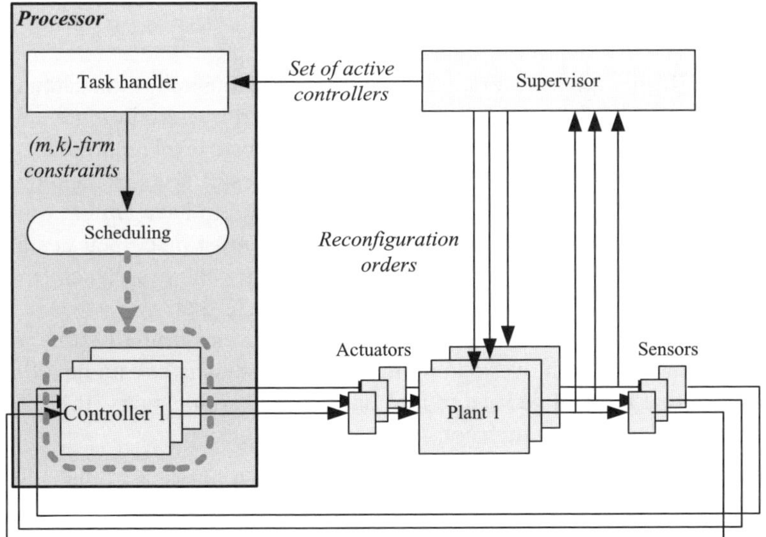

Figure 5.1. *The global scheduling architecture*

short, the activation period and the worst case execution time of the control law. Of course, the schedulability of this task set must be guaranteed in sense of (m_i, k_i)-firm.

5.1.2. *Problem statement*

The global problem of finding, at each instant an optimized scheduling of control tasks can be divided into several sub-problems.

– Firstly, we have to ensure, at least the stability of each controlled system. This will be achieved by the specification of the highest value of k_i in the constraint (m_i, k_i) of each task τ_i sharing the processor that ensures the stability under a $(1, k_i)$-firm constraint.

– A second problem concerns the optimization issue. Two main considerations have to be taken into account in this context: what is the cost function? what are the criteria? For the first question, we will use the (Linear Quadratic Regulator LQR) cost function that is classically applied in the control community. The determination of the value of the parameter m_i in the (m_i, k_i) constraint of each task τ_i will answer the second question.

Solutions to these different problems will be presented in the remaining part of this chapter. But, first of all, let us give more details in the following section on how to schedule tasks under (m, k)-firm constraint.

5.2. Scheduling under (m, k)-firm constraint

The (m, k)-firm model was first proposed by Hamdaoui and Ramanathan in order to precisely characterize the timing constraints required by certain kinds of applications [HAM 94]. It allows the specification of the guarantee level required by real-time applications tolerating deadline miss of certain instances of tasks or messages. More specifically, in this chapter a task or a message τ_i is said to have (m_i, k_i)-firm deadlines if at least m_i out of any k_i consecutive instances must meet their deadlines and if the schedulability analysis of a set of tasks or messages subject to such (m, k)-firm constraints must furnish a deterministic guarantee. Note that, $m_i = k_i = 1$ means that each instance of τ_i is constrained by a firm deadline. Initially, this technique was introduced to deal with overloading in real-time message stream handling, but it was soon seen that it could be used in real-time control applications [RAM 99]. Two problems must be solved for such constraint specifications:

– scheduling policy: how to schedule a set of tasks or messages subject to (m, k)-firm;

– schedulability analysis: for a given scheduling policy, how to guarantee that the constraints are satisfied.

The main solutions to these problems are briefly described in the following.

Two classes of scheduling policies that were developed to take into account the (m, k)-firm constraints were studied: dynamic scheduling and static scheduling.

In the following, we consider a set of tasks or message streams τ_i, each of them being defined by

– C_i, the largest time needed by the instances of τ_i to complete the task instance execution or the message instance transmission;

– D_i, the relative deadline of τ_i, supposed to be the same for each instance;

– and, when τ_i is periodic or sporadic, its period or the minimum inter-arrival h_i;

– m_i and k_i, the parameters of the (m, k)-firm constraint imposed to τ_i;

– the (m_i, k_i)-pattern, Π_i defined as the sequence $\Pi_i(0)$, $\Pi_i(1)$, .., $\Pi_i(k - 1)$ where $\Pi_i(j) = 1$ indicates that the $(j + n.k)^{th}$ $(n \geq 0)$) instance of τ_i has to meet its deadline, $\Pi_i(j) = 0$ indicates that it is not mandatory that the $(j + n.k)^{th}$ instance of τ_i meets its deadline, and consequently, $\sum_{j=n}^{n+k_i-1} \Pi_i(j) = m_i$ for $n > 0$.

5.2.1. *Dynamic scheduling policy under (m,k)-firm constraints*

In [HAM 95], the (Distance Based Priority DBP) algorithm is proposed for non-pre-emptive tasks. It implements a dynamic scheduling that is based, at each scheduling point (e.g. arrival instant or departure instant of a task instance), on the distance to a failure state of each task or message. The failure of τ_i is defined as the situation where more than $k_i - m_i$ instances among the k_i last ones have missed their deadline. The highest priority is given to the task that has the lowest distance to its failure. A FIFO strategy is applied when the distance to the failure is the same for several tasks. The schedulability analysis, based on Markov chain, proposed in [HAM 95] provides a probabilistic guarantee. Further studies have been proposed for improving this approach. In particular, in [LI 06], a sufficient condition is specified for a set of periodic or sporadic messages, scheduled using DBP and earliest deadline first (EDF).

5.2.2. *Static scheduling policy under (m,k)-firm constraints and schedulability issue*

If these dynamic policies provide a good quality of service for a set of streams, they appear quite costly in the context of control applications sharing processors or network resources. To deal with such applications, [RAM 99] considered a system composed of several control applications. The controllers are implemented as pre-emptive tasks running on two processors. When a failure occurs at one processor, all the tasks that were allocated there migrate to the other one, leading to a possible overload situation. In this case, a static scheduling based on (m, k)-firm constraints is proposed. The mandatory instances of each task are scheduled according to the fixed priority of the task, while the lowest priority is given to the optional instances. Given a couple (m_i, k_i) for each τ_i, the instance number a ($a \geq 0$) is mandatory if $a = \left\lfloor \left\lceil \frac{a.m_i}{k_i} \right\rceil \times \frac{k_i}{m_i} \right\rfloor$. This classification has been improved in [QUA 00] by a global approach that determines on-line the mandatory instances of all the tasks τ_i subject to a (m_i, k_i)-firm constraint. As this problem has been proved NP-hard, a heuristic was proposed. Furthermore, [QUA 00] provides, on the one hand, an algorithm for evaluating of an upper bound of each τ_i response time for a "deeply-red" pattern (a "deeply-red" pattern is such that $\Pi_i(a) = 1$ if $0 \leq a < m_i$ and $\Pi_i(a) = 0$ in the other case) and, on the other hand, it demonstrated that if for each task τ_i, the response time is less than the deadline for a (m_i, k_i) "deeply-red" pattern, each τ_i is schedulable for any (m_i, k_i) pattern.

5.2.3. *Static scheduling under (m, k)-constraints and mechanical words*

In [JIA 05], a new method using the properties of the mechanical words is developed for the schedulability analysis of a set of non-pre-emptive tasks under (m_i, k_i)-firm constraints. First, it proved that the static patterns defined in the literature can be characterized in the form of the mechanical words leading to a largely simplified schedulability assessment. Then, by identifying the defaults of these patterns, it proposed a new way, based on a cellular line, to determine the (m_i, k_i) pattern of each task τ_i. Through intensive simulations, this technique has been demonstrated to achieve an improvement in the schedulable region. Considering $\alpha = \frac{m_i}{k_i}$, if α is rationale, then the mechanical word whose slope is α is (k_i)-periodic; the classification of the instances as mandatory or optional is therefore based on formula (5.1):

$$\Pi_i(a) = \lceil (a+1).\alpha \rceil - \lceil a.\alpha \rceil , \forall 0 \leq a < k_i. \tag{5.1}$$

The upper bound of the response time of a non-pre-emptive task τ_i subject to a (m_i, k_i) constraint is obtained, assuming that the sequence is convergent, by the limit of R_i^q when $q \rightarrow \infty$. R_i^q is given by the classical recurrent equation (5.2):

$$R_i^0 = C_i$$

$$R_i^q = \max_{j>i}(C_j) + \sum_{j<i} \left\lceil \frac{m_j}{k_j} \left\lceil \frac{R_i^{q-1}}{h_j} \right\rceil \right\rceil C_j, \tag{5.2}$$

where $i > j$ means that the priority of the task τ_i is lower than the priority of task τ_j.

If, for each mandatory instance of each task τ_i, $R_i^q + C_i < D_i$, then the system is proved to be schedulable.

Let us consider now a set of pre-emptive tasks scheduled using a fixed priority strategy defined by the rate monotonic algorithm (the larger the period is, the smaller the priority is); in [RAM 99] and [JIA 05], it is proved that it is possible to limit the length of the interference interval of a task τ_i due to a task of higher priority τ_j to the basic period h_i (the basic period is the interval between two instances of a task subject to a hard real-time constraint, meaning a (k, k)-firm one). In this case, we first consider a set of intervals associated with the task τ_i and defined in (5.3). h_j is the basic period of a task τ_j. $i > j$ means $h_j < h_i$ (i.e. the priority of the task τ_i is lower than the priority of task τ_j):

$$S_{i,j} = \left\{ \left\lfloor l\frac{k_j}{m_j} \right\rfloor h_j, \ \forall l \epsilon \mathbb{N} \ | \ \left\lfloor l\frac{k_j}{m_j} \right\rfloor h_j < h_i \right\} \tag{5.3}$$

$$S_i = \bigcup_{j=1}^{i-1} S_{i,j}.$$

Then, the response time of a task may be evaluated by equation (5.4):

$$n_j(t) = \left\lceil \frac{m_j}{k_j} \left\lceil \frac{t}{h_j} \right\rceil \right\rceil$$

$$W_i(t) = C_i + \sum_{j=1}^{i-1} n_j(t)C_j. \tag{5.4}$$

If, for each task τ_i, $\min_{r \epsilon S_i}(\frac{W_i(r)}{r}) \leq 1$, then the (m, k)-firm constraint can be met by all the tasks.

This condition is sufficient in general cases. It is necessary and sufficient if the tasks are synchronous (i.e. the first release time of all the tasks starts at the same time, often at $t = 0$).

5.2.4. *Sufficient condition for schedulability assessment under (m,k)-pattern defined by a mechanical word*

The computation of the sequence W_i for each task τ_i, as it is presented in the recurrence relation (5.4), is non-deterministic and it is difficult, if not impossible, to apply it on-line. Therefore, below, we propose a sufficient condition that ensures the schedulability of a set of tasks under (m, k)-firm constraints. We consider a task set $(\tau_1, \tau_2, ..., \tau_n)$; these tasks are periodic and their periods, named "basic period" in the following are, respectively, $h_1, h_2, ..., h_n$. We assume here that the tasks are scheduled according to a pre emptive fixed priority policy based on the rate-monotonic algorithm (the larger the basic period is, the lower the priority is) and that the (m, k)-pattern of each task is defined by a mechanical word through equation (5.1):

The (m_i, k_i)-firm constraint of each task τ_i is satisfied if

$$C_i + \sum_{j=1}^{i-1} n_{i,j} C_j < h_i, \forall 1 \leq i \leq n, \tag{5.5}$$

where $n_{i,j} = \left\lceil \frac{m_j}{k_j} \left\lceil \frac{h_i}{h_j} \right\rceil \right\rceil$.

In fact, $\sum_{j=1}^{i-1} n_{i,j} C_j$ gives the workload generated by all the tasks whose priority is higher than τ_i before instant h_i (remember that we are dealing with the special case where $h_1 < h_2 \cdots < h_n$). So it is clear that the deadline is met for τ_i if the total workload $C_i + \sum_{j=1}^{i-1} n_{i,j} C_j$ can be finished within $[0, h_i]$.

Note that the above schedulability condition is sufficient and necessary when the basic period of a task is a multiple of the basic period of all the tasks with higher priority. In the other cases, it degenerates to a sufficient condition.

5.2.5. *Systematic dropping policy in control applications*

This chapter introduces an approach based on the (m, k)-firm model in order to schedule a set of tasks or messages sharing a common resource. The proposed scheduling principles can be seen as a particular case of the period adjustment technique. More specifically we consider that the period of activities sharing a resource (tasks implementing the control law or samples transmitted by a sensor on a network) can be chosen among a limited number of multiples of the basic sampling period. For example, if the basic period for sampling the plant state is h_i and if the (m_i, k_i)-pattern, Π_i of an activity τ_i (task or message) is $\Pi_i = [10100110]$, then, the actual period will be h_i or $2.h_i$ or $3.h_i$. In fact, in this chapter, we will consider that each optional instance of τ_i is dropped systematically.

As introduced in [SET 96] and improved in [EKE 00], the determination of (m_i, k_i)-pattern, Π_i for each activity τ_i is relevant to an optimization problem for which the cost function is an indicator of the system performance. In the following, we propose a co-design approach dealing with both points: the scheduling parameters and the control parameters. In short, the technique that will be presented in the following sections aims to determine an optimal configuration, or more exactly one that is sub-optimal in practice, of m_i, k_i, Π_i for each activity τ_i and of the gain γ_i of the controller implemented by the task.

5.3. Stability analysis of a multidimensional system

The purpose of this section is to present how to guarantee at least the stability condition of control systems. In fact, we are interested by systems whose structure is presented in Figure 5.2.

5.3.1. *Generic model*

The plant is defined as a linear continuous-time system whose evolution is modeled by equation (5.6):

$$dx = Ax\,dt + Bu\,dt + dv_c, \qquad (5.6)$$

where x is the state vector of the system and u is the output of the controller $x \in \mathbb{R}^p$, $u \in \mathbb{R}^q$. The parameters A and B are two matrices whose dimensions are, respectively, (p, p) and (p, q). v_c is a white noise whose covariance is $R_c dt$. The dimension of the constant matrix R_c is (p, p).

The plant state is sampled periodically; the sampling period is noted h. The jth sampled plant state vector transmitted at times jh is consumed by the linear discrete controller defined in equation (5.7); it is noted x_j in the following:

$$u_j = -Lx_j \quad i = 0, 1, 2.... \qquad (5.7)$$

We assume that this controller is implemented as a task. As the purpose is to share the processor among several controller tasks and therefore to decrease the processor bandwidth consumed by this task, in case of an overload situation, several of its instances are rejected according to a (m, k)-firm model and under the constraint that the stability of the system has to be ensured.

Each time an instance is executed, it produces a command u_j. This command is maintained until a new command is produced. The use of the (Zero-Order Holder

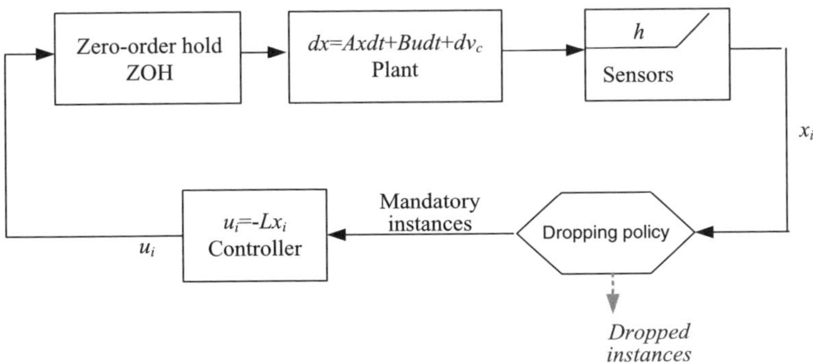

Figure 5.2. *Control system architecture*

ZOH) and the periodicity of the systematic instances dropping defined by the (m, k)-pattern provide a discrete time behavior of the system modeled by equation (5.8):

$$x_{j+1} = \Phi_j x_j + \Gamma_j u_j + v_j, \; j = 0, 1, 2..., \tag{5.8}$$

where Φ_j and Γ_j are given by

$$\Phi_j = e^{A f_j}$$

$$\Gamma_j = \int_0^{f_j h} e^{As} ds B.$$

The value of f_j is the distance, in terms of the number of basic sampling periods, between two updates of the command. For example, if, for a given controller task, the (m, k)-pattern is $\Pi = [1001000]$, we get $f_0 = 3$ and $f_1 = 4$.

v_j is a discrete white noise with a zero mean and

$$Ev_j v_j^T = R_1 \left(f_j h \right) = \int_0^{f_j h} e^{As} R_c e^{A^T s} ds. \tag{5.9}$$

As formerly written, the task instances are periodically rejected according to a given (m, k)-pattern, $f_j + f_{j+1} + f_{j+m-1} = k$ and $f_j = f_{j+m}$, $\forall j > 0$, the period of the system defined in (5.8) is m, meaning also $\Phi_j = \Phi_{j+m}$, $\Gamma_j = \Gamma_{j+m}$.

5.3.2. *Example of multidimensional system*

As an example, we propose to study the control of a cart that can move in one direction guided by a rail. The purpose of the control strategy is to drive the cart to a given position. We suppose that the friction parameters are negligible.

An initial reference position is defined and the position of the cart, d is measured with regard to this reference. $\dot{d}$ notes the speed of the cart. Therefore, the state vector of the plant is $x^T = [d, \dot{d}]$. The parameters p and q of the generic model are, respectively, equal to 2 and 1. The simplified model of the plant is given by the following equation (5.10):

$$\ddot{d} = -k_1 \dot{d} + k_2 u. \tag{5.10}$$

An identification of the system furnished the value of the parameters k_1 and k_2:

$$k_1 = 12.6559 \text{ and } k_2 = 1.9243$$

The continuous model of the system according to the state space representation is given by the generic equation (5.6). In this example, $A = \begin{bmatrix} 0 & 1 \\ 0 & -k_1 \end{bmatrix}$, $B = \begin{bmatrix} 0 \\ k_2 \end{bmatrix}$ and $v_c = 0$. The output of the cart is periodically sampled (period h). The closed-loop model is given by equation (5.11):

$$x_{j+1} = \Phi(h)\,x_j + \Gamma(h)\,u_j,\qquad(5.11)$$

with $\Phi(h) = \begin{bmatrix} 1 & \frac{1-e^{-k_1 h}}{k_1} \\ 0 & e^{-k_1 h} \end{bmatrix}$ and $\Gamma(h) = \frac{k_2}{k_1}\begin{bmatrix} h - \frac{1-e^{-k_1 h}}{k_1} \\ 1 - e^{-k_1 h} \end{bmatrix}$.

The command elaborated by the controller is

$$u_j = L(x_{\mathrm{ref}} - x_j),\qquad(5.12)$$

where L is the gain $L = \begin{bmatrix} k_c & k_d \end{bmatrix}$ and x_{ref} is the targeted reference defined as $x_{\mathrm{ref}} = \begin{bmatrix} d_{\mathrm{ref}} & 0 \end{bmatrix}^T$ (d_{ref} is the reference of the position on the rail where the cart has to stop).

5.3.2.1. *Sampling period definition*

The basic sampling period h_{basic} is determined according to the empirical "*rule of thumb*" formulated in [ÅST 97]: $0.2 < \omega_0 < 0.6$, where ω_0 is the natural frequency of the system ($\omega_0 = 20$, for the cart considered in this example). The period h_{basic} has to be chosen in $[0.01, 0.03]$. A study of this system shows that its performance in terms of rise time and overshot is optimal for $h_{\mathrm{basic}} = 0.01$s. So we will use this value as the basic sampling period.

5.3.2.2. *Controller parameters*

We note $k_{c,0.01}$ and $k_{d,0.01}$, the parameters of the controller obtained for a basic sampling period $h_{\mathrm{basic}} = 0.01$s. Their values are evaluated by solving the Linear Quadratic Regulator (LQR) problem proposed in [ÅST 97]. The final results are $k_{c,0.01} = 121$ and $k_{d,0.01} = 6.5$.

5.3.3. *Stability condition*

The configuration of a dropping policy based on the (m, k)-firm model needs to identify the values of k, m, and the (m, k)-pattern. The first problem to solve concerns the stability of the system, and the question is which parameter of the constraint is critical for ensuring this property. This identification relies on the intuitive idea that is: if a system is stable for a given $(1, k)$-firm policy, it will be stable for any dropping policy based on a (m, k)-firm. Therefore, we propose to evaluate, for each task, $k_{\max}$, the greatest value of k, so that the system is guaranteed to be stable for each constraint (m, k) with $k \le k_{\max}$ and $m \le k$.

The analysis of equation (5.8), shows that on the one hand, a system subject to a (m, k)-firm constraint can be considered as a system with sampling periods varying according to a regular form specified by the (m, k)-pattern and, on the other hand, a system subject to a $(1, k)$-firm constraint is equivalent to a system controlled under a sampling period equal to k times the basic period. Therefore, the determination of $k_{\max}$ is equivalent to the determination of the maximal value of h that ensures the system stability.

In particular, if we study the example proposed in section 5.3.2, let us note $\Psi(h) = \Phi(h) - \Gamma(h) L$, given by

$$\Psi(h) = \left[\begin{array}{cc} 1 - k_c \frac{k_2}{k_1} \left(h - \frac{1-e^{-k_1 h}}{k_1} \right) & \frac{1-e^{-k_1 h}}{k_1} - k_d \frac{k_2}{k_1} \left(h_c - \frac{1-e^{-k_1 h}}{h} \right) \\ k_c \frac{k_2}{k_1} \left(1 - e^{-k_1 h} \right) & e^{-k_1 h} - k_d \frac{k_2}{k_1} \left(1 - e^{-k_1 h} \right) \end{array} \right].$$

By applying the Jury criteria [FRE 63], the following three conditions provide a necessary and sufficient condition for the system stability:

$$1.\ a_2\ <\ 1,$$

$$2.\ a_2\ >\ a_1 - 1,$$

$$3.\ a_2\ >\ -a_1 - 1,$$

where $a_1 = \Psi_{1,1} \Psi_{2,2} - \Psi_{1,2} \Psi_{2,1}$, $a_2 = -\Psi_{1,1} - \Psi_{2,2}$; $\Psi_{i,j}$ notes the element placed in line i, column j of the matrix Ψ; a_1 and a_2 are expressed according to the controller parameters (k_c, k_d) and the sampling period h. This defines a domain of admissible t-uple (k_c, k_d, h).

In practice, the determination of the greatest admissible value of the sampling period starts by fixing $k_d = k_{d,0.01}$, which is the value obtained for k_d when $h = h_{\text{basic}} = 0.01\text{s}$ (the basic period). Then we study k_c according to h under the following constraints deduced from Jury's conditions.

So, we need to analyze the function that expresses the maximal value of h for each value of k_c, with k_d being fixed at the value $k_{d,0.01}$. In fact, we study the function $k_{cMax}(h)$, which is evaluated by

$$k_{cMax}(h) = \left\{ \begin{array}{ll} k_{c1}(h) & \text{if } k_{c1}(h) < k_{c\text{Lim}}(h) \\ k_{c2}(h) & \text{if } k_{c1}(h) \geq k_{c\text{Lim}}(h), \end{array} \right.$$

where

$$
k_{cLim}\,(h) \;=\; \frac{4k_1 e^{k_1 h}}{h k_2 \left(e^{k_1 h} - 1\right)}
$$

$$
k_{c1}\,(h) \;=\; \frac{2k_1 \left(k_1 + k_1 e^{k_1 h} + k_2 k_d - k_2 k_d e^{k_1 h}\right)}{k_2 \left(2 - 2e^{k_1 h} + h k_1 + h k_1 e^{k_1 h}\right)}
$$

$$
k_{c2}\,(h) \;=\; \frac{k_1 \left(k_1 + k_2 k_d\right) \left(e^{k_1 h} - 1\right)}{k_2 \left(-1 - h k_1 + e^{k_1 h}\right)},
$$

and we obtain the stability region in a space $(h,\,k_c)$. This region, for the example proposed in 5.3.2, is identified by the gray color in Figure 5.3. It represents, for each point k_c, all the admissible values of h. For example, for $k_c = k_{c,0.01}$, the maximal value of h, $h_{\max}$, ensuring the stability is given by the abscissa of the point P and h can be chosen in the interval $[0.01, h_{\max}]$.

As mentioned before, the period adjustment based on a (m, k)-firm model is equivalent to a regular sequence of time intervals between to consecutive samples, specified by the (m, k)-pattern; each time interval is a multiple of the basic period. Therefore, for the cart system, $k_{\max}$ is determined by $\left\lfloor \frac{h_{\max}}{h_{\mathrm{basic}}} \right\rfloor$ and the maximal sampling period, ensuring the stability and corresponding to a $(1, k_{\max})$ constraint, is equal to $k_{\max} h_{\mathrm{basic}}$.

5.4. Optimized control and scheduling co-design

Once the stability of the system ensured, there is a further step needed to deal with the optimization issue. So, to do this, we first define a cost function used for determining an optimal control (section 5.4.1) and, then, we identify the global optimization problem for a set of closed loops where the algorithms implementing the control laws share one processor (section 5.4.2). Finally, the proposed approach is illustrated by a case study in section 5.4.3.

The optimization approach relies on two phases:

– The first is done off-line. For each task, τ_i, a value of k_i ensuring the stability of the system is fixed according to the method described in section 5.3.3. For each value of $m_{i,j}$ such that $1 \leq m_{i,j} \leq k_i$, a $(m_{i,j}, k_i)$-pattern $\Pi_{i,j}$ is defined based on the mechanical words technique (see section 5.2). Then, for each pattern $\Pi_{i,j}$, the cost function of the system is evaluated. Such a function is proposed in section 5.4.1.

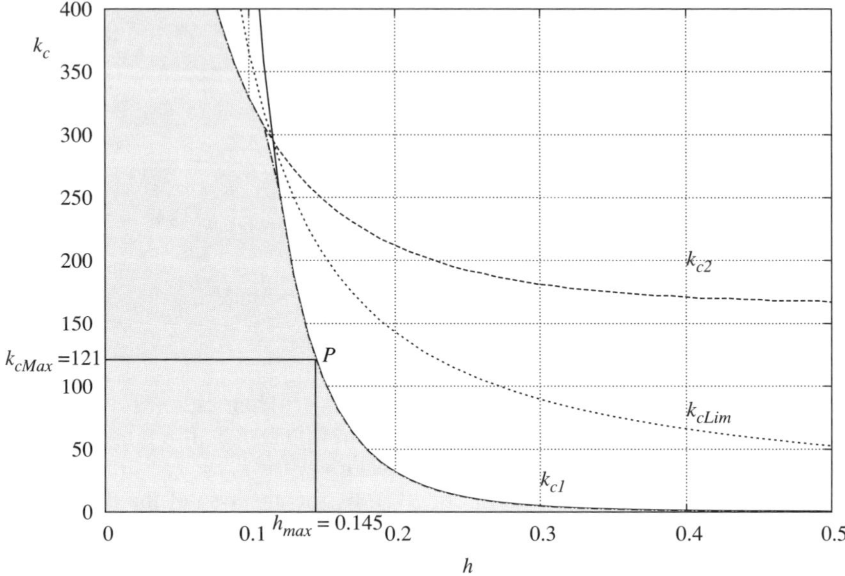

Figure 5.3. *Stability region evaluated on the case study presented in section 5.3.2. k_{c1}, k_{c2}, k_{cLim} and k_{cMax} are functions of the sampling period h. The stability region for the cart system is identified by the gray color*

– The second phase is done by the task handler as illustrated in Figure 5.1. It concerns the global optimization of the system. For each working mode, the task handler has to configure the (m, k)-firm constraints and the scheduling parameters of each task implementing the control law of each system and sharing the same processor.

5.4.1. *Optimal control and individual cost function*

An indicator of the closed-loop performance can be given, among others, by the Least Quadratic (LQ) cost function , which provides a form of "cumulative cost" for an infinite horizon. Applied to a system described by equation (5.6), it is defined by the following formula:

$$J^\infty = \lim_{N \to \infty} \tfrac{1}{N} E(\textstyle\int_0^N x^T(t)Qx(t) + u^T(t)Ru(t)dt),$$

where Q and R are two matrices that respectively weigh the state and the input of the plant.

For a task τ_i that implements the ith controller, given k_i such that the stability is ensured, J_i^∞ is the cost function to minimize. We consider all the numbers of mandatory instances m_i ($1 \leq m_i \leq k_i$), assuming that the corresponding patterns

Π_i are defined using the mechanical words approach; for each possible value of m_i and Π_i we determine the optimal control law, in fact the gain of the controller, which minimizes the cost function. As the intervals between two instances are time varying, according to the pattern, the optimal control law is described by a sequence of gain, $L_{i,p}$ for $p = 1, 2, ..., m_i$. For example, let us consider $k_i = 10$, for $m_i = 3$; the (m_i, k_i)-pattern Π_i is equal to $[1001001000]$, and we need to determine three values of the gain: $L_{i,1}$, $L_{i,2}$, $L_{i,3}$ to apply using the control law when producing the command to the actuator, respectively, at the first, second, and third mandatory instances.

We consider the discrete time-varying model of the plant presented in (5.8). We denote by h_i the basic period of the ith plant controller.

The following discrete form of J_i^∞ is evaluated for each possible value of m_i ($1 \leq m_{i,j} \leq k_i$) and therefore for the corresponding sequences $(f_{i,0}, f_{i,1}, ..., f_{i,m_{i,j}-1})$:

$$J_i(m_{i,j}) \;=\; \frac{1}{m_{i,j}\frac{H_i}{k_i}} \sum_{p=0}^{m_{i,j}\frac{H_i}{k_i h_i}} \left(x_{i,p}^T Q'_{i,p} x_{i,p} + 2 x_{i,p}^T M_{i,p} u_{i,p} + u_{i,p}^T R'_{i,p} u_{i,p} \right)$$

$$+ \frac{1}{m_{i,j}\frac{H_i}{k_i}} \left(E\left(x_{m_i\frac{H_i}{k_i h_i}}^T Q_{i0} x_{m_{i,j}\frac{H_i}{k_i h_i}} \right) + \sum_{p=0}^{m_{i,j}\frac{H_i}{k_i h_i}} \overline{J}_{i,p} \right), \quad (5.13)$$

where H_i is the time horizon of the ith plant, under the condition $\frac{m_{i,j}H_i}{k_i h_i} \epsilon \mathbb{N}^*$, $x_{i,p}$ is the plant state measured at the pth sample and $u_{i,p}$ the corresponding command sent to the actuator, $Q'_{i,p} = \int_0^{f_{i,p}h_i} \Phi_i^T(t) Q \Phi_i(t) dt$, $M_{i,p} = \int_0^{f_{i,p}h_i} \Phi_i^T(t) Q \Gamma_i(t) dt$, $R_{i,p} = \int_0^{f_{i,p}h_i} (\Gamma_i^T(t) Q \Gamma_i(t) + R_i) dt$, $\overline{J}_{i,p} = tr(Q \int_0^{f_{i,p}h_i} R_{ic}(t) dt$ with R_{ic}, the co-variance of v_i, and $\Phi_i(t) = e^{A_i t}$ and $\Gamma_i(t) = \int_0^t e^{A_i t} dt B_i$. The optimal control law that minimizes the cost function 5.13 is given by [ÅST 97] as

$$u_{i,p} = -L_{i,p} x_{i,p} \,, \; p = 0, 1, ..., \tag{5.14}$$

where

$$L_{i,p} = \left(\Gamma_{i,p}^T S_{i,p+1} \Gamma_{i,p} + R'_{i,p} \right)^{-1} \left(\Gamma_{i,p}^T S_{i,p+1} \Phi_{i,p} + M_{i,p}^T \right). \tag{5.15}$$

$S_{i,p}$ is obtained from the recurrent equation

$$S_{i,m\frac{H_i}{k_i}} \;=\; Q'_0$$

$$S_{i,l} \;=\; \Phi_{i,l}^T S_{i,l+1} \Phi_{i,l} + Q_{i,l} - \left(\Gamma_{i,l}^T S_{i,l+1} \Phi_{i,l} + M_{i,l}^T \right)^T$$

$$\left(\Gamma_{i,l}^T S_{i,l+1} \Gamma_{i,l} + R'_{i,l} \right)^{-1} \left(\Gamma_{i,l}^T S_{i,l+1} \Phi_{i,l} + M_{i,l} \right)$$

$$\text{for } 0 \leq l \leq m_{i,j}\frac{H_{i,j}}{k_i h_i}. \tag{5.16}$$

Taking into account the periodicity of the pattern, $\Phi_{i,p} = \Phi_{i,p+m_{i,j}}$ and $\Gamma_{i,p} = \Gamma_{i,p+m_{i,j}}$. Consequently, the solution of equation (5.16) is also periodic with a period $m_{i,j}$ when calculated on a sufficiently long time horizon [BIT 91] i.e. $S_{i,l} = S_{i,l+m_{i,j}}$. Then, the gain of the controller $L_{i,p}$ is designed using the steady-state solution of the Ricatti equation (5.15), and its solution is also periodic:

$$L_{i,p} = l_{i,p+m_{i,j}}.$$

With this computation, it is possible, for a given $(m_{i,j}, k_i)$-pattern, to determine the best sequence of $L_{i,p}$ that allows us to partially compensate the task instance dropouts.

When time goes to infinity, $(\lim H_i \to \infty)$, the influence from the initial condition decreases and because $S_{i,l} = S_{i,l+m_{i,j}}$, equation (5.13) may be written as

$$J_i^\infty(m_{i,j}) = \frac{1}{m_{i,j} h_i} \left(\sum_{p=0}^{m_{i,j}-1} tr S_{i,p+1} R_{i,j} + \sum_{p=0}^{m_{i,j}-1} \overline{J}_{i,p} \right). \tag{5.17}$$

5.4.2. *Global optimization*

Let us now consider the problem introduced at the beginning of section 5.4. As mentioned before, the problem is the global optimization of a set of controllers deployed as a set of tasks sharing the same processor. In the last section (5.4.1), we demonstrated, for a given (m, k)-firm constraint and a given (m, k)-pattern, how to determine the sequence of controller gains that compensate for the task instance dropout between two mandatory instances; this set of gains is identified in order to minimize a cost function that represents a cumulative cost and is derived from the LQ function. Using this evaluation, realized off-line, each task τ_i that may be activated in one of the possible working modes, is characterized by several attributes:

– its basic period h_i and its priority P_i,

– the execution time of the task C_i,

– the parameter k_i that ensures the stability of the system under a period $k_i h_i$,

– the number n_i of values $m_{i,j}$ such that a systematic dropping of the task instances can be done following the $(m_{i,j}, k_i)$-firm constraint:

– for each value $m_{i,j}$:
  - the value of $m_{i,j}$,
  - the $(m_{i,j}, k_i)$ pattern, $\Pi_{i,j}$; we recall that it is defined thanks to the mechanical words,

- the list of gains to apply at each instance of the task: $L_{i,1}$, $L_{i,2}$, ...,$L_{i,m_{i,j}}$

- $J_{i,j}$ the value of the cost function obtained by using these gains repeatedly on each interval $[pk_i, (p+1)k_i]$ for $p \geq 0$.

This information is used on-line for the resolution of the global optimization problem. As soon as a new working mode is defined, the task handler knows which plant needs to be controlled and by which controller; at this point, it has to choose, for each active controller, and therefore for each corresponding task τ_i , what the value of the parameter m_i is that has to be applied in order to minimize a global cost function. We denote the number of tasks activated in one working mode by n.

Let us consider a set of tasks τ_i , $1 \leq i \leq n$, described by the parameters given above; the global optimization problem consists in determining $(s_{i,1}, s_{i,2}, .., s_{i,n_i})$ for each task τ_i that minimizes

$$\sum_{i=1}^{n}\sum_{j=1}^{n_i}(-s_{i,j}J_{i,j}), \tag{5.18}$$

with $s_{i,j}\epsilon\{0,1\}$, $\displaystyle\sum_{j=1}^{n_i}s_{i,j} = 1, i = 1,..,n.$

Under the schedulability constraint that has to be met by all tasks τ_i, $i = 1,..,n$, this condition is equivalent to the one formulated in equation (5.5):

$$C_i + \sum_{j=1}^{i-1}\left[\frac{\sum_{p=1}^{n_j}s_{j,p}m_{j,p}}{k_j}\left\lceil\frac{h_i}{h_j}\right\rceil\right]C_j < h_i. \tag{5.19}$$

This optimization problem can be seen as a Multiple-Choice, Multiple Dimension Knapsack (MMKP) problem [MAR 90] that has been proved to be NP-hard. Therefore, solving this problem on-line requires developing an heuristic algorithm ensuring that it can provide a tight sub-optimal solution. In the following, we apply a slightly modified version of the computationally cheaper algorithm HEU proposed in [KHA 02]. For our optimization problem, the proposed algorithm is

5.4.3. *Case study*

In this section, we apply the method presented above to a case study. Let us consider four cart systems, $cart_1$, $cart_2$, $cart_3$, and $cart_4$, similar to the one presented in section 5.3.2. The control of these cart systems can be active or not depending on

---

Algorithm 5.1: Modified computationally cheaper heuristic

1) to find a feasible solution first, that is to say, select $m_{i,j}$ for each τ_i while satisfying the constraints given in (5.19);

for this purpose, the algorithm HEU is modified by always setting the value of $m_{i,j}$ of each task τ_i to be equal to 1 (if the solution is infeasible in this case, no other solution will be feasible);

2) and then to iteratively improve the solution by replacing, for eachτ_i, the current value of $m_{i,j}$ by another value corresponding to a better performance while keeping the constraints (5.19) satisfied;

if no such solution can be found, the algorithm tries an iterative improvement of the solution which;

a) first replaces $m_{i,j}$ for one task τ_i, which is not schedulable with the current value of $m_{i,j}$;

b) and then replace the value of $m_{i',j}$ for all tasks $\tau_{i'}$ ($i' \neq i$) by a value providing a worse performance;

the original algorithm HEU tries to find, after the first step, a better solution requiring less resource consumption which, however, does not exist in our model, therefore, this property also help us to delete an unprofitable search procedure in HEU;

3) The iteration finishes when no other feasible solution can be found;

---

the working mode chosen by the supervisor (see Figure 5.1). The tasks implementing each controller share the same processor.

A Matlab/Simulink model of the system is specified. The scheduling policy is implemented using the toolbox TrueTime [CER 03]. The system is then analyzed by tracing an indicator of the control performance during the simulation of this model running on a given scenario. This indicator is given, for each controlled cart system, by function (5.20), which is evaluated at each simulation step:

$$J_i(t) = \int_0^t \left(x_i^T(s) Q_i x_i(s) + u_i^T(s) R_i u_i(s) \right) ds, \tag{5.20}$$

where $Q_i = \begin{bmatrix} 1 & 0 \\ 0 & 0 \end{bmatrix}$ and $R_i = 0.00006$ for each cart$_i$, $i = 1, 2, 3, 4$.

Furthermore, we can observe the state of each task instance during the simulation (running, pre-empted, not activated).

5.4.3.1. *Plants and controllers*

The generic continuous model of the cart system, cart_i for $1 \le i \le 4$ (see equation 5.6) is given by

$$dx = \begin{bmatrix} 0 & 1 \\ 0 & \frac{-11.4662}{M_i} \end{bmatrix} xdt + \begin{bmatrix} 0 \\ \frac{1.7434}{M_i} \end{bmatrix} udt + dv_c, \tag{5.21}$$

where M_i is the mass of cart_i:

$$M_1 = 1.5, \;\; M_2 = 1.2, \;\; M_3 = 0.9, \;\; M_4 = 0.6.$$

The controller of each cart_i is noted as $controller_i$, the task which implements the $controller_i$ is τ_i, and its basic period is h_i:

$$h_1 = 0.007, \;\; h_2 = 0.0085, \;\; h_3 = 0.01, \;\; h_4 = 0.0115.$$

5.4.3.2. *Scheduling parameters*

The tasks τ_i are scheduled according to a fixed pre-emptive priority policy under an implicit deadline constraint for their mandatory instances ($D_i = h_i$). The priorities of the tasks are defined thanks to their basic period (the larger the period is, the lower the priority is). So, in any working mode, there is the following relation between the task priorities:

$$\text{priority}(\tau_1) > \text{priority}(\tau_2) > \text{priority}(\tau_3) > \text{priority}(\tau_4)$$

Let us now fix the parameters of the constraint (m, k)-firm for each task. The value of the parameter k_i is defined in order to ensure the stability of the system under a period equal to $k_i h_i$. In this case study, we identified the following value of k_i:

$$k_1 = 5, \;\; k_2 = 8, \;\; k_3 = 10, \;\; k_4 = 1.$$

We consider, in the considered experiments, that the four tasks have the same execution time:

$$C_1 = C_2 = C_3 = C_4 = 3 \text{ ms}$$

and that the execution time of the task handler is $C_{th} = 2,5$ ms. Furthermore, its priority is higher in the system.

5.4.3.3. *Optimal controller*

The controller of cart_i is defined by $u_i = L_i x_i$, where the gain L_i is evaluated for each interval between two consecutive mandatory instances according to the (m_i, k_i)-firm strategy used for this plant. As detailed in section 5.4.2, the value of the gain is calculated in order to optimize the control performance during this interval. The cost function, to minimize, in this case study is the discrete form of

$$J_i = \lim_{N \to \infty} \int_0^N \left(x^T(t) Q_i x(t) + u^T(t) R_i u(t) \right) dt, \tag{5.22}$$

where $Q_i = \begin{bmatrix} 1 & 0 \\ 0 & 0 \end{bmatrix}$ and $R_i = 0.00006$ for each cart_i, $i - 1, 2, 3, 4$.

5.4.3.4. *Simulation scenario*

The system is observed along a scenario that introduces three working modes and two types of working mode switchings:

– at time $t = 0s$, the first two cart systems (cart_1 and cart_2) are to be controlled; therefore, only two tasks are running: τ_1 and τ_2 implementing, respectively, the Controller_1 and Controller_2; the set of tasks to be scheduled is $\{\tau_1, \tau_2\}$;

– at time $t = 1s$, the fourth cart system ($cart_4$) has to be controlled; so, τ_4 implementing the $Controller_4$ is activated; the set of tasks to be scheduled is $\{\tau_1, \tau_2, \tau_4\}$;

– at time $t = 2s$, the third cart system ($cart_3$) has to be controlled; so, τ_3 implementing the $Controller_3$ is activated; the set of tasks to be scheduled is $\{\tau_1, \tau_2, \tau_3, \tau_4\}$.

Two pre-emptive-fixed priority scheduling policies were modeled:

– *Hard real-time constraints.* We consider that all the task instances are mandatory; in this case, the gain of each controller is constant and determined in order to optimize the cost function (5.22) for the basic sampling period of each system.

– *Adaptive system.* In this case, we implement a systematic dropout of the non-mandatory instances according to the (m, k)-firm constraints specified for each task; the gain of the controller is adapted to each inter-sample interval length in order to optimize the performance of each system; the value of k is constant for a given system in each working mode, while the value of m is evaluated for each system at the beginning of each working mode in order to optimize the global cost function proposed in (5.18) under the schedulability condition (5.19).

5.4.3.5. *Simulation results for hard real-time constraints*

The control performance of each system as well as the evolution of the task state are illustrated in Figures 5.4–5.6.

– In the interval $[0, 1[$, two tasks are periodically activated: τ_1, with the period $h_1 = 0.007$ s, and τ_2, with the period $h_2 = 0.0085$ s; all the instances of both tasks meet their deadline.

– At time $t_1 = 1$ s, the supervisor decides to include the control of $cart_4$ leading to the activation of the task τ_4; its period is $h_4 = 0.0115$ s; so, it will be activated successively at 1 s, 1.0015 s, 1.013 s, 1.0245 s, etc.; we can observe in Figure 5.4 that no instance of τ_4 meets its deadline; the starting time of several instances is delayed and the completion of all the instances are after their deadline; as the priorities of τ_1 and τ_2 are higher than that of τ_4, the activation of τ_4 has no impact on the scheduling of the two other tasks that meet always their deadlines;

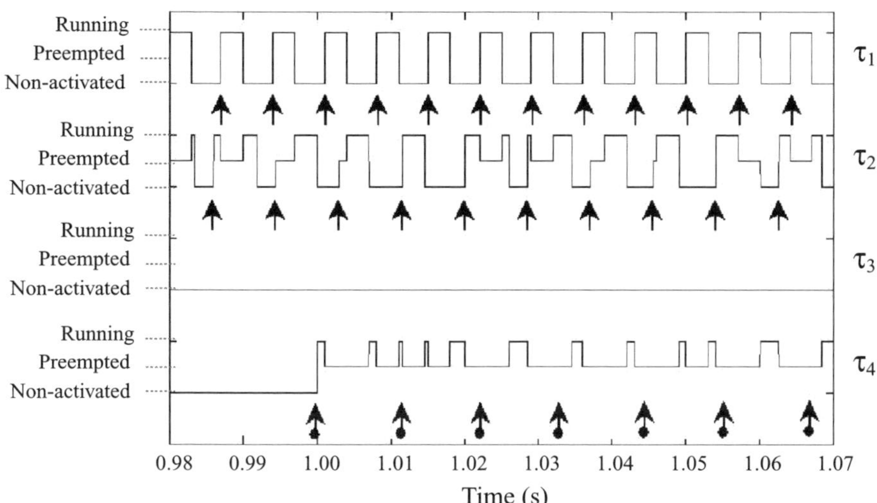

Figure 5.4. *Task evolution under hard real-time constraints strategy. The detailed time interval is* $[0.98, 1.07]$ *and it includes a working mode switch at time* $t_1 = 1$ s*: before* t_1*, only two tasks are activated (*τ_1 *and* τ_2*); at time* t_1*, the plant* cart$_4$ *has to be controlled leading to the activation of the task* τ_4

– At time $t_2 = 2$ s, the supervisor includes the control of cart$_3$ leading to the activation of the task τ_3 at time 2 s, 2.0115 s, 2.023 s, etc.; we can observe in Figure 5.5, that no instance of this new task meets its deadline; furthermore, the interference of the three tasks of higher priority, τ_1, τ_2, and τ_3, makes the processor unavailable for the task τ_4, leading all the instances of this task to fail to run.

An analysis of Figure 5.6 shows how the performance of the system varies, as evaluated by function (5.20). We can note that for cart$_4$, the performance is acceptable between $t = 1$ s, up to $t = 2$ s, which is before the activation of task τ_3. Then the performance of cart$_4$ diverges. This is due to the interference of tasks τ_1, τ_2, and τ_3, whose priorities are higher than that of τ_4. As mentioned before, this task will never run. Finally, the performance of cart$_3$ is always acceptable despite the instances of the corresponding control task τ_3 never meeting their deadline.

5.4.3.6. *Simulation results for* (m, k)-*firm constraints*

The control performance of each system as well as the evolution of the task state are illustrated in the Figures 5.7–5.9.

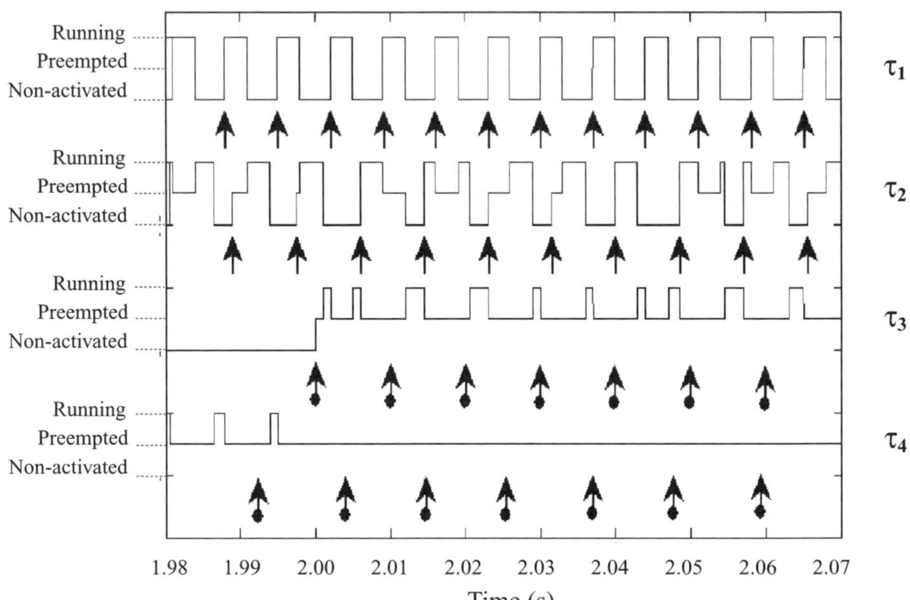

Figure 5.5. *Task evolution under hard real-time constraints strategy. The detailed time interval is* $[1.98, 2.07]$ *and includes a working mode switch at time* $t_2 = 2$ s: *before* t_2, *three tasks are activated* (τ_1, τ_2 *and* τ_4)*; at time* t_2, $cart_3$ *has to be controlled leading to the activation of the task* τ_3

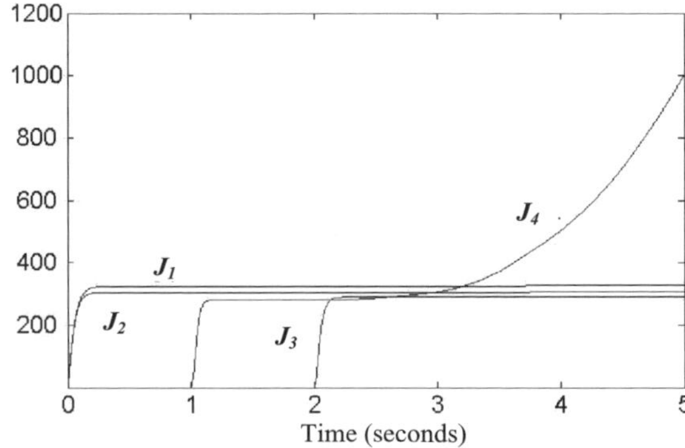

Figure 5.6. *Control performance under hard real time constraints strategy*

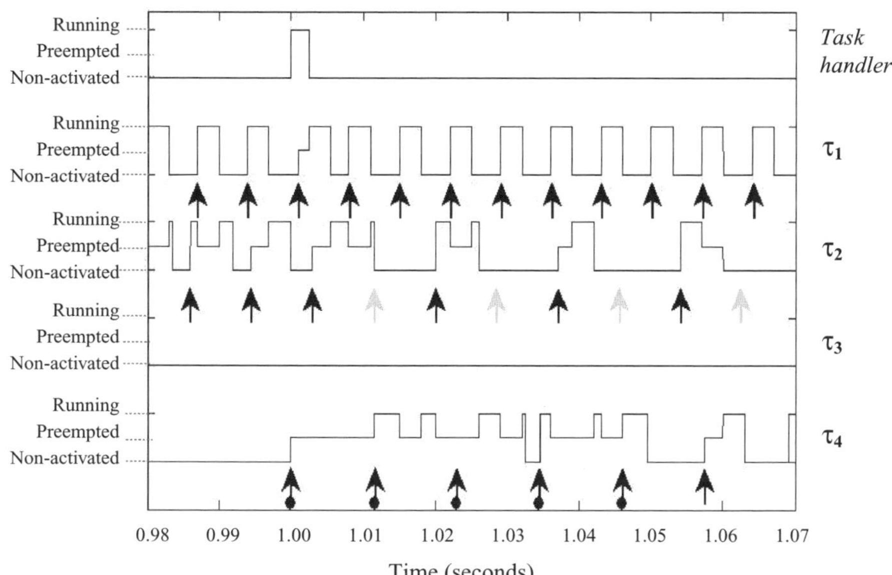

Figure 5.7. *Task evolution under (m, k)-firm strategy. The detailed time interval is $[0.98, 1.07]$ and includes a working mode switch at time $t_1 = 1$ s: before t_1, only two tasks are activated (τ_1 and τ_2); at time t_1, $cart_4$ has to be controlled, leading to the activation of task τ_4*

– In the interval $[0, 1[$, two tasks are activated periodically: τ_1, with the period $h_1 = 0.007$ s, and τ_2, with the period $h_2 = 0.0085$; these tasks are schedulable under hard real-time constraints as seen in section 5.4.3.5; moreover, in this working mode, the global cost function is optimized for $m_1 = k_1$ and $m_2 = k_2$, so all the instances of τ_1 and τ_2 are mandatory.

– At time $t_1 = 1$ s, the supervisor decides to include the control of $cart_4$ leading to the activation of task τ_4; in Figure 5.7, at this time, the task handler is activated; as its priority is higher in the system, its completion is at time $t = t_1 + C_{th} = 1.0.0025$ s; at time t_1, the optimization of the global cost function (5.18), realized by the task handler, furnishes the value of $m_1 = 5$ and $m_2 = 4$ providing the rules for the dropping policy: $(5, 5)$-firm for τ_1 and $(4, 8)$-firm for τ_2 under the (m_2, k_2)-pattern $\Pi_2 = [10101010]$; as $k_4 = 1$, all the instances of τ_4 are mandatory; Figure 5.7 shows that all the mandatory instances of τ_1 and τ_2 meet their deadline, while no instance of

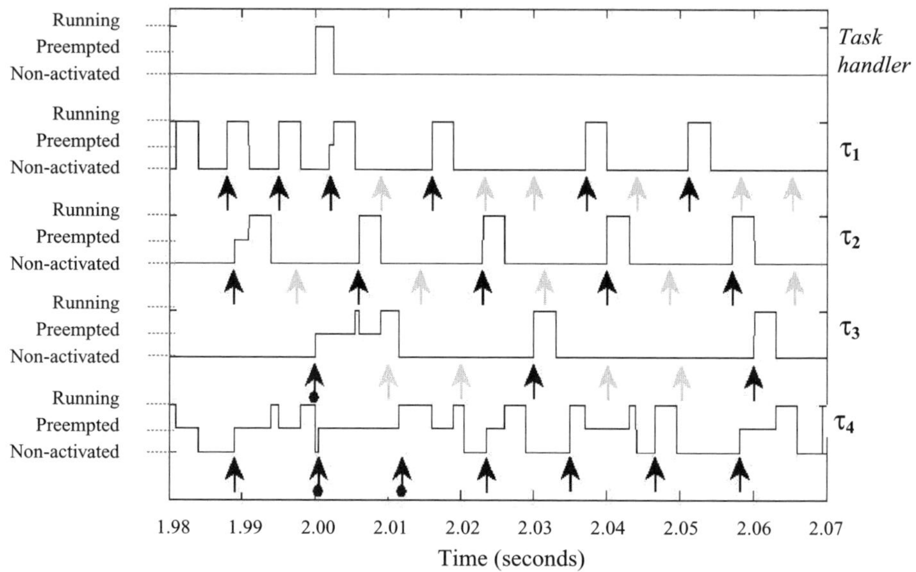

Figure 5.8. *Task evolution under (m, k)-firm strategy. The detailed time interval is $[1.98, 2.07]$ and includes a working mode switch at time $t_2 = 2\,$s: before t_2, three tasks are activated (τ_1, τ_2 and τ_4); at time t_2, $cart_3$ has to be controlled, leading to the activation of the task τ_3*

τ_4 does; nevertheless, all its instances run to completion, under a delayed starting time and delayed completion time.

– At time $t_2 = 2$ s, the supervisor includes the control of $cart_3$, so a new working mode that integrates the former tasks τ_1, τ_2, and τ_4 and the new task τ_3 is started; the task handler has to redefine the optimal configuration of the task activation rules; in this case, function (5.18) is minimized for the following (m_i, k_i)-firm constraints:

- $(m_1, k_1) = (2, 5)$ and $\Pi_1 = [10100]$,
- $(m_2, k_2) = (4, 8)$ and $\Pi_1 = [10101010]$,
- $(m_3, k_3) = (3, 10)$ and $\Pi_1 = [1001001000]$,
- $(m_4, k_4) = (1, 1)$.

We can observe, in Figure 5.8, that the two first instances of τ_4 and the first instance of τ_3 , activated at the beginning of this new working mode, do not meet their deadline because of a transient overload due to the execution of the task handler; after that point, all the mandatory instances of the four tasks meet their deadline.

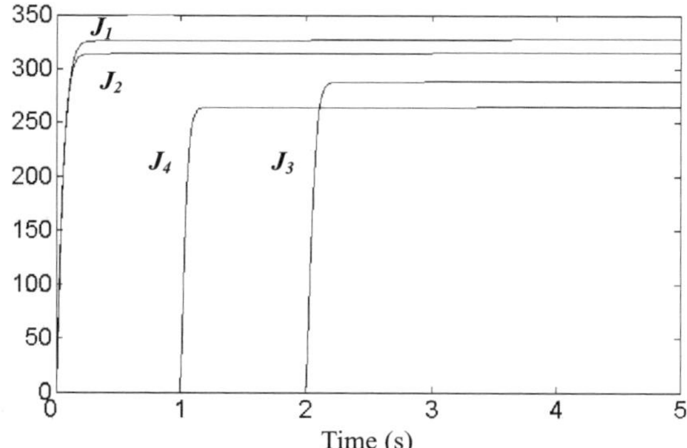

Figure 5.9. *Control performance under* (m, k)-*firm strategy*

Figure 5.9 illustrates the evolution of the cost function, provided by formula (5.20). For each cart system, the performance becomes quickly stable and remains constant in each working mode.

The optimization algorithm of the global cost function (5.18) under the schedulability constraints (5.19) for a set of control tasks sharing the same processor allows for applying a best dropout policy of several well-identified instances for each tasks; the consequence is a wider availability of the processor and the possibility for each task to run its mandatory instances to completion before the deadline (the relative deadline is the basic period of each task). Furthermore, the controller itself is optimized and an adaptive gain suited to the (m_i, k_i)-pattern is defined.

5.5. Plant-state-triggered control and scheduling adaptation and optimization

Let us consider in this section, more precisely, the different situations of the states of the plants that may be controlled:

– the plant is not controlled; this situation occurs when the plant does not exist for the overall system (plant controlled only during certain time interval, plant deactivated because its output is out of a given domain);

– it is controlled and its state is a steady state;

– it is controlled, but when this control is just activated, its state is a transient state.

We have to take into account the last two cases and, in particular, we must adapt the cost functions in order to deal with both situations. For this purpose, we propose taking into consideration, depending on the situation of the plant, an infinite-horizon or a finite-horizon cost function in order to find the optimal configuration of each m_i and of each controller gain L_i.

5.5.1. *Closed-loop stability of switching systems*

A change in the value of m_i, for a given k_i, produces a sampling period variation, and then we consider a Discrete Time Switched System (DTSS) description. To adapt the control law parameters to this variation, we use the design process presented in section 5.4.1. But, as was shown in [SCH 02], controllers designed with optimal-LQ techniques may suffer from instability under certain switching sequences, i.e. when the sampling period changes. Because of this undesirable result, [SCH 02] adopts a linear matrix inequalities (LMI) framework to design stable optimal controllers. We propose using a LMI framework to find a common quadratic Lyapunov function (CQLF); then the asymptotic stability is guaranteed for any (m_i, k_i)-firm sequence for each plant, proving the stability of the control.

Firstly, for a fixed control task τ_i, k_i, we consider each possible value of the number of mandatory instances, m_i. For each of these, we compute the corresponding m_i controller parameters $\mathcal{L}_{i,m_i} = \left\{ L_{i,m_i}^0, L_{i,m_i}^1, ..., L_{i,m_i}^{m_i-1} \right\}$ by using equation (5.15).

Secondly, we consider, for the control task τ_i, the set of open-loop discrete time models (5.6), $\Theta_i = \left\{ (\Phi_{i,1}), (\Phi_{i,2}), ..., (\Phi_{i,k_i}) \right\}$ and evaluate the k_i periods, taking into account the possible interruptions in a planned sequence at any time.

By using the elements in both sets $\mathcal{L}_{i,m_i}$ and Θ_i, we can establish a new set of $m_i.k_i$ closed-loop models (5.7) without noise, $A_{i,p}^l = \Phi_{i,l} + \Gamma_{i,l} L_{m_i}^p$, where l varies between 1 and k_i $((\Phi_{i,l}, \Gamma_{i,l}) \, \epsilon \Theta_i)$ and p between 0 and $m_i - 1$ $((L_i^p \epsilon \mathcal{L}_{i,m_i}))$.

In order to prove the stability of the DTTS [LIB 99], for k_i and each possible value of m_i, we need to find a CQLF for the set of matrices $A_{n,p}^i$, where $n = 1, ..., k_i$ and $p = 0, ..., m_{i-1}$. Then, we can define a set of inequalities:

$$\left(A_{n,p}^i\right)^T P A_{n,p}^i - P < 0, \ \forall n = 1, ..., k_i, \ \forall p = 0, ..., m_{i-1}. \tag{5.23}$$

Note that the identification of a CQLF is a sufficient condition. In order to solve the problem and find the matrix $P = P^T > 0$, we use the LMI toolbox from the Matlab tool.

5.5.2. *On-line plant state detection*

As mentioned in the previous section, three plant states are identified: non-activated plant, steady plant state (or near), and transient plant state. Reaching or leaving the first situation for a plant modifies the value of the number of control tasks n. The dead-band approach presented in [OTA 02] is used to distinguish the steady and the transient states of a plant. Each controlled plant has a state, which asymptotically tracks the reference r, and is supervised by the supervision task. Let y_i be the state of the plant i. y_i can be a subset of the plant state vector x_i defined in section 5.3.1. For instance, taking the previously studied cart system with $x = [position\ speed]'$, if we want the position to follow the reference position r, the variable *position* is then the parameter of the interest, and we can define $y = [1\ 0] * x$. The following condition on two successive samples (n^{Te} and $(n+1)^{NH}$ ones) is set up:

$$|y_i(h_i + Nd_i) - y_i(nh_i)| < \min\{\delta|y(nh_i)|, \text{th}\},$$

where th is a threshold to prevent false identifications due to noises, δ is a parameter to adjust for each plant, h_i is the detection period implemented by the supervision task for plant i. If the condition is verified, then the plant is considered to be in steady state; otherwise, it is in a transient state. The advantage of this plant state detection mechanism is that it depends on the actual evolution and it detects, in the same way, reference changes or/and non-modeled perturbations.

5.5.3. *Global optimization of control tasks taking into account the plant state*

As in section 5.4, we deal with several working modes where, in each mode, a number of control tasks have to share one processor. In this section, we take into account that the working modes are identified, on the one hand, by a number of plants to be controlled, meaning, a number of given control tasks and, on the other hand, by the situation in which the plant is found. This situation corresponds to the two above-mentioned cases: the plant is in a transient state (due to a new reference or noises) or is in steady state. Intuitively, the constraints, in terms of performances, that have to be applied to the control are not the same for these two situations. Therefore, we have to consider a more complex cost function than the one presented in section 5.4.

We suppose that the value k_i of each τ_i has been carefully chosen and is constant during the execution of application. The value of m_i has to be chosen in $[1..k_i]$ on-line by the task handler. We note n the number of tasks activated in one working mode. For each control task τ_i, each possible value of m_i is associated with two values G_{i,m_i} and G'_{i,m_i} corresponding to the control performance, respectively, in a transient situation and in a steady one. Supposing that a lower value of G_{i,m_i} or G'_{i,m_i} represents a better control performance, in the event of a situation change of the plant states, the aim of the task handler is to find, for each τ_i, a value so that the sum of G_{i,m_i} or G'_{i,m_i} (according to the situation into which the plants fall) for $j\epsilon[1\ldots k_i]$ and $i\epsilon[1\ldots n]$ is minimized the subject to the task schedulability condition (5.19). This is formally formulated as the following optimization problem.

Considering in a set of tasks τ_i, $1 \le i \le n$, described by h_i, their basic period (considered as their relative deadline, i.e. $D_i = h_i$), C_i, their worst case execution time, k_i the parameter of their (m, k)-constraint, n_i, the number, and the list of possible values for the parameter m in the (m, k) constraint; the global optimization problem consists of determining $(s_{i,1}, s_{i,2}, \ldots, s_{i,n_i})$ for each task τ_i that minimizes

$$\sum_{i=1}^{n} \sum_{j=1}^{n_i} (s_{i,,j} G_{i,,j} I + s_{i,,j} G'_{i,,j} F),$$

with $s_{i,,j} \epsilon \{0, 1\}$, $\displaystyle\sum_{j=1}^{n_i} s_{i,,j} = 1$, $i = 1, \ldots, n$ and such that condition (5.19) is verified.

F and I indicate the current situation: in transient state, $F = 0$ and $I = 1$ while in the steady state $F = 1$ and $I = 0$.

The values of $G_{i,,j}$ and $G'_{i,,j}$ have to reflect the performance offered by each solution. We identified two ways for defining the performance indicator.

– In the first case, $G_{i,,j}$ is defined on a finite horizon by formula (5.13) while $G'_{i,,j}$ is evaluated on an infinite horizon using (5.17):

$$\begin{aligned}
G_{i,,j} &= J_i(m_{i,j}) \\
G'_{i,,j} &= J_i^\infty(m_{i,j}).
\end{aligned} \tag{5.24}$$

We have to note that the optimization problem is to minimize the overall cost of the application. However, the sub-systems with lower costs may suffer from greater control performance degradation due to a low value of m_i. That is to say, the task handler maintains the value of each m_i as high as possible for the sub-systems with greater costs by reducing the value of m_i for the sub-systems with lower costs.

– The second solution is concerned by a cost that represents performance degradation:

$$\begin{aligned}
G_{i,j} &= \frac{J_i(m_{i,j}) - J_i(k_i)}{J_i(k_i)} \\
G'_{i,,j} &= \frac{J_i^\infty(m_{i,j}) - J_i^\infty(k_i)}{J_i^\infty(k_i)},
\end{aligned} \tag{5.25}$$

where $J_i(x)$ and $J_i^\infty(x)$ are defined, respectively, by functions (5.13) and (5.17). In this case, the control performance criteria avoid the problem given for the first solution presented. The control performance degradation of each sub-system is treated equally. On the other hand, the overall cost of application may not be optimal. So, the choice of control performance representation should be identified according to the application requirements.

In both cases, the time horizon H_i for the finite-horizon cost function is an important design parameter, which directly affects the overall control performance, and needs to

	A_i	B_i	$R_{i,c}$ (incremental covariance of $v_{i,c}$)
Plant$_1$	$\begin{bmatrix} 0 & 1 \\ -18 & 0 \end{bmatrix}$	$\begin{bmatrix} 0 \\ 516 \end{bmatrix}$	$\begin{bmatrix} 0.0025 & -0.005 \\ -0.005 & 0.001 \end{bmatrix}$
Plant$_2$	$\begin{bmatrix} 0 & 1 \\ 0 & -12.6558 \end{bmatrix}$	$\begin{bmatrix} 0 \\ 1.9243 \end{bmatrix}$	$\begin{bmatrix} 0.005625 & -0.075 \\ -0.075 & 1 \end{bmatrix}$
Plant$_3$	$\begin{bmatrix} 0 & 1 \\ -22.206 & -0.9424 \end{bmatrix}$	$\begin{bmatrix} 0 \\ 0.48036 \end{bmatrix}$	$\begin{bmatrix} 0 & 0 \\ 0 & 22.2066 \end{bmatrix}$
Plant$_4$	$\begin{bmatrix} 0 & 1 & 0 & 0 \\ 0 & 0 & -14 & 0 \\ 0 & 0 & 0 & 1 \\ 0 & 0 & 28 & 0 \end{bmatrix}$	$\begin{bmatrix} 0 \\ 2 \\ 0 \\ 2 \end{bmatrix}$	$\begin{bmatrix} 0 & 0 & 0 & 0 \\ 0 & 0.0025 & 0 & 0 \\ 0 & 0 & 0 & 0 \\ 0 & 0 & 0 & 0 \end{bmatrix}$

Table 5.1. *Parameters of the plants whose controllers share the same processor and scheduled according to their transient and steady state*

be carefully chosen. We propose opting for H_i as the settling time (defined approximately as three times the rise time). Moreover, the off-line computation of (5.25) is done by considering the typical values of the reference values and by neglecting noise.

As in the solution presented in section 5.4.2, the optimization problem results from the MMKP. To solve this problem on-line, we also propose using the heuristic algorithm (HEU) presented in [KHA 02].

5.5.4. *Case study*

In this section, we illustrate the scheduling approach presented above by studying the control of four plants. $Plant_1$ (resp. $Plant_2$, $Plant_3$, $Plant_4$) corresponds to a harmonic oscillator system, (resp. to a cart system, a pendulum and an inverted pendulum). The controller of $Plant_i$ is denoted by $Controller_i$. Each plant is modeled by the differential equation (5.6) whose parameters are given in the Table 5.1. The controller of the $Plant_i$ is defined by equation (5.7).

The rise time specifications of each plant are, respectively, 0.2, 0.2, 0.3, and 0.5. Then, the basic sampling periods are related to rise time specifications, i.e. $h_1 = 0.02s$ for $Plant_1$, $h_2 = 0.02s$ for $Plant_2$, $h_3 = 0.03s$ for $Plant_3$, and $h_4 = 0.05s$ for $Plant_4$.

We suppose that the first state variable in vector x of each plant is the variable supervised by the supervision component. In other words, the controller tries to keep it tracking the plant state reference asymptotically. The step response target for the cart

	Q_i	R_i
$Controller_1$	$\begin{bmatrix} 5 & 0 & 0 \\ 0 & 0 & 0 \\ 0 & 0 & 25 \end{bmatrix}$	200
$Controller_2$	$\begin{bmatrix} 1.25 & 0 \\ 0 & 0.0085 \end{bmatrix}$	0.0001
$Controller_3$	$\begin{bmatrix} 1 & 0 \\ 0 & 0 \end{bmatrix}$	0.00001
$Controller_4$	$\begin{bmatrix} 1 & 0 & 0 & 0 \\ 0 & 0 & 0 & 0 \\ 0 & 0 & 2 & 0 \\ 0 & 0 & 0 & 0 \end{bmatrix}$	0.001

Table 5.2. *Weights used for each performance evaluation of each controller sharing the same processor and scheduled according to the transient or steady state of the controlled plant*

($Plant_2$) is an over-damped response, while for the others they are under-damped, being the damping coefficient greater than 0.6 (overshoot $< 10\%$). The gain of the controllers is computed for each possible value of $m_{i,j}$, according to the approach proposed in 5.4.1. The design weights, which allow the satisfaction of the above-mentioned rise time and overshoot, are presented in Table 5.2.

Using the LMI control toolbox of Matlab/Simulink tool, the set of inequalities (for the overall set of discrete plants and controllers parameters), has a QCLF, guarantying stability. To allow for fast changes between different $(m_{i,j}, k_i)$-firm constraints at an $m_{i,j}$ adjustment, the controller parameters are calculated off-line and stored in a table.

As applied in section 5.4, to the control task τ_i which implements the controller $Controller_i$ is assigned the rate-monotonic priority, the task with the largest period has the lowest priority; its execution has no influence on the other tasks. Therefore, no task instance classification will be applied to τ_4, or in other words, it is executed under (k, k)-firm constraint. Using the approach proposed in section 5.3.3, the value of k_i is set to, respectively, $k_1 = 6$, $k_2 = 5$, $k_3 = 5$ and $k_4 = 1$. The value of m_i for the plant τ_i may vary within $[1..k_i]$. The worst case execution time of each task is $C_1 = C_2 = C_3 = C_4 = 9$ ms.

The optimal costs as well as the sequence of corresponding gains, associated with the different possible (m_i, k_i)-firm constraints of each task τ_i are evaluated off-line and stored in order to be used on-line by the task handler. Then the Matlab/Simulink model is simulated. The deployment characteristics of the global system, for short, the specific scheduling policy, are done by using TrueTime toolbox [CER 03].

5.5.4.1. *Simulation scenario*

The evolution of the scenario is illustrated in Figure 5.10.

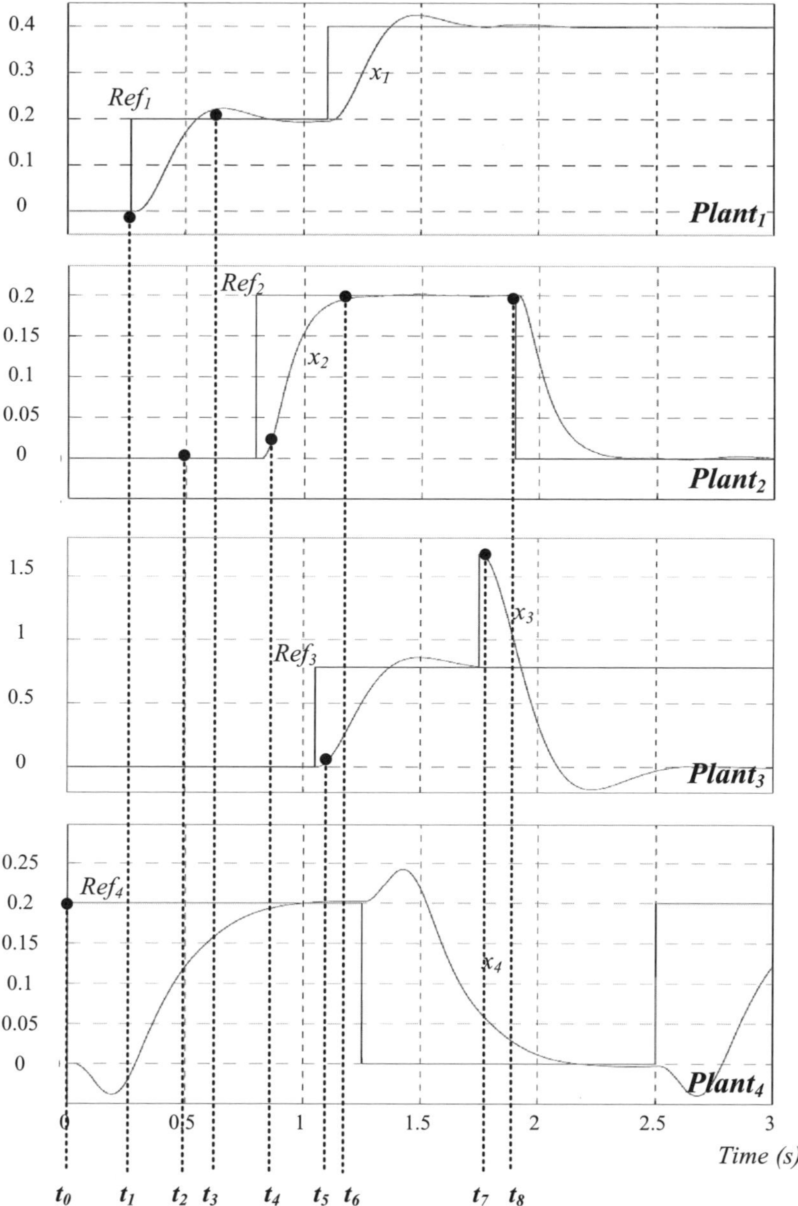

Figure 5.10. *Activation, reference (Ref_i for $Plant_i$) and output (x_i for $Plant_i$) evolution of the four plants whose controllers share the same processor*

	$Plant_1$ ($k_1 = 6$)		$Plant_2$ ($k_2 = 5$)		$Plant_3$ ($k_3 = 5$)		$Plant_4$ ($k_4 = 1$)	
	State	m_1	State	m_2	State	m_3	State	m_4
$t < t_0$	steady	6	-	-	steady	5	-	-
$[t_0 t_1[$	steady	4	-	-	steady	5	steady	1
$[t_1 t_2[$	transient	6	-	-	steady	2	transient	1
$[t_2 t_3[$	transient	3	steady	1	steady	2	transient	1
$[t_3 t_4[$	steady	2	steady	1	steady	5	transient	1
$[t_4 t_5[$	steady	2	transient	2	steady	2	transient	1
$[t_5 t_6[$	steady	2	transient	1	transient	5	steady	1
$[t_6 t_7[$	steady	2	steady	1	transient	5	steady	1
$[t_7 t_8[$	steady	6	steady	5	-	-	transient	1
$t \geq t_8$	steady	2	-	-	-	-	transient	1

Table 5.3. *Values of m_i as they are selected by the task handler at each working mode switch; these values are evaluated for each task activated*

It shows, for each $Plant_i$, if it is controlled or not and, in the first case, the reference that is applied and the output of the plant, respectively, noted in figure Ref_i and x_i for $Plant_i$. Table 5.3 provides, at each significant instant, the value of m_i that is selected by the task handler for each current task. An instant is significant if it leads to a switch between two working modes: a new plant has to be controlled, one plant is no longer activated, one plant goes from steady state to transient state, or a plant reaches its steady state.

These instants are identified on-line by the supervisor.

The sequence of significant instants in the proposed scenario is described below.

– Just before the observation of the system, we suppose that only $Plant_1$ and $Plant_3$ are controlled, while the other plants are not activated and not controlled. These two plants are in a steady state.
In this case, the system is schedulable under (k_1, k_1) and (k_3, k_3) constraints for τ_1 and τ_3.

– At time t_0, $Plant_4$ is activated and, therefore, the set of tasks to schedule is $\{\tau_1, \tau_3, \tau_4\}$; no reference is applied to this new plant. The three plants are in a steady state.
The task handler has to find the optimal configuration of m_i for this set of tasks and to deduce the corresponding sequence of values for L_i, the gain of each controller. For this purpose, it uses the infinite-horizon cost for each plant. The result is $m_1 = 4$ and $m_3 = 5$ (we have to note that m_4 has to be always equal to 1.)

– A transient state is detected by the supervisor at time t_1 for $Plant_1$ (occurrence of a new reference) and $Plant_4$
The task handler looks for an optimal configuration of the (m_i, k_i) constraints by

taking into account the finite-horizon costs for $Plant_1$ and $Plant_4$ and the infinite-horizon cost for $Plant_3$. This results in $m_1 = 6$ and $m_3 = 2$.

– $Plant_2$ is activated at time t_2 and the supervisor identifies it is in a steady state while $Plant_1$ and $Plant_4$ are still in a transient state.
The values of m_i defined by the task handler at this time are, respectively, $m_1 = 3$, $m_2 = 1$ and $m_3 = 2$.

– The supervisor detects a steady state for the $Plant_1$ at time t_3.
The new values evaluated for m_i are $m_1 = 2$, $m_2 = 1$ and $m_3 = 5$.

– At time t_4, the values of m_i have to be modified as the supervisor detects a transient state for $Plant_2$.
Therefore, $m_1 = 2$, $m_2 = 1$ and $m_3 = 5$.

– $Plant_3$ enters in a transient state at t_5.
The adjustment of the values of m_i results in $m_1 = 2$, $m_2 = 1$ and $m_3 = 5$.

– $Plant_2$ goes from transient to steady state at t_6.
This leads the task handler to readjust the m_i parameters, using the infinite-horizon cost for τ_1 and τ_2 and to the finite-horizon cost for τ_3: $m_1 = 2$, $m_2 = 1$, $m_3 = 5$.

– At t_7, an inadmissible perturbation enters in $Plant_3$ whose output reaches $\frac{\pi}{2}$ and consequently, this plant is deactivated, thus reducing the number of tasks to 3. At the same time, $Plant_4$ enters a transient state.
The task handler can therefore increase the values of m_i for the tasks τ_1 and τ_2: $m_1 = 6$ and $m_2 = 1$.

– At the end of the scenario, at time t_8, a transient state is detected by the supervisor for $Plant_2$.
The last values for the m_i parameters that are chosen by the task handler are $m_1 = 2$ and $m_2 = 5$. We recall that during the simulation, parameter m_4 is always equal to 1 (hard real-time constraint).

5.5.4.2. *Observed performance*

In order to analyze the control of the performance degradation due to the (m, k)-firm policy, we evaluate the LQ cost, given by formula (5.20), for each system during the simulation time. Let us note J_i^{adaptive} as this value. On the other hand, during the same simulation time and under the same simulation setup, we calculated the nominal performance of each plant provided by the same formula when each system is controlled by a control task on a separate processor; in this case, the constraint applied is a (k, k) one and the sampling period as well as the activation period of the control task are equal to the basic period of each controller. We note J_i^{nominal}, the value obtained for $Plant_i$. Table 5.4 presents an example of the performance degradation.

Of course, these results depend on the simulation setup, and they are only exposed here to show that using the proposed technique, the degradation of the performance should be kept as small as possible in each situation subject to the task schedulability.

	$J_i^{adaptive}$	$J_i^{nominal}$	$\left\|\dfrac{J_i^{adaptive}-J_i^{nominal}}{J_i^{nominal}}\right\|$
$Plant_1$	395	359	10%
$Plan_2$	52.5	40.04	31.1%
$Plant_3$	199.4	154.6	29.1%
$Plant_4$	26.347	25.31	0.05%

Table 5.4. *Performance degradation of the four plants*

$Plant_4$ suffered the lower cost degradation due to the fact that this system has to respect a hard real-time constraint $m_4 = k_4 = 1$. $Plant_2$ suffered the maximum cost degradation, due to the $Plant_2$ performance indicator, which generates the reduction of m_2 as soon as the other plants require the use of the processor.

5.5.4.3. *Summary*

Through this case study we can see that the on-line adaptation mechanism we proposed can effectively control the degradation of the system performance during the plant state transient period and system overload. In fact, given the current states of the controlled plants, the proposed approach allows us to derive a (m, k)-firm constraint for each control task and the corresponding optimal control gain while still meeting the (m, k)-firm schedulability condition of the total control tasks. Moreover, Table 5.4 shows that comparing to the idle case where the processor has infinite capacity, using our approach does not induce much performance degradation. Notice that, for the simulated scenario, if we do not allow sample data dropping (i.e. in case that all tasks are considered under hard real-time constraint), the processor will be in an overload at time $t = 0.5$ s. As the tasks τ_3 and τ_4 have lower priority, they are no longer executed by the processor and this leads to the instability of $Plant_3$ and $Plant_4$. Figure 5.11 shows that starting from $t = 0.5$ s where the higher priority task τ_2 is released, the processor can no longer execute task τ_3 correctly and could never execute task τ_4. As an example, Figure 5.12 clearly shows that $Plant_3$ is not stable.

5.6. Conclusions

Computing and networking resource sharing is a common trend in NCS for achieving cost-effective solutions. However, an overload situation may occur, either by the dynamic application configuration changes or by the implementation system performance variations.

For dealing with system overload situations, this chapter proposed an approach based on selectively dropping some samples while still guaranteeing the (m, k)-firm schedulability of the control tasks sharing a same processor. By adjusting both the acceptable (m, k) values and the control gains, we have shown, through case studies, that the performance degradation is efficiently controlled. The key point is the use of

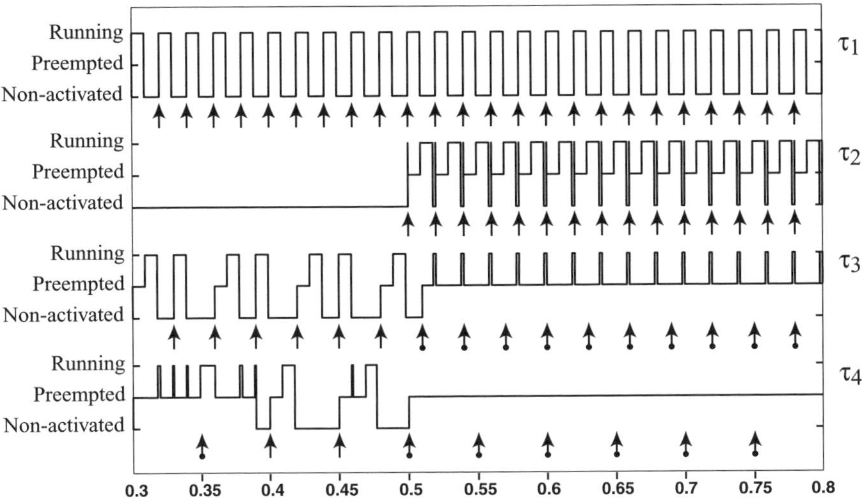

Figure 5.11. *Simulation trace of task execution without selective sample dropping*

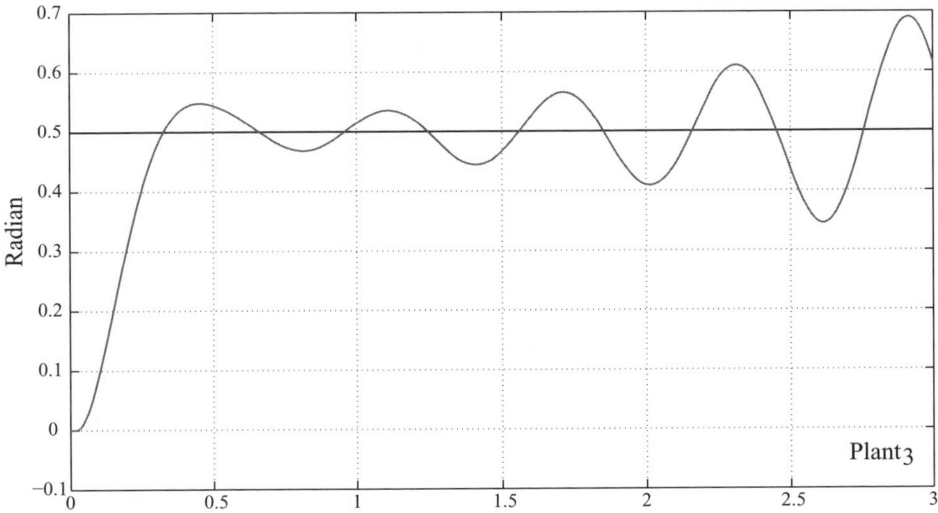

Figure 5.12. *Unstability of $Plant_3$*

an algorithm that allows for finding on-line the optimal values of m and the control gains leading to minimizing a global control performance cost function.

This approach has been further extended to also take into account the transient plant state, where the plant needs more resources to be controlled than when it is in its steady state. The stability of this dynamic parameter switching closed-loop has been discussed, and we showed that the asymptotic stability is guaranteed if one can find a CQLF. This extension makes the resource utilization still more efficient. In fact, the idea behind this is that the shared resource is dynamically allocated to the plant that needs it (when it is in its transient phase), rather than being allocated to the plants that are in their steady-state and need less control.

As indicated in the introduction section of this chapter, the approach presented can be considered as an alternative to the direct sampling period adaptation for dealing with system overload situations. Using such an approach, the resulting system will be more robust as it can auto-adapt to fit with certain system overload situations.

5.7. Bibliography

[ÅST 97] ÅSTRÖM K., AND WITTENMARK B., *Computer-Controlled Systems*, Information and System Sciences Series, Prentice Hall, 3rd edition, 1997.

[BIT 91] BITTANTI S., COLANERI P., AND DE NICOLAO G., The periodic Riccati equation, *The Riccati Equation*, p. 127–162, Springer-Verlag, Berlin, 1991.

[CER 03] CERVIN A., HENRIKSSON D., LINCOLN B., EKER J., AND BERNHARDSSON B., ÅRZÉN K. E., How does control timing affect performance?, *IEEE Control Systems Magazine*, vol. 23, p. 16–30, June 2003.

[EKE 00] EKER J., HAGANDER P., AND ARZEN K.-E., A feedback scheduler for real-time controller tasks, *Control Engineering Practice*, vol. 8, p. 1369–1378, 2000.

[FRE 63] FREEMAN H., *Discret Time Systems*, John Wiley, New York, 1963.

[HAM 94] HAMDAOUI M., AND RAMANATHAN P., A service policy for real-time customers with (m,k)-firm deadlines, *Fault-Tolerant Computing Symposium*, Austin, USA, p. 196–205, April 1994.

[HAM 95] HAMDAOUI M., AND RAMANATHAN P., A dynamic priority assignment technique for streams with (m,k)-firm deadlines, *IEEE Transactions on Computers*, p. 1443–1451, December 1995.

[JIA 05] JIA N., HYON H., AND SONG Y., Ordonnancement sous contraintes (m,k)-firm et combinatoire des mots, *13th International Conference on Real-Time Systems*, Paris, France, April 2005.

[KHA 02] KHAN P., LI K., MANNING E., AND AKBAR M., Solving the knapsack problem for adaptive multimedia system, *Studia Informatica*, vol. 2, p. 161–182, 2002.

[LI 06] LI J., SONG Y.-Q., AND SIMONOT LION F., Providing real-time applications with graceful degradation of QoS and fault tolerance according to (m,k)-firm Model, *IEEE Transactions on Industrial Informatics*, vol. 2, p. 112–119, 2006.

[LIB 99] LIBERZON D., AND MORSE S., Basic problems in stability and design of switched systems, *Control Systems Magazine*, vol. 19, p. 59–70, October 1999.

[MAR 90] MARTELLO S., AND TOTH P., *Knapsack Problems: Algorithms and Computer Implementations*, John Wiley, New York, 1990.

[OTA 02] OTANEZ P., MOYNE J., AND TILBURY D., Using deadbands to reduce communication in networked control systems, *American Control Conference*, Anchorage, USA, May 2002.

[QUA 00] QUAN G., AND HU X., Enhanced fixed-priority scheduling with (m,k)-firm guarantee, *21st IEEE Real-Time Systems Symposium*, Orlando, Florida, USA, p. 79–88, November 2000.

[RAM 99] RAMANATHAN P., Overload management in real-time control applications using (m,k)-firm guarantee, *IEEE Transactions on Parallel and Distributed Systems*, vol. 10, p. 549–559, June 1999.

[SCH 02] SCHINCKLE M., CHEN W.-H., AND A. RANTZER, Optimal control for systems with varying samplin rate, *American Control Conference*, Anchorage, USA, May 2002.

[SET 96] SETO D., LEHOCZKY J. P., SHA L., AND SHIN K. G., On task schedulability in real-time control system, *17th IEEE Real Time Systems Symposium*, Washington, DC, USA, p. 13–21, December 1996.

[SIM 05] SIMON D., AND BENATTAR F., Design of real-time periodic control systems through synchronisation and fixed priorities, *International Journal of Systems Science*, vol. 36, p. 57–76, 2005.

Chapter 6

Fault Detection and Isolation, Fault Tolerant Control

6.1. Introduction

Process automatic control has various objectives. Control performance concerns system stability (which must be guaranteed under any and all conditions), the error between reference and output in the control loops (which has implications, for instance, on product quality or motion precision), robustness against changes in the system's parameters, or the time required to obtain a desired output. It also involves cost in an overall sense (energy or raw material consumption).

A second objective is increasingly being taken into account today by the control, which is improving the *safety* of the system, so as not to present any hazards to the men working close by, to the equipment or to the environment, while guaranteeing control performance. A more general term is *dependability* [LAP 92], used both in systems' engineering and in computer science, and which covers the concepts of safety (the absence of catastrophic consequences), *availability* (readiness for correct service), *reliability* (continuity of correct service), *integrity* (the absence of improper system alteration), and *maintainability* (ability to undergo modifications and repairs).

Integrated automation is, therefore, not only focused on maintaining certain variables at their set-point value, but is also concerned from an overall perspective about the system and its various operating modes. These are the *normal operating mode(s)*

Chapter written by Christophe AUBRUN, Cédric BERBRA, Sylviane GENTIL, Suzanne LESECQ and Dominique SAUTER.

and start-up or shut-down modes, which evidently are taken into account in the initial design of the control system. However, various *failure modes* are also considered, which correspond to the various states that a failure or a malfunction can produce on the process.

Performances of closed-loop controlled systems can be altered by the occurrence of abrupt or incipient faults, which can cause serious damage to the system. A way to prevent system deteriorations is to develop controllers capable of accommodating faults. Associated with rapidly increasing demands for higher system performance, product quality, productivity, and cost efficiency, *fault diagnosis* (FD) and *fault tolerant control* (FTC) have become key issues in product development and system design, and therefore have received much attention in the academic community as well as in industry. Increasing plant dependability may have an even greater impact on improving economic efficiency than control performance.

This chapter will first deal with the different concepts used in diagnosis and fault-tolerant control. Diagnosis aims at deciding if a system, made up of various interconnected physical components, is in a normal state or not (in which case it is said to be faulty). This is known as the *fault detection* (FD) step. After that, based on a certain number of observed signals one tries to identify a list of faulty components, which is called the *fault isolation* (FI) step. Consequently, the control must be adapted to the faulty situation before the faulty component(s) can be repaired (FTC).

Fault detection and isolation (FDI) and FTC algorithms are implemented as tasks in the real-time computer, exactly as the control tasks mentioned in the previous chapters. They will be briefly evoked in the second section of this chapter. In the third section, the problems raised by the use of a network for transmitting the information necessary for diagnosis and FTC will be examined. After that, a number of pragmatic solutions will be discussed, before presenting a number of more theoretical approaches, some of which are still undergoing research.

6.2. FDI and FTC

6.2.1. *Introduction to diagnosis*

A *fault* is defined as an unpermitted deviation of at least one characteristic property of a system from the usual condition. It may initiate a *failure*, defined as a permanent interruption of a system's ability to perform a required function under specific operating conditions [ISE 06]. A fault is revealed thanks to *fault indicators*. A fault indicator is a signal elaborated with variables measured on the system. When an undesired behavior is detected, confusion must be avoided between an unforeseen fault on the process or its instrumentation, a non-robust regulator being at its stability limit, or a non-measurable disturbance effect. The latter, for instance, is not desired, but is

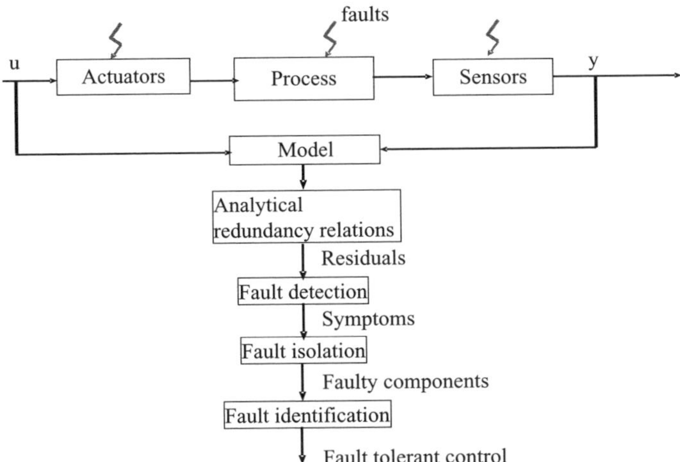

Figure 6.1. *Model-based diagnostic steps*

part of the normal functioning of a complex system. This discrimination is probably the main difficulty in the diagnosis.

Diagnosis is broken down into several steps (Figure 6.1). Fault *detection* consists in deciding whether a fault indicator is significant and consequently generating the corresponding *symptom*, generally a Boolean quantity. Fault *isolation* consists of deducing from a set of symptoms a set of faulty physical components. This set must be as small as possible (minimal diagnosis). Following this, component maintenance can be programmed. Fault *identification* consists in determining the size and time behavior of a fault. It can help maintain the faulty process in operation provided that the failure is not crucial or is precociously detected.

Online monitoring can be based simply on variables overshooting fixed thresholds, which gives limited results because this method is not at all relevant for dynamic transients. Material redundancy may also be used to guarantee system safety. In this case, the various components (valves, pumps, etc.) are duplicated. The same can be done with the instrumentation. An elementary means of obtaining a reliable measurement is to use three sensors, using their measurement mean provided that they are similar. If one of them is very different from the others, the corresponding sensor is suspected to be failing, and its measurement is discarded.

However, these solutions are obviously very costly, which is why now advanced solutions are based on *analytical redundancy*, which means using the system's mathematical model for diagnosis. In what follows, this model is the normal operating model and diagnosis consists in simultaneously testing the consistency of the measurements with various sub-models, corresponding to different sets of components.

This decomposition into elementary fault indicators, also named *residuals*, depends above all on the instrumentation and makes up an important phase of the diagnosis, the design of the *residual generator*, which is carried out off-line. With a set of well-designed symptoms, a fault can not only be detected but also isolated.

Analytical redundancy requires knowledge of the influence of faults on the system's model. It is obvious that accurately representing all the possible faults of an industrial facility could be very time consuming. Consequently, the methods that will be presented here are based on the normal behavior model, where faults are introduced very simply. For instance, an extra signal is added in some place in the block diagram. This is known as an additive fault. An additive fault on a sensor allows for the modeling of an off-set (constant signal) or a drift (ramp signal). In the same way, an additive signal on an actuator can be used to model a fault (wrong valve opening, wrong motor positioning). An unmeasurable disturbance can also represent a fault (leakage in a tank). Other faults are better represented through the modification of model parameters. These are called multiplicative faults.

Several families of diagnostic methods exist, based on a single signal modeling [BAS 93], on the knowledge of the process' history (expert system or pattern recognition) [VEN 03b; VEN 03c], or on the dynamical models of sub-systems [VEN 03a], which, stemming from the classical control models, will be discussed next.

6.2.2. *Quantitative model-based residuals*

The models representing the normal behavior of various sub-systems are assumed to have been obtained. These models may be expressed as transfer functions or state equations, either linear or nonlinear. The models are used to generate residuals, the interpretations of which make up the symptoms of the system's faulty state. To be accurate fault indicators, the residuals have to respect the following properties:

1) no fault $\Longrightarrow r(t) = 0$

2) fault $\Longrightarrow r(t) \neq 0$.

The contra-positive of (1) leads to

$$r(t) \neq 0 \Longrightarrow \text{fault}$$

that is the base of diagnostic reasoning. Property (2) ensures the detection capacity of the indicator. This expression is theoretically clear but difficult to guarantee in practice. Checking that a residual value is not zero results in converting a residual numerical value into a symptom. In practice, the model is never perfectly accurate and measurements are noisy, and therefore residuals are never null. The difference has to be made between "small" residual values, representative of an unfaulty situation, and "high" values representative of a fault. This evaluation is further complicated by the fact that residual sensitivities to distinct faults may be very different. The simplest way to solve this problem is by comparing the residual value to a fixed threshold (*a priori* fixed by an expert). Choosing the threshold value is by no means an easy task. The thresholds could be adaptive (they change on-line with the experimental conditions). Statistical methods have also been proposed: a change in the residual mean or standard deviation is sought. Fuzzy decision making is another way to process the residual: based on the fact that the zero residual concept is vague, it can be described by a fuzzy subset.

Let y be the output of a sub-system, u its input, f the faults, and d the unmeasurable disturbances:

$$y(t) = g(u, f, d, t). \tag{6.1}$$

With y_m being the output measure and u_m being the input one, the model leads to

$$h(y_m, u_m, f, d, t) = 0. \tag{6.2}$$

Ideally, this equation should be decoupled with the disturbances d and then decomposed into two parts:

$$h(y_m, u_m, f, t) = -he(y_m, u_m, f, t) + hc(y_m, u_m, t). \tag{6.3}$$

This allows for the defining of a residual r in such a way that $r = hc$ is its computational form, depending only on measurable signals, and he is its evaluation form, showing how it depends on the various signals, including faults. r is evaluated on-line. Let us consider an example where G_u is the system transfer function, and f_u and f_y, respectively, represent an additive actuator fault and an additive sensor fault:

$$y_m(s) = y(s) + f_y(s) = G_u(s)(u_m + f_u)(s) + f_y(s) \tag{6.4}$$

$$r = \underbrace{y_m(s) - G_u(s)u_m(s)}_{hc} = \underbrace{G_u(s)f_u(s) + f_y(s)}_{he}. \tag{6.5}$$

In order to isolate faults, various residuals are generated, sensitive to some faults and insensitive to the other ones:

$$r_j = he_j(u_m, y_m, f, t)/f_k \neq 0 \Rightarrow r_j \neq 0. \tag{6.6}$$

Using r_j as the indicator of fault, f_k requires that a fault f_k leads to a significant value of r_j and ideally that r_j be decoupled with d and with other faults $f_{i,i \neq k}$. This is not always possible. Therefore, an incidence table is built (Table 6.1). The columns in this table represent faults and the rows represent the residuals. Each column represents a *fault signature,* and the table is sometimes called an *incidence matrix.* The matrix is filled with the theoretical values of the Boolean symptoms deduced from the evaluation form of the residuals. A "1" in the element ij of the table means that residual r_i is sensitive to faults f_j, while zero means the contrary. If the columns in this table are

	f_1	f_2	f_3
r_1	1	1	0
r_2	1	0	1
r_3	0	1	1

Table 6.1. *Signature table*

different, then the faults can be isolated. If two columns are similar, the two corresponding faults cannot be isolated with the designed residuals. In this case, the model
equations have to be rearranged in order to obtain optimal fault isolation.

On-line, residuals are generally evaluated at each sampling time. The symptom
vector is then generated. A particular fault is isolated when this vector is similar to a
fault signature. Using the signature table implies occurrences of single faults and that
all faults have been listed in advance. Otherwise, one is confronted with an unknown
situation.

To generate residuals, two methods will be described in the following subsections.
The first one is based on parity relations, while the second is based on state observers.

6.2.2.1. *Parity relations*

The idea is to rearrange the model equations in order to obtain optimal fault isolation. The whole set of model equations are used and combined in various manners,
with each combination eliminating some signals and generating a residual. These analytical redundancy equations (ARE) must be based only on measured values. We have
to find the combinations that lead to independent columns in the incidence matrix.

In Figure 6.2, u_i signals represent inputs and y_i intermediary signal measurement
points, while M_i symbolizes the sub-models. Fault indicators can be based on either
elementary relations or their combinations:

$$r_1(u_1, y_1), r_2(y_1, y_2), r_3(y_3, y_4), r_4(u_2, y_1, y_3), r_5(u_2, y_1, y_4), r_6(u_2, y_3, y_2). \quad (6.7)$$

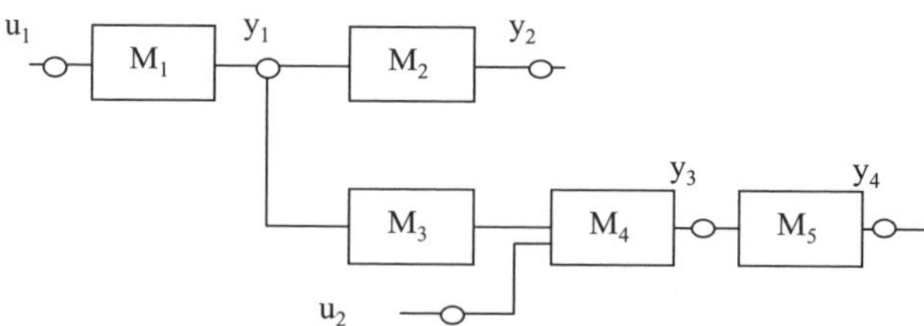

Figure 6.2. *Complex system decomposition for ARE generation*

Consider the example of a transfer matrix model with p inputs and m outputs

$$\begin{bmatrix} y_1 \\ \dots \\ y_m \end{bmatrix} = \begin{bmatrix} G_{11} & \dots & G_{1p} \\ & \dots & \\ G_{m1} & \dots & G_{mp} \end{bmatrix} \begin{bmatrix} u_1 \\ \dots \\ u_p \end{bmatrix}, \tag{6.8}$$

where G_{ij} represents the transfer between output y_i and input u_j. The polynomial form of this model is

$$A_{11}(s)y_1(s) = B_{11}(s)u_1(s) + \cdots + B_{1p}(s)u_p(s). \tag{6.9}$$

From (6.9), the evaluation form of one residual is deduced

$$r_1 = A_{11}(s)y_{m1}(s) - B_{11}(s)u_{m1}(s) - \cdots - B_{1p}(s)u_{mp}(s), \tag{6.10}$$

where the index m stands for "measured value". Adding to model 6.9 the fault model, the dependence of r_1 on a sensor fault on y_1 or on actuator faults on all the inputs is shown easily.

In order to structure the residual space, a projection matrix W can be used to obtain new residuals

$$r' = Wr. \tag{6.11}$$

Matrix W is used to make some residuals sensitive to a subset of faults and insensitive to the complementary subset [GER 98]. The same technique can be applied to a state model. This method does not require many hypotheses on the model, which could be nonlinear.

6.2.2.2. *Observers bank*

Observers are algorithms based on the system's state model, the objective of which is state reconstruction (state following). In order to simplify the following explanations, the model is assumed to be linear, but a nonlinear observer may be synthesized too:

$$\dot{x} = Ax + Bu \qquad y = Cx, \tag{6.12}$$

where $x \in \Re^n$ is the state vector, and y and u are assumed to be scalar to simplify the explanations. The corresponding observer equation is

$$\dot{\widehat{x}} = A\widehat{x} + Bu + L(y - \widehat{y}) \tag{6.13}$$

$$\widehat{y} = C\widehat{x}. \tag{6.14}$$

It can be noticed (equation (6.13)) that the observer is made of a prediction part, based on the model simulation, and a correction part, based on the difference between the real output y and the output computed by the observer $\widehat{y}$. From equation (6.13), the following general equation for state observation is deduced:

$$\dot{\widehat{x}} = (A - LC)\widehat{x} + Bu + Ly. \tag{6.15}$$

The observation error ε can be computed combining equations (6.15) and (6.12):

$$\dot{\widehat{x}} = (A - LC)\widehat{x} + Bu + LCx \tag{6.16}$$

$$\varepsilon = x - \widehat{x} \tag{6.17}$$

$$\dot{\varepsilon} = (A - LC)\varepsilon. \tag{6.18}$$

From this last equation, the Laplace transform of the observation error is found

$$\varepsilon = (sI_n - A + LC)^{-1}(x_0 - \widehat{x}_0), \tag{6.19}$$

where I_n denotes the n-dimensional identity matrix. After any observer initialization $\widehat{x}_0$, if the observer is stable ((A-LC) eigenvalues with negative real parts), the observed state $\widehat{x}$ converges toward the true system state x.

The error between the measured output and the observer output can be used as residual

$$y - \widehat{y} = C(x - \widehat{x}) = \widehat{\varepsilon}_y \tag{6.20}$$

$$\widehat{\varepsilon}_y = C(sI - A + LC)^{-1}(x_0 - \widehat{x}_0), \tag{6.21}$$

where $\widehat{\varepsilon}_y$ has the properties necessary to define a residual, as explained in section 6.2.2, and more generally

$$r = T\widehat{\varepsilon}_y \tag{6.22}$$

can be used as residual, where T is used to obtain isolation properties.

A discrete formulation is obtained similarly for sampled-data systems with sampling period h:

$$\Phi = \exp(Ah) \qquad \Gamma = \int_0^h \exp(As)B\,ds \tag{6.23}$$

$$x_{k+1} = \Phi x_k + \Gamma u_k \qquad y_k = Cx_k \tag{6.24}$$

$$\widehat{x}_{k+1} = \Phi\widehat{x}_k + \Gamma u_k + L(y_k - \widehat{y}_k) \tag{6.25}$$

$$\widehat{y}_k = C\widehat{x}_k \tag{6.26}$$

$$\widehat{x}_{k+1} = (\Phi - LC)\widehat{x}_k + \Gamma u_k + Ly_k \tag{6.27}$$

$$r_k = T(y_k - \widehat{y}_k). \tag{6.28}$$

An unmeasured disturbance can make x different from $\widehat{x}$ and thus y different from $\widehat{y}$. An interesting characteristic of observers is that they can be computed in such a way that the observer state is decoupled from unknown disturbances [PAT 00]. Residuals can be systematically structured when using observers. Two main methods have been proposed: dedicated or generalized observer banks. When it comes to dedicated schemes and N actuator faults, N observers are designed so that observer i uses all the outputs but only input u_i (this requires the system to be observable with this input only). The other $(N-1)$ inputs are considered as unknown inputs. This observer is obviously sensitive only to the actuator i fault, provided there are no sensor faults. In the dedicated scheme for N sensor fault isolation, N observers are designed so that observer i uses all the inputs but only output y_i. Then, this observer is only sensitive to the sensor i fault, provided that there are no actuator faults. In this way, the incidence matrix contains only one "1" in each line. In the case of generalized schemes and actuator faults, N observers are designed so that observer i uses all the inputs but u_i. This observer is obviously sensitive to all actuator faults but the one discarded. For sensor faults, N observers are designed, such that observer i uses all the outputs but y_i, which makes it sensitive to all sensor faults but the one discarded. In this way, the incidence matrix contains only one "0" in each line.

6.2.3. *Example*

A simple example is now described in order to illustrate the concepts defined above. It illustrates the behavior of a direct current motor when sensor faults occur, and various ways to design residuals.

6.2.3.1. *The system-residual generation*

Figure 6.3 shows the system and its controller together with the diagnostic block. The motor is simulated with a continuous-time block defined in Simulink. The controller is a PI defined by a discrete Simulink block. The sampling period is $h = 30$ ms. The controller is tuned to compensate the system's dominant pole. The various parameters are defined in Table 6.2. The diagnostic algorithm is fed by the following measurements: voltage at the motor input $u_{\mathrm{mes}}(t)$, current $i_{\mathrm{mes}}(t)$, motor velocity $\omega_{\mathrm{mes}}(t)$.

The system transfer function is given by

$$H(s) = \frac{A}{(1 + \tau_1 s)(1 + \tau_2 s)} \tag{6.29}$$

with $A = 124$ rad s^{-1} V, $\tau_1 = 555$ ms, $\tau_2 = 1.5 10^{-3}$ ms. The two time constants have very different orders of magnitude. The controller transfer function is

$$C(z) = \frac{5.08\,10^{-2}z - 4.78\,10^{-2}}{z - 1}. \tag{6.30}$$

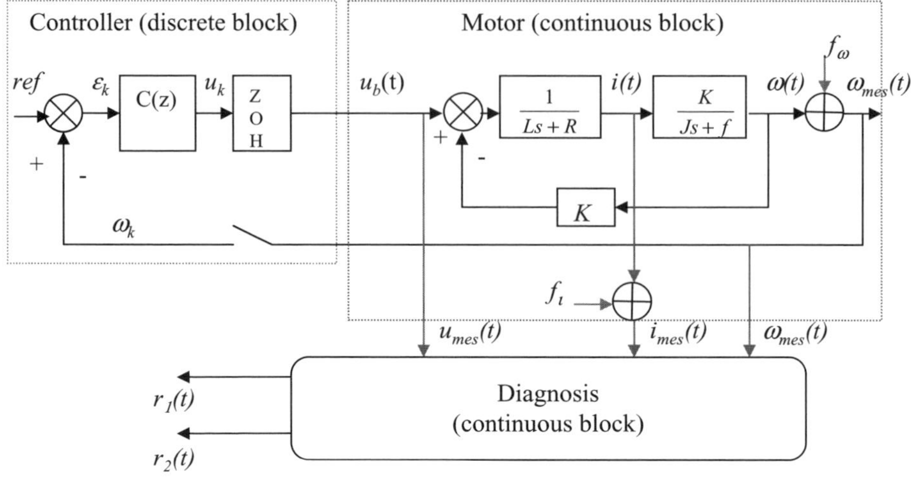

Figure 6.3. *A closed-loop DC motor control and diagnosis*

Residuals are generated using the system model

$$J\frac{d\omega_{\mathrm{mod}}(t)}{dt} + f\omega_{\mathrm{mod}}(t) = Ki_{\mathrm{mes}}(t) \tag{6.31}$$

$$L\frac{di\mathrm{mod}(t)}{dt} + Ri_{\mathrm{mod}}(t) = u_{\mathrm{mes}}(t) - K\omega_{\mathrm{mes}}(t) \tag{6.32}$$

$$r_1(t) = \omega_{\mathrm{mes}}(t) - \omega_{\mathrm{mod}}(t) \tag{6.33}$$

$$r_2(t) = i_{\mathrm{mes}}(t) - i_{\mathrm{mod}}(t). \tag{6.34}$$

$\omega_{\mathrm{mes}}(t)$	Angular velocity measured by the sensor
$\omega_{\mathrm{mod}}(t)$	Angular velocity calculated by the diagnostic algorithm
ω_k	Sampled angular velocity received by the digital controller
$i(t)$	Current of the DC motor
$i_{\mathrm{mes}}(t)$	Current measured by the sensor
$i_{\mathrm{mod}}(t)$	Current calculated by the diagnostic algorithm
u_k	Controller output
$u_b(t)$	Voltage at the output of the zero-order hold
$u_{\mathrm{mes}}(t)$	Voltage measured by the sensor
f_ω, f_i	Sensor faults
$r_1(t), r_2(t)$	Residuals
$R = 0.67\,\Omega$	Resistor
$L = 1\,\mathrm{mH}$	Self-inductance
$J = 3.35\ 10^{-5}\mathrm{kg\ m^2}$	Inertia
$f = 2.29\ 10^{-5}\mathrm{N\ ms}$	Friction coefficient
$K = 5\ 10^{-3}$	Velocity constant and torque constant

Table 6.2. *Parameters of the system in Figure 6.3*

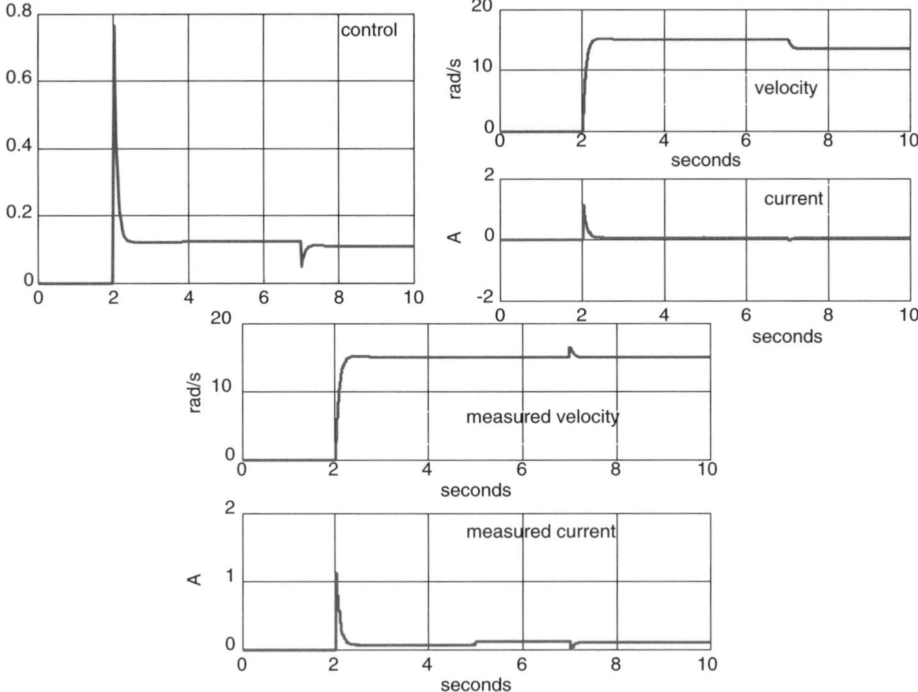

Figure 6.4. *Motor with faults*

A reference change from 0 to 15 rad s^{-1} has been introduced at time $t = 2$ s. A fault of amplitude 1.5 rad s^{-1} (a 10% of the nominal value) has been introduced on the velocity sensor at time $t = 7$ s and a fault of amplitude 0.005 A (7% of the nominal mode) has been introduced on the current sensor at time $t = 5$ s. The simulation results are found in Figure 6.4. It is observed that both faults f_i and f_ω have an effect on the current and velocity. The control is sensitive only to the fault f_ω, because the measured current is not used in the loop. Following a transient state, the measured velocity is again equal to the set-point thanks to the regulator, but the real velocity is smaller due to the fault. This can be detected because the measured velocity is no longer consistent with the voltage at the motor input (Figure 6.5).

6.2.3.2. *Observer-based residuals*

The results presented in the previous subsection are purely theoretical. In practice, the diagnosis is implemented to the computer, and so the algorithm for computing the residuals must be discretized. This is not as obvious as the well-known discretization of a controller associated with a zero-order hold (ZOH). In this particular case, the solution of the discrete equation, obtained through the z-transfer function of the holder associated with the system transfer function, is equal to the continuous time behavior

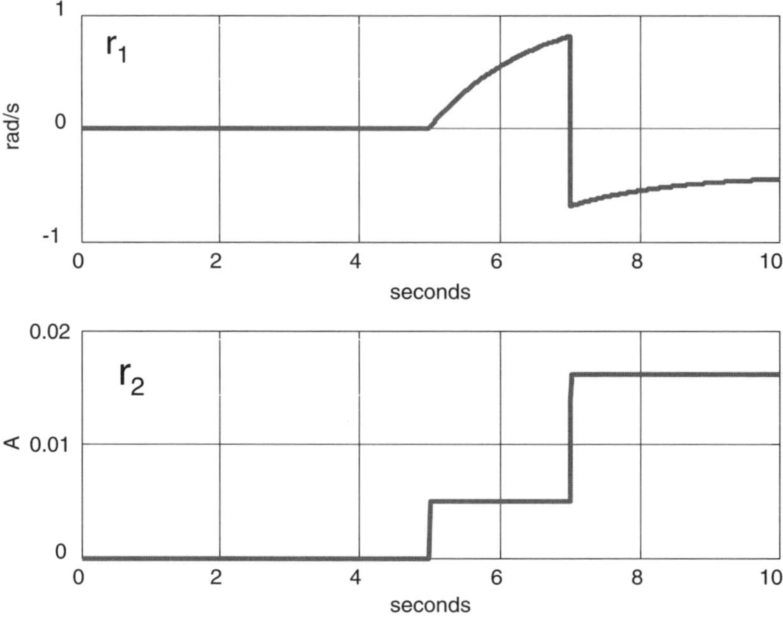

Figure 6.5. *Motor continuous model residuals*

of the system at the sampling time. The same is true for state representation computed with equation (6.23). Now, signals in the analytical redundancy equations are not hold and vary all throughout the sampling period. Thus, the discrete equations are only an approximation of the differential equations. If the sampling period is very short compared to the residuals dynamics, this approximation can be rather precise, but if this is not the case, many false alarms can be generated, only due to computation errors. This has been fully detailed in [BER 07] and will not be detailed here. The state model of the system is easier to discretize, because its input $u_b(t)$ is constant during the sampling period, and an observer can be designed to generate the residuals. The continuous state model is given by

$$x = \begin{bmatrix} i \\ \omega \end{bmatrix}, \quad A = \begin{bmatrix} -670 & -5 \\ 149.25 & -0.6836 \end{bmatrix}, \quad B = \begin{bmatrix} 1000 \\ 0 \end{bmatrix}. \tag{6.35}$$

The discrete state model (with $h = 30$ ms) is

$$\Phi = \begin{bmatrix} -0.0016 & -0.0071 \\ 0.212 & 0.949 \end{bmatrix}, \quad \Gamma = \begin{bmatrix} 1.449 \\ 6.2 \end{bmatrix}, \tag{6.36}$$

and an observer could be

$$\widehat{x} = \begin{bmatrix} \widehat{i} \\ \widehat{\omega} \end{bmatrix}, \quad \Phi - LC = \begin{bmatrix} 0 & 0 \\ 0.212 & 0 \end{bmatrix}, \quad L = \begin{bmatrix} -0.0016 & -0.0071 \\ 0 & 0.949 \end{bmatrix}. \tag{6.37}$$

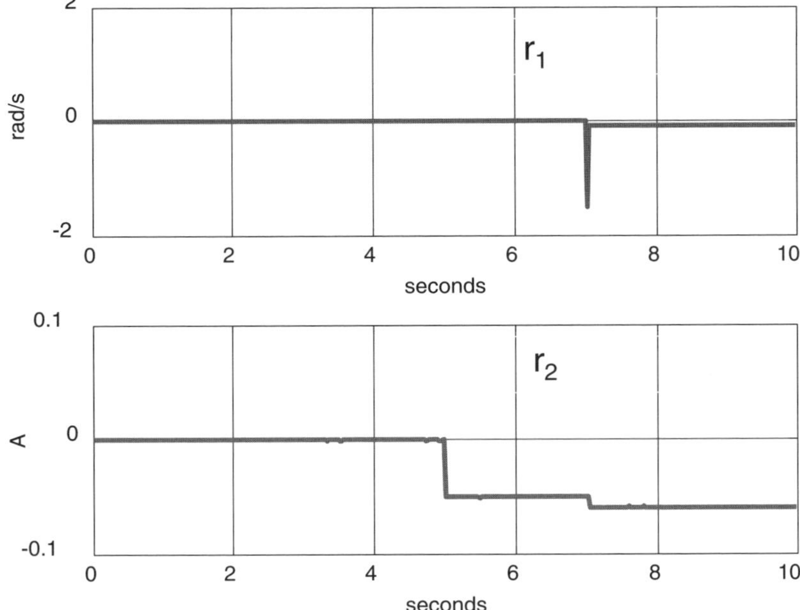

Figure 6.6. *Observer-based residuals for the motor*

The observer here has been designed to have zero eigenvalues, which makes it very fast. Now, the residual $r_1 = (\omega_{\mathrm{mes}} - \widehat{\omega})$ is insensitive to f_i, while the residual $r_2 = (i_{\mathrm{mes}} - \widehat{i})$ is sensitive to both faults (Figure 6.6).

6.2.4. *Diagnostic summary*

Monitoring and controlling an industrial facility require advanced tools, the basic concepts of which have been introduced in this section. All the diagnostic techniques are based on real signals emitted from the process instrumentation, which feed an analytical redundancy relation. When the models are discretized, the general form of such a relation is

$$r_k = f(y_k^1, y_{k-1}^1, \ldots y_{k-n1}^1, y_k^2, \ldots y_{k-n2}^2, \ldots u_k^1, u_{k-1}^1, \ldots u_{k-m1}^1, u_k^2, \ldots u_{k-m2}^2 \ldots),$$

$$(6.38)$$

where k stands for the sampling time, and r_k is such that when there is no fault $r_k <$ threshold. The observers constitute a particular case of equation (6.38) where the residual depends on values measured at time k and $(k-1)$ only.

Section 6.3 explains how using a network to transmit these data deteriorates the computation results.

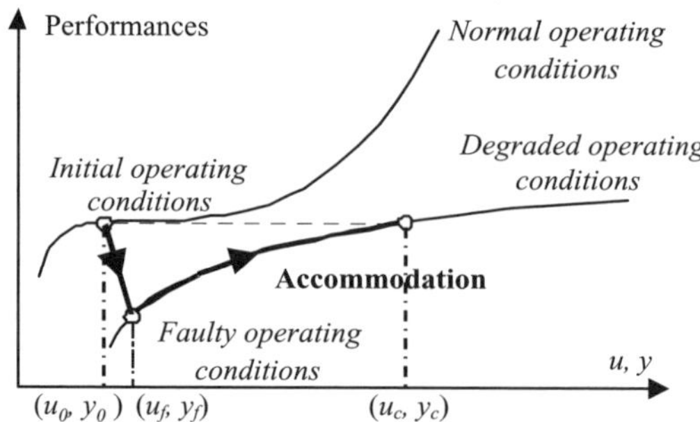

Figure 6.7. *Fault tolerance principle*

6.2.5. *Introduction to FTC*

A number of methods allowing for system diagnosis have been outlined above. Diagnosis means first detecting that the system in not in its normal operating mode, and then isolating the sub-system – and possibly the particular component – which is faulty. This knowledge can be used to take online control decisions, when the fault is detected early on. Prior to any decisions, worker safety must first be ensured, then production loss must be limited, and these are accomplished through the use of an appropriate control.

The task to be tackled in achieving fault tolerance is the design of a controller having a structure enabling it to guarantee satisfactory performance, not only when all control components are operational, but also when instruments are operating in a faulty mode. In this context, the aim of this section is to introduce the main concepts of FTC, and to take stock of recent methods for fault handling.

In the past few years, a number of FTC approaches have been reported, but most of them were developed for particular applications, mainly relevant to flight control or aerospace [MAY 91; BAN 99]. FTC has not reached its full maturity, however, and still remains an open methodology [PAT 97; MAH 03; BLA 03]. FTC concepts have been extensively treated in [BLA 06]. That is to say that, within the framework of FTC, there are many ways to achieve fault tolerance in the design of a controller.

Currently, FTC concepts can be separated into *passive* and *active* approaches. The key difference between these two approaches is that the active FTC system includes an FDI system, and the fault handling is carried out based on the fault information that the FDI system delivers, while in a passive FTC system, the system components and controllers are designed in such a way that they are robust to possible faults to a certain degree.

The passive approach makes use of robust control techniques to ensure that a closed-loop system remains insensitive to certain faults. When redundant actuators are available, methods dealing with this approach are also called *reliable control methods* [JOS 87; VEI 92; ZHA 98]. In the active approach, a new set of control parameters is determined in such a way that the faulty system reaches the nominal system performance. The principle of active approaches is illustrated in Figure 6.7: after the fault has occurred, the system deviates from its nominal operating point – defined by its input/output variables (u_0, y_0) – to a faulty one (u_f, y_f). The goal of FTC is to determine a new control law that takes the degraded system parameters into account and then drives the system to a new operating point (u_c, y_c), while maintaining the main performances (such as stability and accuracy) as close to the initial performances as possible. It is therefore important to accurately define the degraded modes that are acceptable with regard to the required performances, since after the occurrence of faults conventional feedback control design may result in unsatisfactory performances such as tracking errors, instability, and so on.

When the exact model of the failed system is known, the control system can be accommodated so that system performances are recovered and the new system behaves as initially specified. [GAO 91], [GAO 92], and [MOR 90] suggest a basic approach based on what they called the *pseudoinverse method* which has been revisited by [STA 08]. In practice, however, the faults are unanticipated and the model of the impaired system is not available.

To overcome the limitations of conventional feedback control, new controllers have been developed with accommodation capabilities – or *tolerance* – to faults. These fault tolerant controllers fall into four different categories:

– Adaptive control seems to be the most natural approach to fault tolerance: when the effects of faults appear as parameter changes and are identified online, the control law is reconfigured automatically based on new parameters [BOD 97; OCH 91; RAU 95]. [WU 98] consider a loss of effectiveness in actuators and suggest using an augmented state Kalman Filter to estimate both the fault-free state and the faulty parameters. The estimated fault-free state is then used to feed the controller. These approaches have the advantage of not requiring that the faults be categorized *a priori*, although the design of robust identification and control algorithms presents significant challenges.

– Integrated approaches represent another trend [NET 88]. They consist of integrating fault monitoring and control procedures. In this case, the possible actuator or sensor faults are represented by signals and are estimated by the same algorithm that computes the control law [MUR 96; TYL 94]. The faults are identified first, and then the controller is built to be insensitive to them, but the operator may be made aware of possible faults thanks to the alarm monitoring function.

– The FTC problem can also be formulated as a multi-objective problem based on the assumption that, like the uncertainties, the effects of faults can be expressed by means of *linear fractional transformation* (LFT). Following this methodology, a linear matrix inequality formulation for fault tolerant controller synthesis has been introduced by [CHE 98]. Another approach based on convex optimization has also been considered: a linear quadratic controller is used and the reconfiguration is achieved by choosing new values of the weighting matrices in the performance index in a manner appropriate to offset the effect of faults [LOO 85; SAU 98].

– Finally, another way to achieve FTC relies on supervised control where an FDI unit provides information about the location and time occurrence of any fault. Faults are compensated via an appropriate control law triggered according to the diagnosis of the system. This can be achieved using gain scheduling [JIA 98] or compensation via additive input design [NOU 00; THE 98]. Methods combining model-based and knowledge or heuristic techniques were also successfully used to tune the controller [AUB 93; KWO 95; PAT 97].

The ability of a plant to function safely and efficiently depends to a large extent on effective communication and information sharing between the interconnected units. With the significant growth in computing and networking abilities in recent times, as well as the rapid advances in actuator/sensor technologies, there has been an increased reliance on distributed computing and process operations across computer networks. A critical issue, however, that must always be considered in the design of any networked control system (NCS) is its robustness with respect to failure situations. By network failure, a total breakdown in the communication between the control system components is meant, as a result of, for example, some sort of physical malfunction in the networking devices or severe overload of the network resources that causes it to shut down.

The following section will examine the consequences of using a network on FDI and FTC.

6.3. Networked-induced effects

An NCS is a feedback control system in which information is transmitted among the system components in the form of data flows through a network, as illustrated in Figure 6.9. The functionality of a typical NCS is established by the use of four basic elements in the control loop: sensors to collect information, controllers to provide control signals, actuators to perform the control, and the communication network to enable the exchange of information. In order to guarantee a safe operation, some specific supervision functions are added to perform FDI and FTC. In practice, the network links in the NCS are usually unreliable, because control signals sent by the controller and plant measurements sent by the sensors may be lost or corrupted by noise during transmission.

6.3.1. *Example of network-induced drawbacks*

Consider the example of the diagnosis of the direct current motor (section 6.2.3), which is simulated using Simulink as above, together with the TRUETIME toolbox for the network [OHL 07]. A CAN network (1 Mbits s^{-1}, frame size 64 bits) is introduced between the controller and the actuator and between the sensors and the controller–diagnoser. The sensor task is time triggered (run at each multiple of the sampling period). The controller and the actuator tasks are event triggered (by the arrival of the measured velocity and the control signal, respectively). The diagnostic task is event-triggered (by the arrival of both measurements and of the new control signal, whose dates are not verified). When the network is dedicated to this application, the results are identical to those presented in section 6.2.3. Now, if the network is assumed to be overloaded, which can be simulated introducing some data loss probability (0.5 for all transmitted data in Figure 6.8), the functioning is genuinely disturbed. High residual amplitudes appear, for instance, when the reference is changed, even though there is no fault. This is because, during the transient phase, measurement values are very different from one sampling period to another. Computing with an obsolete value (a value that has not been updated because it has not been received) makes

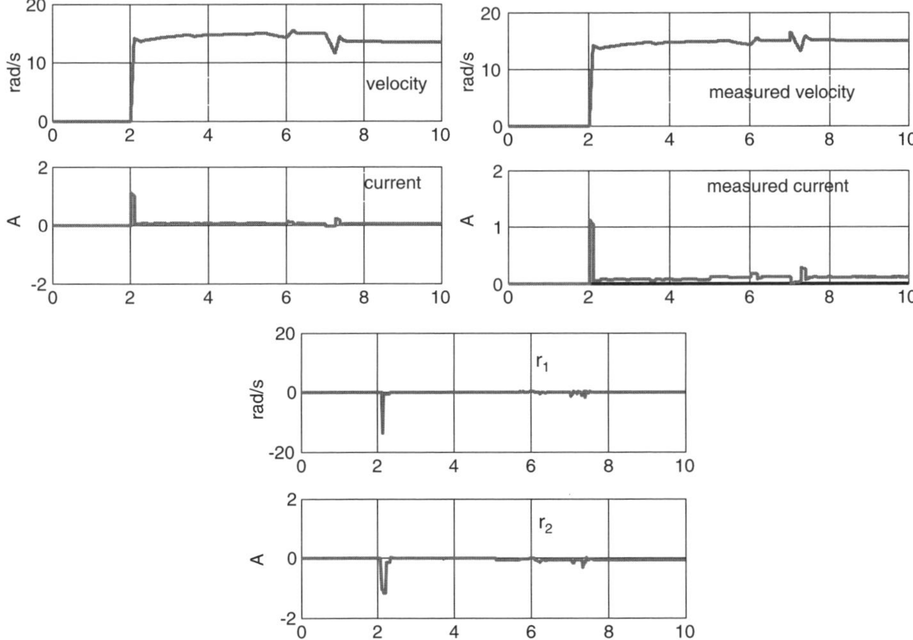

Figure 6.8. *Variables and residuals using a CAN network with a 0.5 data loss probability*

an important difference, interpreted as a fault, unless the threshold is increased. But this increase could further result in missed detection. This example illustrates the drawbacks of using a network if the way the diagnosis is processed is not modified to take the network into account.

6.3.2. *Modeling data dropouts*

The immediate result of complete network failure is the loss of all the associated sensor and actuator signals, which results in the plant switching from a closed-loop to an open-loop operating mode. For stable plants, this may lead only to performance degradation, while for unstable plants it may lead to instabilities. Note that packet losses are inevitable in any well-functioning network. However, such losses typically occur intermittently and do not cause a breakdown in communication. Here, we consider an NCS where the controller sends control signals to the actuator through a packet-dropping network as illustrated in Figure 6.9.

On the basis of the fact that the FDI system and the controller are remotely located at the same place, packet dropout of sensor measurements is assumed to be known to the FDI system, whereas packet dropout of control commands is assumed not to be. It is further assumed that the network satisfies an (User Datagram Protocol) UDP-type protocol, which means that the controller does not receive any acknowledgment when a packet has been received by the actuator. In other words, the controller does not know if the packet was dropped or not. Therefore, the following discrete time model for the NCS is obtained from the continuous time model equations (6.12) and (6.23)

$$\begin{cases} x_{k+1} = \Phi x_k + \Gamma u_k + \Xi f_k \\ y_k = C x_k, \end{cases} \tag{6.39}$$

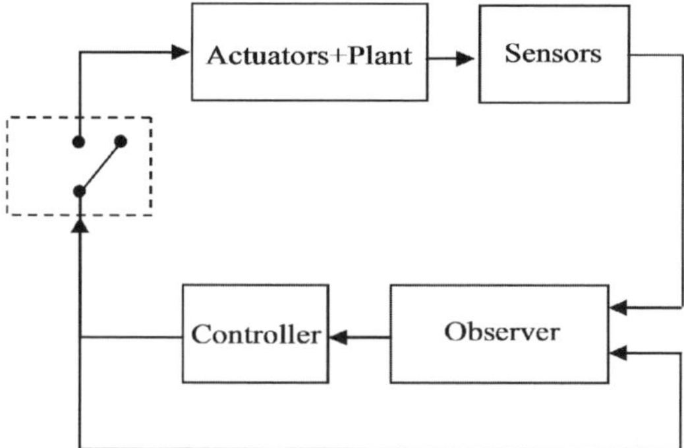

Figure 6.9. *A feedback control system with packet dropouts*

where $x_k \in \Re^n$ is the state vector, $y_k \in \Re^m$ the output observation vector, $u_k \in \Re^p$ the input vector, $f_k \in \Re^q$ the fault vector, Ξ is the faults distribution matrix on the state vector. It is further assumed that under intermittent communication, when the packet containing the control signal is dropped, the control signal sent to the actuator is kept equal to the previous one.

Introduce $\bar{u}_k = [\bar{u}_1 \ \bar{u}_2 \cdots \bar{u}_p]^T$ with $\bar{u}_{i,k} = \delta_{i,k} u_{i,k}$ and $\delta_{i,k} = -1$ when the packet containing $u_{i,k}$ is dropped; otherwise, $\delta_{i,k} = 0$. Then the model of the system with unreliable data transmission can be expressed as

$$\begin{cases} x_{k+1} = \Phi x_k + \Gamma u_k + \Gamma \bar{u}_k + \Xi f_k \\ y_k = C x_k. \end{cases} \tag{6.40}$$

It is to be noticed that the model in (6.40) can be rewritten as

$$\begin{cases} x_{k+1} = \Phi x_k + \Gamma u_k + H n_k \\ y_k = C x_k, \end{cases} \tag{6.41}$$

where $H = [\Gamma \ \Xi], \Gamma = [b_1, \ldots, b_p], \Xi = [e_1, \ldots, e_q]$ and

$$n_k = [\bar{u}_{1,k}, \bar{u}_{2,k}, \ldots, \bar{u}_{p,k} \ f_{1,k} \ f_{2,k}, \ldots, f_{q,k}]^T.$$

For data acquisition, it is supposed that the sensor is time triggered with a constant sampling time h.

Data dropouts will impact any standard observer-based residual generator. Consider for system (6.39), the state observer

$$\begin{cases} \hat{x}_{k+1} = \Phi \hat{x}_k + \Gamma u_k + L(y_k - C \hat{x}_k) \\ \hat{y}_k = C \hat{x}_k. \end{cases} \tag{6.42}$$

Under faulty conditions but considering unreliable data communication, from (6.41) and (6.42), the estimation error $\varepsilon_k = x_k - \hat{x}_k$ and the output residual of the state observer propagate as

$$\begin{cases} \varepsilon_{k+1} = (\Phi - LC)\varepsilon_k + H n_k \\ r_k = T C \varepsilon_k. \end{cases} \tag{6.43}$$

Suppose that the ith control signal is not transmitted to the actuator at time instant r, and consider the following index:

$$\rho_i = \min \{\nu = 1, 2, \ldots : \quad C G^\nu b_i \neq 0\}. \tag{6.44}$$

Due to the additive effects of dropouts and faults, the output residual can now be expressed as

$$r_k = \tilde{r}_k + \Phi^u_{k,p} \lfloor \bar{u}^i_p \cdots \bar{u}^i_{k-s} \cdots \bar{u}^i_{k-1} \rfloor + \Phi^f_{k,p} \lfloor f^i_r \cdots f^i_{k-s} \cdots f^i_{k-1} \rfloor, \tag{6.45}$$

with

$$\Phi^*_{k,p} = LC\Phi^*_{k-1,\rho_i}\,b_i$$
$$\Phi_{k-1,k-j} = G_{k-1}G_{k-2}\cdots G_{k-j},$$
(6.46)

where $\tilde{r}_k$ is the output residual if there is no packet dropped and if there is no fault and $G = \Phi - LC$.

Note. *Network breakdowns.* Unlike data dropouts, the losses resulting from network failure are more prolonged and sustained over time. In addition, unlike the problems of sensor or controller failure in the classical (hardwired) control architecture, where the failure effects may be confined to a specific control loop, the effects of network failure are typically more severe in that they cause a breakdown in the sensor-controller–actuator communication for multiple loops simultaneously. Depending on the extent to which network architecture is centralized, thousands of sensors and actuators could be connected through the same network and thus suffer from the consequences of plant-wide communication failure.

6.3.3. *Modeling network delays*

If it is assumed that the system is controlled over a network, then we have to take into account the sensor to controller τ_{sc} delays and controller to actuator delays τ_{ca}. For data acquisition, it is supposed that the sensor is time triggered with a constant sampling period h. By event-triggered controller or actuator, we mean that the calculation of the new control or actuator signal begins as soon as the new control or actuator information arrives, as is represented in Figure 6.10.

Taking into account the network-induced delay and the control input (zero hold is assumed) over a sampling interval $[kh, (k+1)h,]$ we have

$$u(t) = \begin{cases} u_{k-1}, & t \in [kh, kh + \tau_k] \\ u_k, & t \in [kh + \tau_k, (k+1)h]. \end{cases}$$
(6.47)

Note that the delays, in general, cannot be considered constant and known. Network induced delays may vary, depending on the network traffic, medium access protocol, and the hardware. Under actuator or component faults, it is supposed that the plant is described by the following equation:

$$\{x_{k+1} = \Phi x_k + \Gamma_0 u_k + \Gamma_1 u_{k-1} + \Xi f_k, \quad y_k = Cx_k + Ff_k,$$
(6.48)

where matrices Γ_0 and Γ_1 are defined as follows:

$$\Gamma_0 = \int_0^{h-\tau_k} e^{As}B\,ds \quad \Gamma_1 = \int_{h-\tau_k}^{h} e^{As}B\,ds,$$
(6.49)

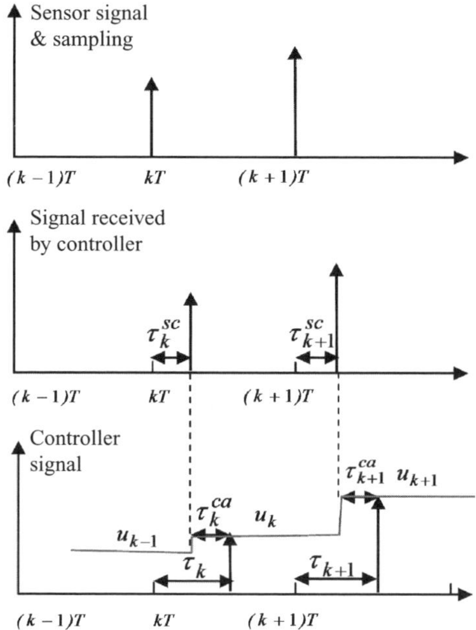

Figure 6.10. *Delay timing diagram for a network control system NCS*

and F is the faults distribution matrix on the outputs observation vector. If we introduce the control increment $\Delta u_k = u_k - u_{k-1}$, the model in (6.48) can also be rewritten as

$$\begin{cases} x_{k+1} = \Phi x_k + \Gamma u_k - \Gamma_1(\tau_k)\Delta u_k + \Xi f_k \\ y_k = C x_k + F f_k. \end{cases} \tag{6.50}$$

The FDI system introduced earlier in this section was designed *a priori* to achieve certain characteristic performances without considering the time delays. Let us now examine the effects of network-induced delays on the residual generator given by (6.42). From equation (6.43), the estimation error $\varepsilon_k = x_k - \hat{x}_k$ and the residual vector r_k propagate as

$$\begin{cases} \varepsilon_{k+1} = (\Phi - LC)\varepsilon_k - \Gamma_1(\tau_k)\Delta u_k + (\Xi - LF)f_k \\ r_k = TC\varepsilon_k. \end{cases} \tag{6.51}$$

Therefore, it appears that the robustness of the fault diagnosis system against network induced delays depends on the amplitude of the unknown term $-\Gamma_1(\tau_k)\Delta u_k$.

6.4. Pragmatic solutions

Diagnostic algorithms can be run at each sampling time or on demand. In the event that they are run on demand, the measurements are not taken permanently, but only during certain periods, when a sequence of values (which can be of several thousands) is captured and transmitted. This sequence makes up a burst of values transmitted from the sensors to the diagnoser. This can be the case if a component is functioning from time to time or if for instance the diagnosis is based on a frequency analysis made on a given time window. Whatever the case, the timing requirements placed on the measurements are very stringent: they must be regularly sampled.

6.4.1. *Data synchronization*

6.4.1.1. *Clock synchronization*

Timing considerations have led to means for enforcing the timing requirements in control systems implemented with technologies such as network communication, local computing and distributed objects. One such technique is the use of system components that contain real-time clocks, all of which are synchronized to one another within the system synchronization technology. The IEEE 1588 standard addresses clock synchronization requirements, defining a protocol enabling precise synchronization in measurement and control systems.[1] The protocol enables heterogenous systems that include clocks of various inherent precision, resolution and stability to synchronize. The protocol supports systemwide synchronization accuracy in the sub-microsecond range with minimal network and local clock computing resources. The default behavior of the protocol allows simple installation and operation. Since the creation of the 1588 standard, clocks in all system components are synchronized to a specified uncertainty, data or actions of system components based on these clocks are also synchronized according to the application's specifications.

The "1588 clock" is typically used in an actuator component as a mechanism to generate the actuation trigger by comparing the time of the "1588 clock" to a specified "trigger time" provided to the component as part of the application. Sensor components in a system supporting the IEEE 1588 standard will typically include a "1588 clock" and a small microprocessor. Such components are often termed "smart transducers" in that they perform processing on the raw data and add additional information about the measurement, such as the time of measurement.

The time-stamp object that enables the user of CANopen systems to adjust a unique network time should also be mentioned here. The time stamp is mapped to one single CAN frame with a data length code of 6 bytes. These six data bytes provide the "Time of Day" information, which is given as milliseconds after midnight and days since January 1st, 1984.

In conclusion, if the network transmits perfectly synchronized values, as described above, there is no delay problem. If it transmits at least variable values with the

---

1. http://ieee1588.nist.gov

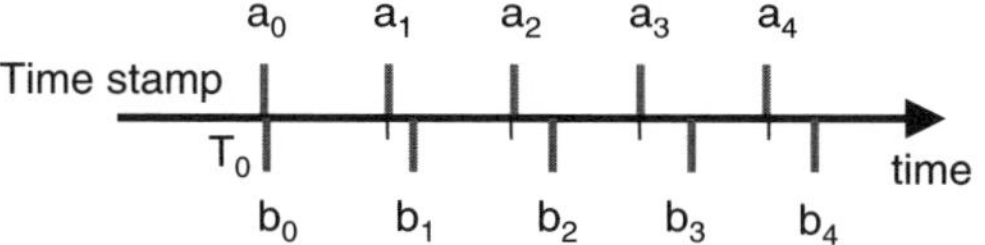

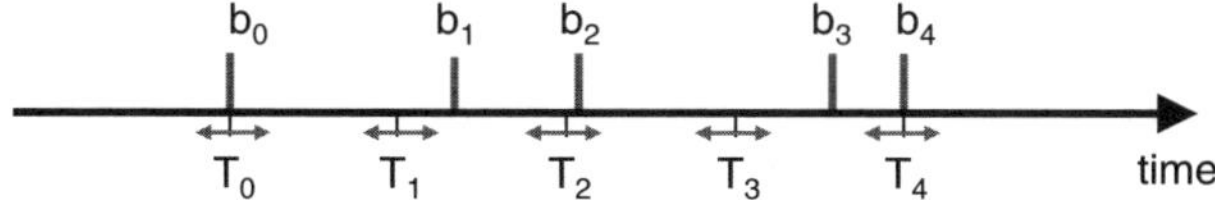

Figure 6.11. *Data synchronization. Top: a_i ideal data; b_i received data; bottom: data reconstruction*

exact date at which they were sampled, then ideally, perfectly sampled data can be reconstituted. This is described in the following section.

6.4.1.2. *Data reconstruction*

When the diagnostic algorithm receives a single datum or a sequence of data arriving within a burst or a packet, its first task is to arrange them with increasing dates thanks to the time stamps. Then, it must examine the dates of all the data embedded in one residual computation (see equation 6.38) and compare them to the sampling period multiples. If the dates are very similar (within an accepted period margin), the residual can be calculated; otherwise, the data need to be modified.

Figure 6.11 shows an example. Values b_0, b_2, and b_4 will be stored directly; they are within the acceptable period margin around T_0, T_2, and T_4. A value of T_1 will be obtained by linear interpolation of the values b_0 and b_1. A value for T_3 will not be obtained. The time difference $T(b_3) - T_2$ is greater than an acceptable value. The data are considered lost. In the proposed procedure, it is assumed that the signal does not

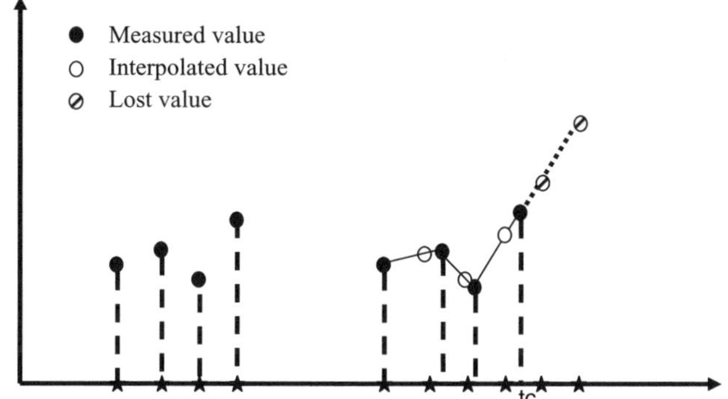

Figure 6.12. *Data reconstruction*

vary a lot during the sampling period h, which is the case if Shannon's sampling theorem is respected. A linear approximation is, therefore, sufficient to describe the signal variation during the time h. In this case, complicated interpolation or extrapolation procedures such as spline methods are not necessary.

Let tc be the current time. Figure 6.12 shows a first data flow that has been perfectly sampled and a second flow that has not. Four data are received in the same packet, represented by black points. Linear interpolation gives us the white points, the values of which will be used in the residual computation instead of the second, third, and fourth measured values. The same procedure is used if a single datum is received in the current sampling period but the residual needs a history of this variable. The history is synchronized using linear interpolation. After tc, it is assumed in Figure 6.12 that two data are lost. An extrapolation could be done, but this may be particularly dangerous in a diagnostic procedure, because extrapolation supposes that the signal behavior does not change a lot, which implies that no fault is present. It seems better in this case to block the diagnostic algorithm, which will be explained in section 6.4.2.

6.4.1.3. *Example*

What exactly is meant by "acceptable period margin"? A simple example will allow us to discuss this point. Let us consider a system step response that needs to be transmitted. Let y_0 and y_f be the initial and final values of the signal. The rise time tm is defined as the time during which the response evolves from 10% to 90% of its final value. A reasonable choice for the sampling period is $h \in [\frac{tm}{4}, \frac{tm}{10}]$. Choose $h = \frac{tm}{4}$. The signal variation during one sampling period during the transient is $0.2(y_f - y_0)$. If there is a desynchronization of Δh, the amplitude error is $0.2(y_f - y_0) * \frac{\Delta h}{h}$. This error can be compared to the noise standard deviation σ. If it has the same order of magnitude, then it is acceptable. The acceptable desynchronization is therefore

$$\Delta h = \frac{\sigma h}{0.2(y_f - y_0)}. \tag{6.52}$$

Approximating $\frac{\sigma}{(y_f - y_0)}$ by the noise-to-signal ratio B/S gives us

$$\frac{\Delta h}{h} = \frac{5}{S/B}. \tag{6.53}$$

With a smaller sampling time ($h = \frac{tm}{10}$), the result would be

$$\frac{\Delta h}{h} = \frac{12.5}{S/B}. \tag{6.54}$$

In conclusion, for $h = \frac{tm}{4}$ (respectively $h = \frac{tm}{10}$) and for a very small signal-to-noise ratio (1% for instance), Δh must be smaller than 5% of the sampling period (respectively 12.5%), but for a signal-to-noise ratio of 10%, it could be half of this period (respectively 125%). Given a signal-to-noise ratio of 10%, a sampling time equal to $h = \frac{tm}{4}$ and a rise time equal to 60 s, a reasonable period margin could be 7.5 s, while for a much more rapid system with a rise time equal to 500 ms, the margin could not be higher than 62.5 ms.

6.4.2. *Data loss and diagnostic blocking*

In this section, equation (6.38) is examined when one or several data in this equation are lost. For control algorithms, when the new control value computed by the controller is not transmitted to the actuator task, the best solution is to maintain the previous control value. If the system is in a permanent state, then this has no influence; if the system is in a transitory state, it is equivalent to introducing a delay equal to the sampling period, which will have a negative influence only if the system is not robust, is at its stability limit, or is extremely under-sampled.

The situation is very different when considering diagnostic algorithms. Computing equation (6.38) with old value(s) of some input or output signal will result in a residual value different from zero (in practice, it will be greater than the threshold) and thus induce false alarms. This effect will last as long as these data are needed in the equation, and, as a result, several sampling periods ($ni + 1$ if data y^i will be lost (see equation (6.46) for explanations relative to observers). In most cases, it is better not to deliver a diagnosis than to generate false alarms, which could lead to a regulator reconfiguration, or perhaps even worse, to an emergency shut down! Diagnosis is an event-triggered task. Consequently, operators could decide to run the diagnostic algorithm only once all the data have been received. A more subtle procedure is proposed here: data loss is now considered as a fault to detect and take into account in the diagnostic procedure. In the case of a network functioning like CAN, data are emitted and possibly re-emitted during a packet lifetime, which is equal to the sampling period. Following the sampling period, a new control has been computed and must be applied, or new data have been acquired, corresponding to the actual process state. It is more convenient not to overload the network transmitting the obsolete data; therefore, when data have not been received within the sampling period, it is considered permanently lost.

The data loss can be managed at the level of a generalized signature table. The list of necessary data is known for each residual (equation 6.38). A fault indicator r^i_{network} is set to "1" if at least one datum in the residual r_i history is lost. This makes up a symptom dedicated to the network's functioning, with the network being considered as a component of the system, with faults f^i_{network} to be diagnosed. The symptoms corresponding to each residual are introduced in the signature table. As an example, consider a system subject to three faults f_1, f_2, and f_3. Two residuals have been designed r_1 and r_2. When there is no data loss, $r^i_{\text{network}} = 0$, the first three columns of Table 6.3 allow for discriminations between the three faults. When one of the network

	f_1	f_2	f_3	$f_{network}^1 \wedge (f_1 \vee f_2)$	$f_{network}^2 \wedge (f_1 \vee f_3)$	$f_{network}^1 \wedge f_{network}^2$
r_1	1	0	1	Φ	1	Φ
r_2	1	1	0	1	Φ	Φ
$r_{network}^1$	0	0	0	1	0	1
$r_{network}^2$	0	0	0	0	1	1

Table 6.3. *Signature table with data loss*

faults is present, $r_{network}^i = 1$, the corresponding residual r_i is not relevant. Some partial information may be deduced from the value of the other residuals. When both $r_{network}^i = 1$, diagnosis is not possible.

6.5. Advanced techniques

6.5.1. *Residual generation with transmission delay*

In [YE 04], the authors propose a time-varying parity-space-based residual generator where robustness is achieved thanks to a decoupling procedure. Consider the system model described by (6.39). When the sampling period h is sufficiently small compared to the dynamics of the system, with the Taylor approximation of e^{As} we obtain

$$\Gamma_1(\tau_k) = \int_{h-\tau_k}^{h} e^{As} B ds = A^{-1}\left[1 - e^{A\tau_k}\right] e^{Ah} B \approx \Phi B \tau_k. \tag{6.55}$$

Assume also that the network delay can be broken down into one deterministic part and one stochastic part

$$\tau_k = \bar{\tau} + \Delta \tau_k, \tag{6.56}$$

where the stochastic part belongs to a zero-mean Gaussian white sequence with a known variance $\text{var}(\Delta \tau_k) = \sigma^2$. From (6.39), and with the introduction of system and sensor noises, we obtain

$$\begin{cases} x_{k+1} = \Phi x_k + \Gamma u_k - \Phi B \bar{\tau} \Delta u_{k-1} - \Phi B \Delta \tau_k \Delta u_{k-1} + \Omega d_k + \Xi f_k \\ y_k = C x_k + F f_k + w_k. \end{cases} \tag{6.57}$$

The NCS model can then be written as

$$\begin{cases} x_{k+1} = \Phi x_k + [\Gamma \ \ \Phi B \bar{\tau}] \begin{bmatrix} u_k \\ \Delta u_k \end{bmatrix} + [\Omega \ \ \Phi B \Delta u_k] \begin{bmatrix} d_k \\ \Delta \tau_k \end{bmatrix} + \Xi f_k \\ y_k = C x_k + F f_k + w_k. \end{cases} \tag{6.58}$$

With $\tilde{\Gamma} = [\Gamma \ \ \Phi B \bar{\tau}]$, $\tilde{u}_k = \begin{bmatrix} u_k \\ \Delta u_k \end{bmatrix}$, $G_k = [\Omega \ \ \Phi B \Delta u_k]$, $v_k = \begin{bmatrix} d_k \\ \Delta \tau_k \end{bmatrix}$, we obtain

$$\begin{cases} x_{k+1} = \Phi x_k + \tilde{\Gamma} \tilde{u}_k + G_k v_k + \Xi f_k \\ y_k = C x_k + F f_k + w_k. \end{cases} \tag{6.59}$$

with the following statistical properties:

[i.] $\{v_k\}$ and $\{w_k\}$ are two discrete Gaussian white noises such that

$$E\left[v_k\right] = 0_{d+1}, \quad E\left[w_k\right] = 0 \quad, \quad E\left[d_k\right] = 0_{d+1}$$

$$E\left[w_k w_j^T\right] = Q.\delta_{kj} \ , \ E\left[v_k v_j^T\right] = \tilde{R}.\delta_{kj} \ i \neq j$$

$$\tilde{R} = \begin{bmatrix} R & 0 \\ 0 & \sigma^2 \end{bmatrix}.$$

[ii.] The measurement noise sequence $\{w_k\}$ and the system noise sequence $\{v_j\}$ are not correlated $E\left[w_k v_j^T\right] = 0 \quad \forall k, \ j$.

[iii.] The initial state x_0 is a random Gaussian variable with mean m_0 and covariance matrix P_0:

$$E\left[x_0\right] = m_0 \ , \quad E\left[(x_0 - m_0)(x_0 - m_0)^T\right] = P_0.$$

[iv.] The initial state x_0 and noises v_k and w_k are not correlated

$$E\left[x_0 \, v_k^T\right] = 0, \quad E\left[x_0 \, w_k^T\right] = 0.$$

With the NCS model given in (6.58), consider the adaptive Kalman filter given by the recursive algorithm for residual generation

$$\begin{cases} \hat{x}_{k+1/k} = \Phi\hat{x}_{k/k-1} + \tilde{\Gamma}\tilde{u}_k + \Phi K_k(y_k - C\hat{x}_{k/k-1}) \\ K_k = P_{k/k-1}C^T(CP_{k/k-1}C^T + R)^{-1} \\ P_{k+1/k} = \Phi P_{k/k-1}\Phi^T + G_k(\Delta u_k)\tilde{R}G_k^T(\Delta u_k) - \Phi K_k CP_{k/k-1}\Phi^T. \end{cases} \qquad (6.60)$$

The innovation sequence $\gamma_k = y_k - C\hat{x}_k$ can thus be used for FDI.

6.5.2. *Adaptive thresholding*

Threshold schemes attempt to analyze residuals signals by comparing their values against a threshold and deciding whether or not a fault has occurred using some decision logic. Although threshold methods are the most common in the literature [STO 03; CAS 05], these methods have always imposed significant trade-offs between false alarms and missed detections. The work presented in this section proposes a methodology for the implementation of dynamic threshold strategies. The threshold must be adapted in order to minimize false alarms as well as missed detections in the case residuals are affected by network communication effects such as transmission delays of information dropout.

6.5.2.1. *Optimization-based approach for threshold selection*

The approach proposed for threshold adaptation is based on the optimization of a performance index related to the maximum amplitude of the residual with respect to the possible variation of the delays inside a bounded region. In [SAU 06], the following multiple input system is considered:

$$\dot{x}(t) = Ax(t) + \sum_{i=1}^{m} B_i u_i(t) + Ef(t), \quad y(t) = Cx(t), \tag{6.61}$$

where the control input is

$$\begin{cases} u_i(t, \tau_i) = u_i(kh - 1), & kh \le t \le kh + \tau_i \\ u_i(t, \tau_i) = u_i(kh), & kh + \tau_i \le t \le (k+1)h. \end{cases} \tag{6.62}$$

A state observer for the system described in (6.61) is given by

$$\begin{cases} \dot{\hat{x}}(t) = A\hat{x}(t) + \sum_{i=1}^{m} B_i u_i(t) + L(y(t) - \hat{y}(t)) \\ \hat{y}(t) = C\hat{x}(t), \end{cases} \tag{6.63}$$

with L being defined by $\Phi - LC = \exp\left\{(A - LC)h\right\}$. Since the control input applied to the observer is $u_i(t, 0) = u_i(kh)$, $kh \le t \le (k+1)h$, the estimation error $\varepsilon(t) = x(t) - \hat{x}(t)$ and the residual propagate as

$$\begin{cases} \dot{\varepsilon}(t) = (A - LC)\varepsilon(t) + \sum_{i=1}^{m} B_i(u_i(t, \tau_i) - u_i(t, 0)) \\ r(t) = TC\varepsilon(t). \end{cases} \tag{6.64}$$

The minimum threshold for fault detection can be considered as a performance index to be maximized for the dynamic system described by (6.64). Thus, the optimization problem is to find the inputs $u_i^*(t)$ on the time interval $[kh, (k+1)h]$ so that the performance index $\Psi(t = kh) = |r(kh)|$ is maximized. For this purpose, let us introduce the Hamiltonian

$$H = \lambda^T[(A - LC)\varepsilon(t) + \sum_{i=1}^{m} B_i(u_i(t, \tau_i) - u_i(t, 0))], \tag{6.65}$$

where λ is the co-state vector. The solution of the optimization problem satisfies the state and co-state equations

$$\dot{\varepsilon}^*(t) \quad = \quad \frac{\partial H}{\partial \lambda^*} = (A - LC)\varepsilon^*(t) + \sum_{i=1}^{n} b_i(u_i^*(t) - u_i(t, 0)) \tag{6.66}$$

$$\dot{\lambda}^*(t) \quad = \quad -\frac{\partial H}{\partial \varepsilon^*} = -(A - LC)\lambda^*(t), \tag{6.67}$$

where * stands for "optimum" and $u_i^*(t) = u_i(t, \tau_i^*)$ is given by the application of the Pontryagin minimum principle

$$\tau_i^* = \underset{0 \le \tau_i \le \tau_i^+}{\mathrm{Arg}(\mathrm{Min}(H))}. \tag{6.68}$$

In addition to the initial conditions $\varepsilon(kh)$, the terminal boundary conditions for this fixed terminal time problem are

$$\left[\frac{\partial \Psi}{\partial \varepsilon} - \lambda(t) \right]_{t=(k+1)h} = 0, \tag{6.69}$$

which gives

$$\lambda((k+1)h) = TC \begin{bmatrix} \mathrm{sgn}(\varepsilon_1(k+1)h \\ \vdots \\ \mathrm{sgn}(\varepsilon_n(k+1)h \end{bmatrix}. \tag{6.70}$$

Thus, the optimization consists of solving state and co-state equations with boundary conditions while simultaneously selecting the values τ_i^* that maximize H. The optimization problem is very simple since over the time frame of the optimization window only one commutation of the control input is applied. Therefore, the solution to (6.65) is given by

$$\tau^* = \underset{0 \le \tau \le \tau^+}{\mathrm{Arg}\left(\mathrm{Min}\left(\lambda^T \sum_{i=1}^{m} B_i \Delta u_i(t, \tau_i)\right)\right)}. \tag{6.71}$$

Obviously, the optimum for τ_i^* (either τ_i^+ or 0) depends solely on the sign of the terms in $\lambda^T \sum_{i=1}^{m} B_i u_i(t, \tau_i)$, which requires to know the sign of $\lambda(t)$. This can be achieved when solving (6.67). Since the inputs are constant over the considered time interval, the optimization can be iterated over several sampling intervals.

6.5.2.2. *Network calculus-based thresholding*

The interval for the admissible delays is calculated by using the network calculus theory. In the event of unexpected changes in the network architecture (such as a component breakdown, change in traffic load, or broken links), the network behavior is modified and transmission delays may vary. In this case, the threshold is adjusted according to the network characteristic variations. With this in mind, a method based on network calculus theory is presented in order to determine upper-bound values $\overline{\tau_{\mathrm{ca}}}$ and $\overline{\tau_{\mathrm{sc}}}$, i.e. the control and measurement delays. Those upper-bounds will be used to adapt the FDI residual to the delays.

The upper-bound delay estimation algorithm applies ideas from the network calculus theory ([CRU 91; GEO 05]). The network architecture considered corresponds to

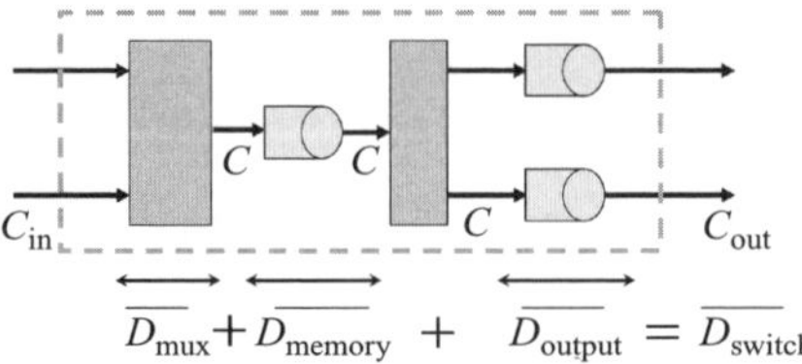

Figure 6.13. *Model of a two-port switch in a full-duplex mode based on shared memory and a cut-through management*

a switched Ethernet architecture (linked to the IEEE 802.1D standard). The approach consists in modeling switches as a combination of basic components: multiplexers, demultiplexers and FIFO queues, as shown in Figure 6.13.

The first step in switch modeling consists of determining an upper-bound delay for the crossing of each of the basic components. The upper-bound delay over the switch is then the sum of the upper-bound delays over the basic components

$$\overline{D_{\text{switch}}} = \overline{D_{\text{mux}}} + \overline{D_{\text{queue}}} + \overline{D_{\text{output}}}, \tag{6.72}$$

where $\bar{D}$ represents the upper-bound value of the delays.

Maximum time for crossing one Ethernet switch. In the mathematical analysis, the traffic arriving at the switch, both periodic and aperiodic, is modeled as a *leaky bucket controller*. Data will arrive at the leaky rate only if the level of the bucket is less than the maximum bucket size. In the network calculus theory the traffic models are represented as arrival curves, and, with the assumption that the traffic follows the leaky bucket model and that the incoming rate is limited by the port capacity, these curves are adjusted and have the following shape:

$$b(t) = \min(\ C_{\text{in}}t \quad \sigma + \rho t\ ), \tag{6.73}$$

where σ is the maximum amount of data that can arrive in a burst, ρ is an upper bound of the average traffic flow rate, and C_{in} is the capacity of the input port. By the same token, service curves are used to represent the minimal data processing activity of the components. Typical arrival and service curves are shown in Figure 6.14.

The approach used in analyzing the upper-bound delay for crossing a two-input multiplexer will now be summarized [GEO 05]. The approach is based on the evolution of a specific parameter, the backlog. The backlog is the number of bits waiting in the component, and it is a measure of congestion level at the component. For the arrival curves in Figure 6.14, the upper-bound backlog occurs at time t when the following curve reaches a maximum:

$$b_1(t) + b_2(t + L/C_2) - C_{\text{out}}t, \tag{6.74}$$

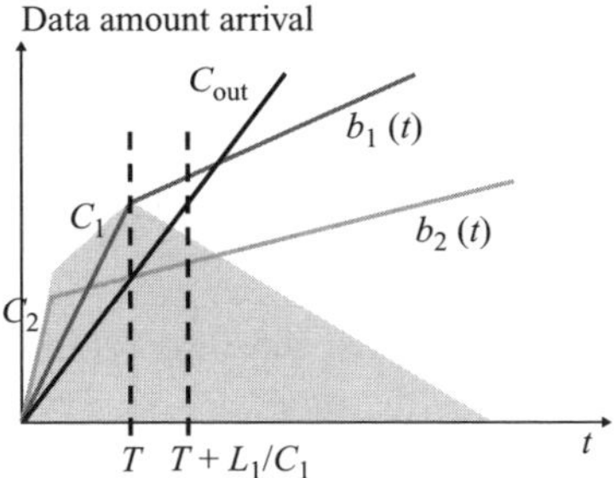

Figure 6.14. *Arrival and service curves and backlog evolution inside the two-input FIFO multiplexer*

where b_1 and b_2 are the arrival curves of streams 1 and 2 at time t, L is the maximum length of the frames, C_2 is the capacity of the import port 2, and C_{out} is the capacity of the output link. When the upper-bound backlog over the component is known, the upper-bound delay over the component is obtained by dividing the maximum backlog value by the capacity of the output link of the multiplexer.

In a FIFO m-inputs multiplexer, the delay for any incoming bit from the stream i is upper-bounded by

$$\overline{D_{\text{mux},i}} = \frac{1}{C_{\text{out}}} \min_k \overline{B_{\text{mux},k}}, \tag{6.75}$$

where $\overline{B_{\text{mux},k}}$ is an upper bound of the backlog in the bursty periods v_k, so that $1 \leq k \leq m$.

For $k = i$, the bursty period is defined by $v_i = \sigma_i/(C_i - \rho_i)$ and the backlog is upper-bounded by

$$\overline{B_{\text{mux},i}} = \sum_{z=1;z\neq i}^{m} \left(\sigma_z + \rho_z \left(v_i + \frac{L_z}{C_z} \right) \right) + v_i (C_i - C_{\text{out}}), \tag{6.76}$$

where σ_i is the burstiness of stream i, ρ_i is the average rate of arrival of the data of stream i, L_i is the maximum length of the frames of stream i, and C_i is the capacity of the import port i. For $k \neq i$ so that $1 \leq k \leq m$, we have $v_k = \sigma_k/(C_k - \rho_k) - L_k/C_k$ and

$$\overline{B_{\text{mux},i}} = \sum_{z=1;z\neq k}^{m} \left(\sigma_z + \rho_z \left(v_k + \frac{L_z}{C_z} \right) \right) + v_k (C_k - C_{\text{out}}) \\ -\rho_i \frac{L_i}{C_i} + L_k. \tag{6.77}$$

For the FIFO queue, the delay of any byte is upper-bounded by

$$\overline{D_{queue}} = \frac{1}{C_{\text{out}}} \frac{(C_{\text{in}} - C_{\text{out}})}{C_{\text{in}} - \rho_{\text{in}}} \sigma_{\text{in}}. \tag{6.78}$$

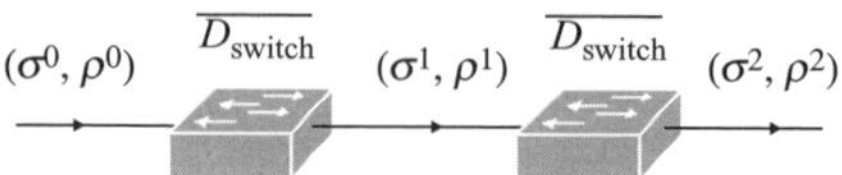

Figure 6.15. *Burstiness along a switched Ethernet network*

For the demultiplexer, it is assumed that the time required to route the output port is relatively negligible compared to the other delays, i.e. the demultiplexer does not generate delays.

Maximum end-to-end delays for crossing a switched Ethernet network. The computation of the upper-bound end-to-end delays requires that special attention be paid to the input parameters of previous equations. The maximum delay value $\overline{D}$ depends on the leaky bucket parameters: the maximum amount of traffic σ that can arrive in a burst, and the upper bound of the average rate of the traffic flow ρ. In order to calculate the maximum delay over the network, it is therefore necessary that the envelope (σ, ρ) be known at every point in the network. However, as shown in Figure 6.15, only the initial arrival curve values (σ^0, ρ^0) are usually known, and so the values for other arrival curves have to be determined.

To calculate all the arrival curve values, the following equations may be used:

$$\begin{aligned} \sigma_{\text{out}} &= \sigma_{\text{in}} + \rho_{\text{in}} D \\ \rho_{\text{out}} &= \rho_{\text{in}}. \end{aligned} \tag{6.79}$$

For example, with the arrival curve (σ^1, ρ^1) shown in Figure 6.14, the envelope after the first switch is

$$\left(\sigma^1, \rho^1\right) = \left(\sigma^0 + \rho^0 \overline{D}_{\text{switch}}, \rho^0\right). \tag{6.80}$$

The last part of the method used to obtain the upper-bounded delay estimate is the resolution of the burstiness characteristic of each flow at each point in the network. First, the burstiness values are determined by solving the equation system

$$\begin{bmatrix} a_{11} & a_{12} & \cdots & a_{1n} \\ a_{21} & a_{22} & \cdots & a_{2n} \\ \vdots & & & \vdots \\ a_{n2} & a_{n2} & \cdots & a_{nn} \end{bmatrix} \begin{bmatrix} \sigma_1 \\ \sigma_2 \\ \vdots \\ \sigma_n \end{bmatrix} = \begin{bmatrix} b_1 \\ b_2 \\ \vdots \\ b_n \end{bmatrix}. \tag{6.81}$$

Once the above equation has been solved, the upper-bound end-to-end delays are obtained from

$$\overline{D}_i = \frac{\sigma_i^N - \sigma_i^o}{\rho_i}, \tag{6.82}$$

where N is the number of crossed switches.

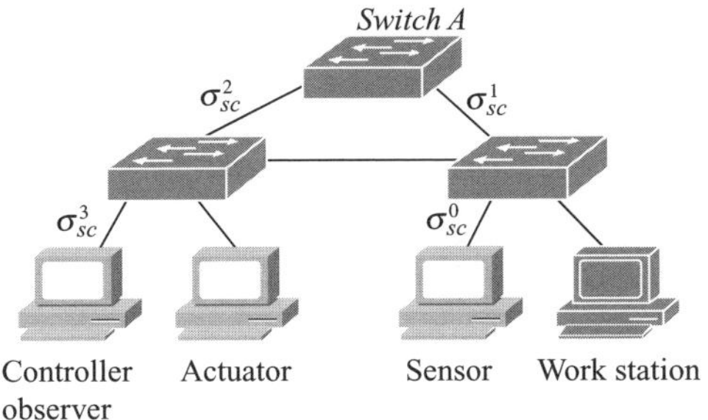

Figure 6.16. *A redundant switched architecture*

The network shown in Figure 6.16 interconnects the controller, actuator, and the sensor using a redundant switched Ethernet architecture, so that if a link between two switches breaks down, the network will be able to carry on the communications. The architecture here is also shared with applications other than the control of the system, so that a workstation is also linked to the network.

Traffic arrivals are modeled as follows: periodical exchanges from the sensor to the controller/observer are constrained by the arrival curve $b_{\mathrm{sc}}(t) = \sigma_{\mathrm{sc}} + \rho_{\mathrm{sc}}t$ and exchanges from the controller to the actuator by $b_{\mathrm{ca}}(t) = \sigma_{\mathrm{ca}} + \rho_{\mathrm{ca}}t$. At the same time, the workstation sends frames to the controller. Traffic here is constrained by the arrival curve $b_w(t) = \sigma_w + \rho_w t$.

Now consider the delay supported by the frames sent by the sensor. Delays depend on the network topology, and consequently the communication path. In a switched Ethernet network, the Spanning Tree Protocol is used to define an active topology in which the loops are eliminated. Firstly, it is assumed that a hierarchical active topology is defined, so that the measures will pass through switch A.

The determination of an upper-bound $\overline{\tau_{\mathrm{sc}}}$ consists of writing equation (6.81). To do this for each flow, it is necessary to write the expression of the output burstiness for each switch and for each switch basic component, as defined in Figure 6.13. Formulas are obtained according to equations (6.75), (6.78), and (6.79). Then the upper bound is obtained with the following expression:

$$\overline{\tau_{\mathrm{sc}}} = \frac{\sigma_{\mathrm{sc}}^3 - \sigma_{\mathrm{sc}}^o}{\rho_{\mathrm{sc}}}. \tag{6.83}$$

The principle will be the same for the delay supported by the control frames. This approach enables network faults to be taken into account. Indeed, in the event of a link failure between two switches, the Spanning Tree Protocol will define a new active topology and new communications paths. By applying the previous analysis once again, a new upper-bound $\overline{\tau_{\mathrm{sc}}}$ may be determined. This will be useful to control the adaptation of the FDI algorithms to the network evolution.

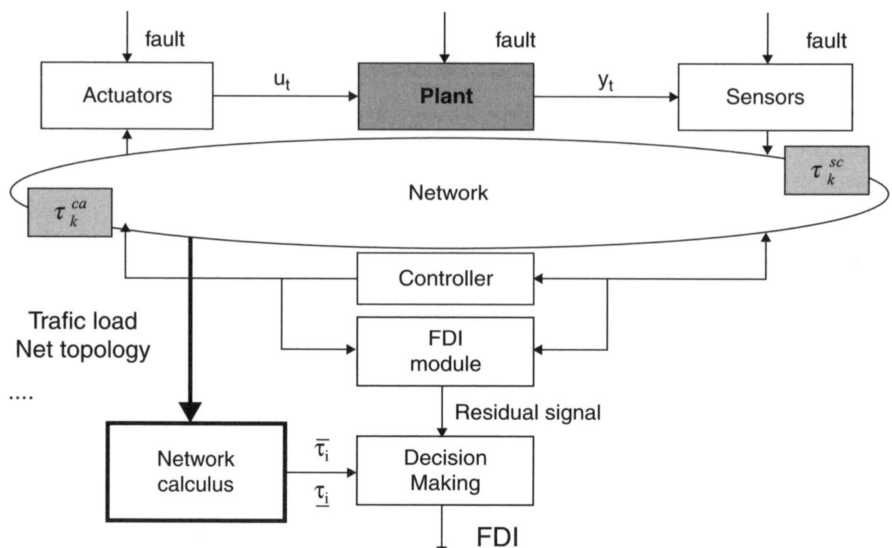

Figure 6.17. *A global FDI scheme with an adaptive threshold*

The global FDI scheme for the NCS is presented in Figure 6.17. The block entitled *Network Calculus* provides the upper bound value of the networked delay to the decision making module. Thus, the residual are evaluated according to the network behavior.

6.5.3. *Fault isolation filter design in the presence of packet dropouts*

In the case of packet dropouts, the subsequent problem to deal with is to design a linear filter the inputs of which are the system measurable outputs and control inputs which generate residual signals that are sensitive to fault and not affected by data dropouts.

The objectives of the filter are (i) to construct a fault isolation filter (FIF) for FDI of multiple faults, (ii) to design a free parameter ensuring that the energy ratio between useful and disturbance signals ω_k defined on the fault indicators is maximized. In what follows, actuators – or component faults – and data dropouts are considered as unknown inputs, which are considered as faults. The following definitions are needed to specify the properties of the detection filter.

DEFINITION.– *The NCS 6.48 is said to have fault detectability index* $\rho = \{\rho_1, \rho_2 \ldots, \rho_q\}$ *if* $\rho_i = \min\{\nu : C\Phi^{\nu-1} f_i \neq 0, \nu = 1, 2 \ldots\}$.

DEFINITION.– *If NCS has finite detectability indexes, the fault detectability matrix Ψ is defined as $\Psi = CD$, with $D = [\Phi^{\rho_1-1}f_1 \ \cdots \ \Phi^{\rho_i-1}f_i \cdots \Phi^{\rho_q-1}f_q]$.*

The following FIF is now introduced as the residual generator for the NCS system described by (6.48):

$$\begin{cases} \hat{x}_{k+1} = \Phi\hat{x}_k + \Gamma u_k + L(y_k - C\hat{x}_k), \\ r_k = T(y_k - C\hat{x}_k). \end{cases} \tag{6.84}$$

Let $G_{fr}(z)$ be the transfer function from f_k to the output residual r_k. Then the following theorems are presented to design L and T so that

$$\begin{aligned} G_{fr}(z) &= TC(zI - (\Phi - LC))^{-1}F \\ &= \operatorname{diag}\left\{ z^{-\rho_1}, \ldots, z^{-\rho_s} \right\}, \end{aligned} \tag{6.85}$$

where (6.85) makes possible the isolation of multiple faults and dropouts.

THEOREM 1. *For the NCS system (6.48) and the residual generator (6.84), the transfer function from faults to residual exhibits a diagonal structure if the following conditions are met:*

$$\begin{cases} (\Phi - LC)\Psi = 0 \\ TC = I. \end{cases} \tag{6.86}$$

THEOREM 2. *Given the condition $rank(D) = s$, the solutions of (6.85) can be parameterized as $K = \omega\Pi + \bar{K}_k\Sigma$, $L = \Pi$ with $\Sigma = \beta(I - \Psi\Pi)$ and $\Pi = \Psi^+$, $\omega = AD$, where $K_k \in \Re^{n\times m - p - q}$ represents the free parameters to be designed, Ψ^+ is the pseudoinverse [32] of Ψ and β is an arbitrary matrix chosen so that $rank(\Sigma) = m - s$.*

From the previous theorem, the FIF is rewritten with the free parameter $\bar{K}_k$ as

$$\begin{cases} \hat{x}_{k+1} = \Phi\hat{x}_k + \Gamma u_k + \omega\alpha_k + \bar{K}_k \sum(y_k - C\hat{x}) \\ \alpha_k = \Pi(y_k - C\hat{x}_k), \end{cases} \tag{6.87}$$

where α_k is a deadbeat filter of fault n_k, given by

$$\alpha_k = \tilde{\alpha}_k + [f^1_{k-\rho_1} \ \cdots \ f^i_{k-\rho_i} \ \cdots \ f^s_{k-\rho_s}]^T. \tag{6.88}$$

The state estimation errors without faults propagates as

$$\tilde{\varepsilon}_{k+1} = (\Phi - LC)\tilde{\varepsilon}_k + w_k - Kv_k, \tag{6.89}$$

where the fault $n^i_{k-\rho_i}$ of detectability index $k - \rho_i$ directly affects the reduced output residual r_k with a time delay equal to its detectability index. r_k can also be viewed as a stochastic deadbeat observer of the fault magnitudes. Furthermore, it is easy to show that $\alpha_k = \Pi(y_k - C\hat{x}_k)$ is decoupled from the faults, while $\gamma_k = \Sigma(y_k - C\hat{x}_k)$ is sensitive to the faults since $\Pi D = I$, and $\Pi\Sigma = 0$. The FIF design can then be used in different ways, depending on $rank(D)$.

Fault and data dropout isolation: when the condition $rank(\Psi) = p + q$ is verified, fault and dropout isolation can be achieved with a residual generator (6.59).

Fault diagnosis robust versus data dropouts: if the conditions $rank(C\Psi_f) = q$ and $rank(C\Psi_{\bar{u}}) = p$ are not satisfied, then multiple isolation is no longer possible. The free parameter $\bar{K}_k$ is independent of the multiple-fault isolation because any $\bar{K}_k$ ensures (6.85), if we suppose that the actuator and component faults are the unknown inputs to be isolated, while data dropouts represent disturbances. The transfer function from $\bar{u}_k$ to $\tilde{\alpha}_k$ is given by

$$G_{\bar{u}\tilde{\alpha}}(z) = \Pi C(zI - (\Phi - LC))^{-1}\Gamma. \tag{6.90}$$

The introduction of $\bar{K}_k$ gives an extra degree of freedom to satisfy some other design requirements. One remaining element would be to design the free parameters $\bar{K}_k$, satisfying the constraints as

$$\|G_{\bar{u}\tilde{\alpha}}(z)\|_\infty := \sup_{\frac{\|\tilde{\alpha}\|_2}{\|\bar{u}\|_2}} < \gamma. \tag{6.91}$$

Control accommodation to dropout: if we consider that $rank(C\Psi_{\bar{u}}) = p$ and that there are no faults, then data dropouts can be detected and isolated. With

$$\alpha_k = \tilde{\alpha}_k + [\bar{u}^1_{k-\rho_1} \quad \cdots \quad \bar{u}^i_{k-\rho_i} \quad \cdots \quad \bar{u}^p_{k-\rho_p}]^T \tag{6.92}$$

the fault free state estimate is sent to the controller, while an additive control signal u^{ad} is used to compensate for the fault effect on the system. Therefore, the total control law applied to the system is given by

$$u_k = -K\hat{x}_k + u^{\mathrm{ad}}_k. \tag{6.93}$$

The additional control law u^{ad} must be computed so that the faulty system is as close to the nominal one as possible. In other words, u^{ad} must satisfy

$$\Gamma u^{\mathrm{ad}}_k + \Gamma \bar{u}_k = 0. \tag{6.94}$$

Using the estimation of the fault magnitude described in the previous section, the solution for (6.94) can be obtained by the following relation if matrix B is full row rank:

$$u^{\mathrm{ad}_i}_k = -\hat{\delta}\bar{u}^i_{k-\rho_i}, \tag{6.95}$$

where $\hat{\delta}$ is an estimate of the data dropout indicator.

6.5.4. *Estimation and diagnosis with data loss*

As stated above, the observer bank is an appealing technique among the FDI community. For network-controlled systems, two observer bank schemes can be implemented. The first one supposes that a centralized computation unit is implemented while the second scheme makes use of sensor nodes with enough computation capabilities to locally estimate the state [SHI 08]. In the first case, raw data are transmitted to a unique computational unit, leading to a centralized observer bank. In the second case, the observers are distributed over the system. In this case, each observer can only make use of local data, or else data exchanges may be necessary. When data are transmitted, these can be delayed or even lost. The observer must therefore tackle this problem.

This section considers the Kalman Filter implementation [VER 07] when measurements are transmitted through a network. Previous works (see for instance [SIN 04] [HES 07] [SCH 07] [HUA 07] [SUN 08] [EPS 08] and references therein) suppose that the output vector $y_k \in \Re^m$ is sent in a unique packet that may be lost. Therefore, when a packet is dropped, all the measurements at time k are lost. However, depending on the topology of the system, on the protocol used, etc., measurements can be sent independently (or at least in several sets). Thus, each output (or set of outputs) is likely to be lost. Let us now assume that each measurement y_k^i is to be sent independently, and so the Kalman Filter is modified in order to deal with data loss.

6.5.4.1. *Problem formulation*

Consider the linear discrete-time varying stochastic system

$$x_{k+1} \ = \ \Phi_k x_k + v_k \tag{6.96}$$

$$y_k \ = \ C_k x_k + \nu_k, \tag{6.97}$$

where $\Phi_k \in \Re^{n \times n}$, $C_k \in \Re^{m \times n}$. v_k and ν_k are two mutually independent sequences of independent and identically distributed (i.i.d.) Gaussian white noises with covariance matrices Q_k and R_k, respectively. Moreover, the pair (Φ_k, C_k) is supposed to be observable and the pair $(\Phi_k, Q_k^{1/2})$ controllable. The packet loss can be modeled with an i.i.d. Bernoulli binary random sequence [SIN 04; SCH 07]. This model is chosen for mathematical tractability. The stability of the Kalman Filter is proved via a modified algebraic Riccati equation (MARE). Other authors (e.g. [XIE 08; HUA 07]) implement a discrete-time binary Markov model. In contrast to the i.i.d. model, the Markov chain captures the temporal correlation of the channel variation. With the

Markov packet loss model, stability results can also be achieved [HUA 07]. In both contexts, the Kalman Filter equations become

$$\widehat{x}_{k+1/k} \;=\; \Phi_k \widehat{x}_{k/k} \tag{6.98}$$

$$P_{k+1/k} \;=\; \Phi_k P_{k/k} \Phi_k^T + Q_k \tag{6.99}$$

$$\widehat{x}_{k+1/k+1} \;=\; \widehat{x}_{k+1/k} + \gamma_{k+1} K_{k+1}(y_{k+1} - C_{k+1}\widehat{x}_{k+1/k}) \tag{6.100}$$

$$P_{k+1/k+1} \;=\; P_{k+1/k} - \gamma_{k+1} K_{k+1} C_{k+1} P_{k+1/k}, \tag{6.101}$$

where Φ^T denotes the transpose of matrix Φ. $\gamma_{k+1} = 1$ (0 respectively) indicates that y_{k+1} has (has not respectively) been received.

$$K_{k+1} = P_{k+1/k} C_{k+1}^T [C_{k+1} P_{k+1/k} C_{k+1}^T + R_{k+1}]^{-1} \tag{6.102}$$

is the gain of the Kalman filter. Matrix P, representing the observer state covariance, is now a random variable because of the randomness of γ. In (6.100), the whole observation vector y is supposed to be lost or received. However, the components of y might not be put in a unique packet, leading to the loss of a subset of the output data. This gives rise to a new Kalman filter with partial data loss which is now presented.

6.5.4.2. *Kalman filter with partial data loss*

The estimation problem is reformulated as follows. In what follows, the lost measurement is replaced with zero and the standard deviation of the associated noise is set to an arbitrarily large value. Consider matrix $M_k \in \Re^{m \times m}$ defined as follows:

$$M_k(i,j) \;=\; 0 \quad \text{if } i \neq j \tag{6.103}$$

$$M_k(i,i) \;=\; 1 \quad \text{if the measurement is present} \tag{6.104}$$

$$M_k(i,i) \;=\; \lambda_i \gg 1 \quad \text{if the measurement is not received.} \tag{6.105}$$

Denote $\bar{R}_k$ the covariance matrix of the noise of the measurement vector when some elements of $y_k = [y_k^1, \cdots, y_k^i, \cdots, y_k^m]^T$ may be lost. $\bar{R}_k$ can be expressed with

$$\bar{R}_k = M_k R_k M_k. \tag{6.106}$$

Therefore, if y_k^i is not received, its associated noise variance becomes $\bar{\sigma}_i^2 = \lambda_i^2 \sigma_i^2$. The Kalman filter formulation can now be derived. Note that there is no change in the filter structure, in contrast to the work in [CAO 09] where the lines in matrix C corresponding to lost data are removed.

Initialization: initial state x_0, initial covariance matrix $P_0 = \lambda I$, where I is the identity matrix and λ is large.

Prediction: the prediction step is unchanged

$$\widehat{x}_{k+1/k} \;=\; \Phi_k \widehat{x}_{k/k} \tag{6.107}$$

$$P_{k+1/k} \;=\; \Phi_k P_{k/k} \Phi_k^T + Q_k. \tag{6.108}$$

Computation of the Kalman filter gain:

$$\bar{K}_{k+1} = P_{(k+1)/k}C_{k+1}^T[C_{k+1}P_{(k+1)/k}C_{k+1}^T + \bar{R}_{k+1}]^{-1} \qquad (6.109)$$

Correction:

$$\widehat{x}_{k+1/k+1} = \widehat{x}_{k+1/k} + \bar{K}_{k+1}(y_{k+1} - C_{k+1}\widehat{x}_{k+1/k}) \qquad (6.110)$$

$$P_{k+1/k+1} = P_{k+1/k} - \bar{K}_{k+1}C_{k+1}P_{k+1/k}. \qquad (6.111)$$

Equation (6.109) can be reformulated using $\bar{C}_{k+1} = T_{k+1}C_{k+1}$, where $T_{k+1} = M_{k+1}^{-1}$:

$$\bar{K}_{k+1} = K_{k+1}T_{k+1}. \qquad (6.112)$$

Therefore,

$$P_{k+1/k+1} = P_{k+1/k} - P_{k+1/k}\bar{C}_{k+1}^T[\bar{C}_{k+1}P_{k+1/k}\bar{C}_{k+1}^T + R_{k+1}]^{-1}\bar{C}_{k+1}P_{k+1/k}. \qquad (6.113)$$

Note that the Kalman filter is prone to serious numerical difficulties that are well documented [HAY 01; VER 07]. For instance, the theoretical properties of matrix P (symmetric definite positive) might be lost during the computation performed using (6.101), (6.111) or (6.113). To numerically ensure the theoretical properties of P, a square-root version of the algorithm must be implemented. Consider $Q_k^{1/2}$, $R_k^{1/2}$, and $P_{k/k}^{1/2}$ are the square root factorization of Q_k, R_k, and $P_{k/k}$, respectively,

$$Q_k = (Q_k^{1/2})^T Q_k^{1/2}, \quad R_k = (R_k^{1/2})^T R_k^{1/2}, \quad P_{k/k} = (P_{k/k}^{1/2})^T P_{k/k}^{1/2}. \qquad (6.114)$$

The factorized version of the Kalman filter given by equations (6.107)–(6.111) is now derived.

Initialization: initial state x_0, initial covariance matrix $P_0 = (\sqrt{\lambda})^2 I$, where I is the identity matrix and λ is large.

Prediction:

$$\widehat{x}_{k+1/k} = \Phi_k\widehat{x}_{k/k} \qquad (6.115)$$

$P_{k+1/k}^{1/2}$ is computed thanks to a QR factorization

$$H\begin{pmatrix} Q_k^{1/2} \\ P_{k/k}^{1/2}\Phi_k^T \end{pmatrix} = \begin{pmatrix} P_{k+1/k}^{1/2} \\ 0 \end{pmatrix}. \qquad (6.116)$$

Computation of the Kalman filter gain: the Kalman gain is obtained with a second QR factorization

$$H\begin{pmatrix} R_{k+1}^{1/2}M_{k+1} & 0 \\ P_{k+1/k}^{1/2}C_{k+1}^T & P_{k+1/k}^{1/2} \end{pmatrix} = \begin{pmatrix} U & Z \\ 0 & P_{k+1/k+1}^{1/2} \end{pmatrix} \quad U\bar{K}_{k+1}^T = Z. \qquad (6.117)$$

Correction:

$$\widehat{x}_{k+1/k+1} = \widehat{x}_{k+1/k} + \bar{K}_{k+1}(y_{k+1} - C_{k+1}\widehat{x}_{k+1/k}). \qquad (6.118)$$

Note that the QR factorization may be partial in (6.117), the aim being to obtain a zero matrix under matrix U. This partial factorization decreases the computational cost.

In the case a filter bank is implemented for diagnostic purpose, the Kalman filter with partial loss as proposed in equations (6.115)–(6.118) can be applied in order to take into account the loss of data.

6.6. Conclusion and perspectives

This chapter has reviewed various concepts that are encountered in FDI and in FTC. The concept of an NCS was then introduced and two drawbacks induced by the network reviewed, namely, packet dropouts, and packet delays. Several approaches to deal with FDI and FTC in the presence of network drawbacks were exposed. These methodologies can be split in pragmatic solutions and advanced techniques.

In the first set of solutions, the problem of data synchronization was presented together with two solutions (clock synchronization, data reconstruction). Moreover, it was shown that packet dropouts imply false alarms. To avoid such false alarms, an indicator was proposed to detect that data had not been received by the diagnostic algorithm, and then freezing the diagnosis.

For advanced techniques, the problem of residual generation in the presence of packet delays induced by the network was first tackled. An adaptive Kalman filter that takes account such drawbacks was presented. Adaptive threshold techniques were then considered. The first one makes use of an optimization approach while the second one is based on network calculus. The last part of this chapter dealt with the problem of packet dropouts. The first solution proposed was the design of a FIF that is insensitive to packet dropouts. The second approach implemented a Kalman filter that is adaptive to data partial loss: the Kalman filter structure remained unchanged and the missing data was replaced with 0, assuming that the amount of lost data was arbitrarily low.

Several techniques presented in this chapter will be exemplified in the next chapter.

6.7. Bibliography

[AUB 93] AUBRUN C., SAUTER D., NOURA H., AND ROBERT M., Fault diagnosis and re-configuration of systems using fuzzy logic: application to a thermal plant, *International Journal of System Sciences*, vol. 24, num. 10, p. 1945–1954, 1993.

[BAN 99] BANDA S., Special issue on reconfigurable flight control, *International Journal of Robust and Nonlinear Control*, vol. 9, num. 14, 1999.

[BAS 93] BASSEVILLE M., AND NIKIFOROV I., *Detection of abrupt changes – Theory and applications*, Prentice Hall, Englewood Cliffs, NJ, Information and System Sciences Series, 1993.

[BER 07] BERBRA C., GENTIL S., LESECQ S., AND THIRIET J.-M., Co-design for a safe networked control DC motor, *3rd IFAC Workshop on networked control systems tolerant to faults Necst*, Nancy, France, June 2007.

[BLA 03] BLANKE M., KINNAERT M., LUNZE J., AND STAROSWIECKI M., *Diagnosis and Fault Tolerant Control*, Springer-Verlag, Berlin, 2003.

[BLA 06] BLANKE M., KINNAERT M., LUNZE J., STAROSWIECKI M., AND SHRODER J., *Diagnosis and Fault-Tolerant Control*, Springer-Verlag, New York, 2006.

[BOD 97] BODSON M., AND GROSZKIEWICZ J., Multivariable adaptive algorithms for reconfigurable flight control, *IEEE Transactions on Control System Technology*, vol. 5, num. 2, p. 217–229, 1997.

[CAO 09] CAO X., CHEN J., GAO C., AND SUN Y., An optimal control method for applications using wireless sensor/actuators networks, *Computer and Electrical Engineering*, vol. 35, num. 5, p. 748–756, 2009.

[CAS 05] CASAVOLA A., FAMULARO D., AND FRANZE G., Deconvolution Scheme for Fault Detection and Isolation of Uncertain Linear Systems – An LMI approach, *Automatica*, vol. 41, num. 8, p. 1463–1472, 2005.

[CHE 98] CHEN J., PATTON R., AND CHEN Z., An LMI approach to fault tolerant control of uncertain systems, *IEEE ISIC/CIRA/ISAS Joint Conference*, Gaithersburg, USA, p. 175–180, November 1998.

[CRU 91] CRUZ R., A calculus for network delay: Part 1: network elements in isolation, *IEEE Transactions on Information Theory*, vol. 37, p. 114–141, 1991.

[EPS 08] EPSTEIN M., SHI L., TIWARI A., AND MURRAY R., Probabilistic performances of state estimation across a lossy network, *Automatica*, vol. 44, p. 3046–3053, 2008.

[GAO 91] GAO Z., AND ANTSAKLIS P., Stability of the pseudo-inverse method for reconfigurable control, *International Journal of Control*, vol. 53, num. 3, p. 717–729, 1991.

[GAO 92] GAO Z., AND ANTSAKLIS P., Reconfigurable control system design via perfect model following, *International Journal of Control*, vol. 56, num. 4, p. 783–798, 1992.

[GEO 05] GEORGES J.-P., DIVOUX T., AND RONDEAU E., Confronting the performances of a switched ethernet network with industrial constraints by using the Network Calculus, *International Journal of Communication Systems*, vol. 18, num. 9, p. 877–903, 2005.

[GER 98] GERTLER J., *Fault Detection and Diagnosis in Engineering Systems*, Marcel Dekker, New York, 1998.

[HAY 01] HAYKIN S., Kalman Filters, *Kalman Filtering and Neural Networks*, John Wiley & Sons Inc., p. 1–22, 2001.

[HES 07] HESPANHA J., NAGHSHTABRIZI P., AND XU Y., A survey of recent results in networked control systems, *Proceedings of the IEEE*, vol. 95, num. 1, p. 138–162, 2007.

[HUA 07] HUANG M., AND DEY S., Stability of Kalman filtering with Markovian packet losses, *Automatica*, vol. 43, num. 4, p. 598–607, 2007.

[ISE 06] ISERMANN R., *Fault diagnosis systems: An Introduction from Fault Detection to Fault Tolerance*, Springer-Verlag, Berlin, 2006.

[JIA 98] JIANG J., AND ZHAO Q., Fault tolerant control systems synthesis using imprecise fault identification and reconfigurable control, *IEEE ISIC/CIRA/ISAS Joint Conference*, Gaithersburg, USA, p. 169–174, November 1998.

[JOS 87] JOSHI S., Design of failure accommodating multi loop LQG type controllers, *IEEE Transactions on Automatic Control*, vol. 32, num. 8, p. 740–741, 1987.

[KWO 95] KWONG W., PASSINO K., LAUKONEN E., AND YURKOVITCH S., Expert supervision of fuzzy learning systems for fault tolerant aircraft control, *Proceedings of the IEEE*, vol. 83, num. 3, 1995.

[LAP 92] LAPRIE J.-C., Ed., *Dependability: Basic Concepts and Terminology*, Springer-Verlag, Berlin, 1992.

[LOO 85] LOOZE D., WEISS J., ETERNO J., AND BARETT N., An automatic redesign approach for restructurable control systems, *IEEE Control System Magazine*, vol. 5, num. 2, p. 16–22, 1985.

[MAH 03] MAHMOUD M., JIANG J., AND ZHANG Y., Active fault tolerant control systems: stochastic analysis and synthesis, vol. 287 of *Lecture notes in control and information sciences*, Springer, Berlin, 2003.

[MAY 91] MAYBECK P., Application of multiple model adaptive algorithms to reconfigurable flight control, *IEEE Transactions on Aerospace and Electronic Systems*, vol. 27, num. 3, p. 470–480, 1991.

[MOR 90] MORSE W., AND OSSMAN K., Model following reconfigurable flight control system for the AFTI/F1-16, *Journal of Guidance*, vol. 13, num. 6, p. 969–976, 1990.

[MUR 96] MURAD G., POSTHETHWAITE I., AND GU D., A robust design approach to integrated controls and diagnostics, *13th IFAC World Congress*, San Francisco, CA, USA, p. 199–204, July 1996.

[NET 88] NETT C., JACOBSON J., AND MILLER A., An integrated approach to control and diagnostics: the 4-parameter controller, *IEEE American Control Conference*, Atlanta, USA, p. 824–835, June 1988.

[NOU 00] NOURA H., SAUTER D., HAMELIN F., AND THEILLIOL D., Fault tolerant control of dynamic systems: application to a winding machine, *IEEE Control System Magazine*, vol. 5, p. 33–49, 2000.

[OCH 91] OCHI Y., AND KANAI K., Design of restructurable flight control systems using feedback linearization, *Journal of Guidance*, vol. 14, num. 5, p. 903–911, 1991.

[OHL 07] OHLIN M., HENRIKSSON D., AND CERVIN A., *TrueTime 1.5 – Reference Manual*, January 2007.

[PAT 97] PATTON R., Fault tolerant control: the 1997 situation, *3rd IFAC Safeprocess Conference*, vol. 2, Kingston upon Hull, UK, p. 1033–1055, August 1997.

[PAT 00] PATTON R., FRANK P., AND CLARK R., *Issues of Fault Diagnosis for Dynamic Systems*, Springer, London, 2000.

[RAU 95] RAUSCH H., Autonomous control reconfiguration, *IEEE Control System Magazine*, vol. 15, num. 6, p. 37–49, 1995.

[SAU 98] SAUTER D., HAMELIN F., AND NOURA H., Fault tolerant control in dynamic systems using convex optimization, *IEEE ISIC/CIRA/ISAS Joint Conference*, Gaithersburg, USA, p. 187–192, November 1998.

[SAU 06] SAUTER D., AND BOUKHOBZA T., Robustness against unknown networked induced delays of observer based, *6th IFAC Symposium Fault Detection and Safety of Technical Processes*, Beijing, China, p. 331–336, August 2006.

[SCH 07] SCHENATO L., SINOPOLI B., FRANCESCHETTI M., POOLLA K., AND SASTRI S., Fundations of control and estimation over lossy networks, *Proceeedings of the IEEE*, vol. 95, num. 1, p. 163–187, 2007.

[SHI 08] SHI L., JOHANSSON K., AND MURRAY R., Estimation over wireless sensor networks: tradeoff between communication, computation and estimation qualities, *17th IFAC World Congress*, Seoul, Korea, July 2008.

[SIN 04] SINOPOLI B., SCHENATO L., FRANCESCHETTI M., POOLLA K., JORDAN M., AND SASTRY S., Kalman filtering with intermittent observations, *IEEE Transactions on Automatic Control*, vol. 49, num. 9, p. 1453–1464, 2004.

[STA 08] STAROSWIECKI M., AND CAZAURANG F., Fault recovery by nominal trajectory tracking, *American Control Conference ACC'08*, Seattle, USA, p. 1070–1075, June 2008.

[STO 03] STOUSTRUP J., NIEMANN H., AND LA COUR HARBO A., Threshold functions for fault detection and isolation, *American Control Conference ACC'03*, vol. 2, Denver, USA, p. 1782–1787, June 2003.

[SUN 08] SUN S., XIE L., XIAO W., AND SOH Y., Optimal linear estimation for systems with multiple packet dropouts, *Automatica*, vol. 44, p. 1333–1342, 2008.

[THE 98] THEILLIOL D., NOURA H., AND SAUTER D., Evaluation of a fault-tolerant control design for actuator faults, *IEEE Conference on Decision and Control*, Tampa, USA, December 1998.

[TYL 94] TYLER M., AND MORARI M., Optimal and robust design of integrated control and diagnostic modules, *IEEE American Control Conference*, Baltimore, USA, p. 2060–2064, June 1994.

[VEI 92] VEILLETTE R., MEDANIC J., AND PERKINS W., Design of reliable control systems, *IEEE Transactions on Automatic Control*, vol. 37, num. 3, p. 290–304, 1992.

[VEN 03a] VENKATASUBRAMANIAN V., RENGASWAMY R., AND KAVURI S., A review of process fault detection and diagnosis. Part 1: quantitative model-based methods, *Computer and Chemical Engineering*, vol. 27, p. 293–311, 2003.

[VEN 03b] VENKATASUBRAMANIAN V., RENGASWAMY R., AND KAVURI S., A review of process fault detection and diagnosis. Part 2: qualitative models and search strategies, *Computers and Chemical Engineering*, vol. 27, p. 313–326, 2003.

[VEN 03c] VENKATASUBRAMANIAN V., RENGASWAMY R., KAVURI S., AND YIN K., A review of process fault detection and diagnosis. Part 3: process history based methods, *Computers and Chemical Engineering*, vol. 27, p. 327–346, 2003.

[VER 07] VERHAEGEN M., AND VERDULT V., *Filtering and System Identification*, Cambridge University Press, Cambridge, UK, 2007.

[WU 98] WU E., ZHANG Y., AND ZHOU K., Control effectiveness estimation using an adaptive Kalman estimator, *IEEE ISIC/CIRA/ISAS Joint Conference*, Gaithersburg, USA, p. 181–186, November 1998.

[XIE 08] XIE L., AND XIE L., Stabilizing sampled-data linear systems with Markovian packet losses and random sampling, *17th IFAC World Congress*, Seoul, Korea, July 2008.

[YE 04] YE H., WANG G., AND DING S., A new parity space approach for fault detection based on stationary wavelet transform, *IEEE Transactions on Automatic Control*, vol. 49, num. 2, p. 281–287, 2004.

[ZHA 98] ZHAO Q., AND JIANG J., Reliable state feedback control system design against actuator failures, *Automatica*, vol. 34, num. 10, p. 1267–1272, 1998.

Chapter 7

Implementation: Control and Diagnosis for an Unmanned Aerial Vehicle

7.1. Introduction

Embedded systems are gaining ground in high-tech industries such as the automotive industry or aeronautics. Most of these systems are network controlled, which raises new control and diagnostic research problems. Analyzing, prototyping, simulating, and guaranteeing the safety of these systems are very challenging topics. Models are needed for the mechatronic continuous system, for the discrete controllers and diagnosers, and for network behavior. Real-time properties (task response times) and the network Quality of Service (QoS) influence the controlled system properties (Quality of Control, QoC).

Miniature unmanned aerial vehicles (UAVs) present unique challenges compared with most robotic applications. While fixed-wing vehicles have extensive applications for military and meteorological purposes due to their range, speed and flight duration, there is a distinct preference for rotor-craft vehicles in indoor and outdoor civilian applications. Thanks to their hover capability, they tend to be useful for many exploration missions such as video supervision of road traffic, surveillance of urban districts, forest fire detection or building inspection. Compared to aircrafts, they represent lightweight, low-cost systems with sensors and actuators that are often subject to faults (offsets, drifts). Consequently, their maintenance is essential, and, at times, challenging. Moreover, even in the presence of faults, the system must still be able to exhibit given properties, such as stability or precision. Hardware redundancy is not

Chapter written by Cédric BERBRA, Sylviane GENTIL, Suzanne LESECQ and Daniel SIMON.

Figure 7.1. *The quadrotor*

conceivable, not only due to cost but also due to constraints on weight, energy, and space. Therefore, a diagnostic module together with a fault-tolerant control (FTC) must be embedded in the UAV to guarantee their properties.

In this chapter, a *quadrotor* is taken as an example of an embedded system whose safety is critical. It consists of a mini helicopter, the four blades of which are actuated by four identical DC motors. It is equipped with an inertial measurement unit (IMU) for system attitude positioning (Figure 7.1). The quadrotor is a partially redundant structure, excellent for applying FDI methods, as well as designing FTC schemes.

Safety critical real-time systems must obey stringent constraints on resource usage such as memory, processing power, and communication, which must all be verified during the design stage. A hard real-time approach is most generally used because it guarantees that all timing constraints are verified. However, it generally results in oversizing the computing power, which is not always compatible with embedded applications. Thus, the soft real-time approach is preferred, since it tolerates occasionally missed deadlines. Interactions between the computing system and the control system must be carefully observed whenever the soft real-time approach is used.

The experiment presented in this chapter aims at illustrating network controlled systems, and is an intermediate platform that precedes the definitive one, and is designed step by step. Each of these steps will be described in succession. The quadrotor will first be classically controlled and diagnosed and the interactions between the controller, the diagnoser, and the system will be studied with Matlab/Simulink, which are standard tools for simulating a controlled physical system (section 7.2). Then, the simulated quadrotor will be controlled and diagnosed through a network, the effects of which will be simulated with the TrueTime toolbox (section 7.3). The next step will be to implement the so-called *hardware-in-the-loop* test bench. In this case, the system will still be numerically simulated, but the control and diagnostic algorithms will be run on the embedded target as real-time tasks, and they will communicate with the numerical simulator through a real network. This design will be made easy thanks to a tool named ORCCAD, which allows for the validation of the proposed real-time architecture (section 7.4). Finally, a number of experiments will be presented to validate some of the new ideas proposed in this book (section 7.5).

7.2. The quadrotor model, control and diagnosis

7.2.1. *The system*

Figure 7.2 schematically depicts the movement of the quadrotor, regarded as a composition of two Planar Vertical Take-Off and Landing (PVTOL), the axes of which are orthogonal, allowing for a movement of six degrees of freedom. Two frames are considered: the *inertial frame* $\mathbf{R}(x_n, y_n, z_n)$ and the *body frame* $\mathbf{B}(x_b, y_b, z_b)$, which is attached to the quadrotor with its origin at the quadrotor's center of mass. In the following paragraphs, the model is focused on the quadrotor's orientation (also named attitude) represented by three angles: yaw, pitch, and roll (ψ, θ, ϕ). Given that the front and rear motors rotate counter-clockwise, while the other ones rotate clockwise, gyroscopic effects and aerodynamic torques tend to be canceled. The throttle input is the sum of the thrusts of each rotor ($f_1 + f_2 + f_3 + f_4$). The pitch movement (θ) is obtained by increasing (or reducing) the velocity of the rear motor while reducing (or increasing) the velocity of the front motor. The roll movement (ϕ) is obtained similarly using the lateral motors. The yaw movement (ψ) is obtained by increasing (or decreasing) the velocity of the front and rear motors while decreasing (or increasing) the velocity of the lateral motors. This can be done while keeping the total thrust T constant, which must satisfy $T \geq mg$, with g representing the earth's gravity.

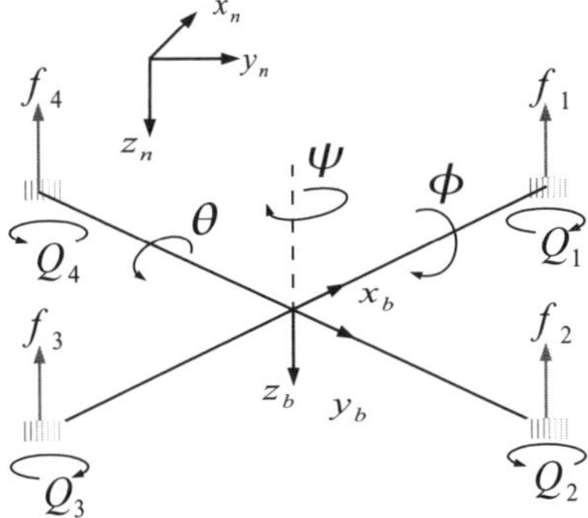

Figure 7.2. *Schematic representation of the forces and frames*

MEMS sensors used in the IMU may suffer several types of faults, such as bias, gain modification, or even loss of information delivered by the sensor. These faults must be detected and isolated early on to guarantee the system's safety. The same applies to the actuators: the malfunctioning of one of the motors must be detected as quickly as possible. The behavior of the quadrotor, subject to different kinds of faults (sensor or actuator faults), has been extensively studied to provide useful information for the design of the diagnosis and FTC [TAN 07b].

7.2.2. *The physical system model*

7.2.2.1. *Introduction to quaternions*

The orientation (attitude) is represented in the computations used for control, observation and diagnosis by a unitary quaternion: $q = [q_0 \ \vec{q}^T]^T$, $\|q\|_2 = 1$ [CHO 92]. A rotation can be represented by a unit vector, $\vec{e}$, known as the Euler axis, and a rotation angle β around this axis. The quaternion q is then defined as

$$q = \begin{pmatrix} \cos\frac{\beta}{2} \\ \vec{e}\sin\frac{\beta}{2} \end{pmatrix} = \begin{pmatrix} q_0 \\ \vec{q} \end{pmatrix} \in \mathbb{H}, \tag{7.1}$$

where

$$\mathbb{H} = \{q \mid q = [q_0 \ \vec{q}]^T, q_0 \in \Re, \vec{q} \in \Re^3, q_0^2 + \vec{q}^T \vec{q} = 1\}. \tag{7.2}$$

$\vec{q} = [q_1 \ q_2 \ q_3]^T$ and q_0 are, respectively, known as the vector and scalar parts of the quaternion.

In attitude control applications, the unitary quaternion can be used to represent the rotation from an inertial coordinate frame $\mathbf{R}(x_n, y_n, z_n)$ located at some point in the space (for instance, the earth North–East-Down frame, NED), to the body coordinate frame $\mathbf{B}$. A coordinate change from $r \in \Re^3$ in the reference frame to $b \in \Re^3$ in the body frame is expressed with the rotation matrix $C(q)$ by

$$b \ = \ C(q)r \tag{7.3}$$

$$C(q) \ = \ (q_0^2 - \vec{q}^T \vec{q})I_3 + 2(\vec{q} \, \vec{q}^T - q_0[\vec{q}^\times]) \tag{7.4}$$

$$C(q) \ = \ \begin{pmatrix} 2(q_0^2 + q_1^2) - 1 & 2(q_1 q_2 + q_0 q_3) & 2(q_1 q_3 - q_0 q_2) \\ 2(q_1 q_2 - q_0 q_3) & 2(q_0^2 + q_2^2) - 1 & 2(q_0 q_1 + q_2 q_3) \\ 2(q_0 q_2 + q_1 q_3) & 2(q_2 q_3 - q_0 q_1) & 2(q_0^2 + q_3^2) - 1 \end{pmatrix} \tag{7.5}$$

with

$$[\vec{q}^\times] = \begin{bmatrix} 0 & -q_3 & q_2 \\ q_3 & 0 & -q_1 \\ -q_2 & q_1 & 0 \end{bmatrix}, \tag{7.6}$$

where q is an efficient attitude representation from a computational point of view, which is of great importance to embedded systems. The main advantage of using q for the attitude representation is to avoid singularities that appear with classical angular representations (Euler angles or Cardan angles). Nevertheless, the figures that will be presented in this chapter will generally provide the angles $(\psi,\ \theta,\ \phi)$ that are physically easier to interpret. Changing q into the angles is a simple nonlinear transformation

$$\psi = \arctan\left(\frac{2(q_0 q_3 + q_1 q_2)}{1 - 2(q_2^2 + q_3^2)}\right) \tag{7.7}$$

$$\theta = \arcsin\left(2(q_0 q_2 - q_1 q_3)\right) \tag{7.8}$$

$$\phi = \arctan\left(\frac{2(q_0 q_1 + q_2 q_3)}{1 - 2(q_1^2 + q_2^2)}\right). \tag{7.9}$$

The mismatch between two attitudes is computed as an error quaternion. If q represents the current attitude quaternion and q_r is the reference quaternion, i.e. the desired orientation, then the error quaternion is given by

$$q_e = q \otimes q_r^{-1} = q \otimes \overline{q}_r. \tag{7.10}$$

Here $\otimes$ denotes the quaternion multiplication [CHO 92] and q^{-1} represents the complementary rotation of quaternion q, because the quaternion is unitary. $\overline{q} = [q_0 - \overrightarrow{q}^{\,T}]$ is the conjugate of q.

7.2.2.2. *The quadrotor model*

With $\omega \in \Re^3$ as the angular velocity of the quadrotor in the body frame $\mathbf{B}$, the dynamic equation for the quaternion is

$$\begin{pmatrix} \dot{q}_0 \\ \dot{\overrightarrow{q}} \end{pmatrix} = \tfrac{1}{2}\begin{pmatrix} -\overrightarrow{q}^{\,T} \\ I_3 q_0 + [\overrightarrow{q}^{\,\times}] \end{pmatrix}\omega \tag{7.11}$$

$$= \tfrac{1}{2}\Xi(q)\omega = \tfrac{1}{2}\Omega(\omega)q,$$

with I_3 being he identity matrix of size 3 and $\Omega(\omega)$ given by

$$\Omega(\omega) = \begin{pmatrix} 0 & -\omega^T \\ \omega & -[\omega^\times] \end{pmatrix}, \tag{7.12}$$

where $[\omega^\times]$ is the self-cross-product defined in equation (7.6). The rotational motion of the quadrotor is expressed by

$$I_f \dot{\omega} = -[\omega^\times] I_f \omega + \tau_a. \tag{7.13}$$

$I_f \in \Re^{3\times3}$ is the symmetric positive definite constant inertia matrix of the quadrotor with respect to frame $\mathbf{B}$. It is assumed to be diagonal

$$I_f = \mathrm{diag}(I_{f_x}, I_{f_y}, I_{f_z}). \tag{7.14}$$

ω_{Mi}^{*}	4-motor velocities set points calculated by the controller and sent to four local PI loops
ω_g	Quadrotor angular velocity measured in the body frame **B**
b_{acc}	Accelerometer measurements in the body frame **B**, $(i = 1, \ldots, 3)$
b_{mag}	Magnetometer measurements in the body frame **B**, $(i = 1, \ldots, 3)$
τ_a	Torque generated by the four rotors, used as control vector
q	Quaternion representing the drone attitude
ϕ, θ, ψ	Euler angles representing the drone attitude
$r_i, (i = 1, \ldots, 6)$	Residuals generated to detect an accelerometer or a magnetometer sensor fault.
$r_i, (i = 7, \ldots, 9)$	Residuals generated to detect a rate gyro sensor fault.

Table 7.1. *Notations*

The gyroscopic torques have been neglected. ω_{Mi} $(i = 1, \ldots, 4)$ are the four motors' velocities. The components of the torque $\tau_a \in \Re^3$ generated by the four rotors are given by

$$
\begin{aligned}
\tau_a^\phi &= d \cdot b \cdot (\omega_{M2}^2 - \omega_{M4}^2), \\
\tau_a^\theta &= d \cdot b \cdot (\omega_{M1}^2 - \omega_{M3}^2), \\
\tau_a^\psi &= k \cdot (\omega_{M1}^2 + \omega_{M3}^2 - \omega_{M2}^2 - \omega_{M4}^2),
\end{aligned}
\tag{7.15}
$$

where d is the distance from the rotors to the quadrotor's center of mass, b and k are two parameters depending on the air density, the radius, the shape, the pitch of the blade and other factors. Finally, the control vector is expressed based on the velocity of the motors

$$
\tau_a = \begin{pmatrix} 0 & db & 0 & -db \\ db & 0 & -db & 0 \\ k & -k & k & -k \end{pmatrix} \begin{pmatrix} \omega_{M1}^2 \\ \omega_{M2}^2 \\ \omega_{M3}^2 \\ \omega_{M4}^2 \end{pmatrix}.
\tag{7.16}
$$

Table 7.2 borrowed from [GUE 08] gives the values of the various parameters used for the Matlab/Simulink simulations presented in this chapter.

d	Distance between a rotor and the center of gravity	0.225 m
b	Parameter for the torque computation	$29.1\ 10^{-5}$
k	Parameter for the torque computation	$1.14\ 10^{-6}$
I_f	Quadrotor inertia matrix	$\mathrm{diag}\{8.28, 8.28, 5.7\}10^{-3}\ \mathrm{kg\ m^2}$
m	Quadrotor mass	0.520 kg
$\vec{\eta}_a$	Accelerometer noise	0.002 (normalized value)
$\vec{\eta}_M$	Magnetometer noise	0.0007 (normalized value)
$\vec{\eta}_G$	Gyrometer noise	$0.01\ \mathrm{rad\ s^{-1}}$

Table 7.2. *Quadrotor parameter values*

7.2.2.3. *The inertial measurement unit (IMU) model*

The estimation of the attitude and of the rotational velocity of the quadrotor is a prerequisite for its attitude control. An IMU is embedded in the quadrotor in order to provide measurements that will be fused so as to estimate the attitude. The IMU consists of three rate gyros $(g1, g2, g3)$, a tri-axis accelerometer $(a1, a2, a3)$, and a tri-axis magnetometer $(m1, m2, m3)$.

1) *Rate gyro modeling.* The angular velocity ω is measured in the body frame $\mathbf{B}$ with the three rate gyros $(g1, g2, g3)$ mounted at right angles. The measurements ω_g delivered by these sensors are usually affected by noise. Theoretically, the integral of ω could give the relative orientation but the presence of noise generates errors that are accumulated over time. The sensor measurements are modeled as

$$\omega_g = \omega + \eta_g, \tag{7.17}$$

where $\omega_g \in \Re^3$ is the vector of the sensor values and $\eta_g \in \Re^3$ is assumed to be zero-mean Gaussian white noise.

2) *Accelerometer modeling.* The tri-axis accelerometer $(a1, a2, a3)$ senses the inertial forces and the gravity in $\mathbf{B}$ ($a1$ is aligned with axis x_b, and so on). The transformation of accelerometer measurements from the inertial frame $\mathbf{R}$ to the body frame $\mathbf{B}$ is computed as follows:

$$b_{\mathrm{acc}} = C(q)(\dot{v} - g) + \eta_{\mathrm{acc},} \tag{7.18}$$

where $b_{\mathrm{acc}} \in \Re^3$ corresponds to the measurements in $\mathbf{B}$ and $\eta_{\mathrm{acc}} \in \Re^3$ is zero-mean Gaussian white noise. From here on, the motion is assumed to be quasi-static so that linear accelerations $\dot{v}$ are neglected. Note that this assumption is fully valid because the quadrotor is controlled so as to obtain hover conditions $(\phi \approx \theta \approx \psi \approx 0)$. Moreover, the Coriolis effect is not taken into account. In this way, accelerometers are only sensitive to the gravitational field $g = [0 \ \ 0 \ \ 9.81]^T$ ms^{-2}. Note that the measurements are normalized. Therefore, $\|b_{\mathrm{acc}}\|_2 \approx \|g\|_2 = 1$.

3) *Magnetometer modeling.* The information provided by the tri-axis magnetometer is added to the inertial measurements. b_{mag1} (respectively b_{mag2}, b_{mag3}) is the measurement along axis x_b (respectively y_b, z_b). The earth's magnetic field is sensed in $\mathbf{B}$. It is defined by

$$b_{\mathrm{mag}} = C(q)h_m + \eta_{\mathrm{mag}}, \tag{7.19}$$

where $h_m = [h_{mx} \ 0 \ h_{mz}]^T = [\frac{1}{2} \ 0 \ \frac{\sqrt{3}}{2}]^T$ and $b_{\mathrm{mag}} \in \Re^3$ are the three components of the magnetic field in $\mathbf{R}$ and $\mathbf{B}$, respectively, and $\eta_{\mathrm{mag}} \in \Re^3$ is zero-mean Gaussian white noise.

Remark. Note that the accelerometer and magnetometer measurements are modeled by *static nonlinear equations* that depend on constant known vectors g and h_m and on matrix $C(q)$, which is a nonlinear function of q.

7.2.3. *The attitude control*

The control and observation of the quadrotor have been extensively studied. A nonlinear control law [GUE 08] has been designed, combined with a nonlinear observer to perform nonlinear control of the quadrotor. A linear approximation of the model has also been obtained: it has been used to design a linear quadratic (LQ) control and an extended Kalman filter (EKF).

7.2.3.1. *Nonlinear control*

The measured rotational velocity ω_g and the quaternion estimation $\widehat{q}$ are used in a feedback loop to compute a bounded attitude control, from which the required velocity of each motor is deduced (7.16). The attitude reference is given by a quaternion reference q_{ref} (Figure 7.3). The controller implemented for the attitude stabilization is detailed in [GUE 08] and supplies for each motor the velocity set point ω^*_{Mi}. This velocity set point feeds a local PI control loop. The nonlinear controller takes into account the saturation of the control signals, which provides a good robustness against disturbances, as can be observed in Figure 7.5.

7.2.3.2. *Linear quadratic control*

In this section, a linearized model is obtained for the quadrotor. The attitude representation makes use of angles, leading to a very simple model. This linear approximation is employed to develop a simple LQ controller [GUE 09]. The kinematic and dynamic equations of the quadrotor are given by

$$(\dot{\phi}, \dot{\theta}, \dot{\psi})^T = M\omega. \tag{7.20}$$

Matrix M is given by

$$M = \begin{pmatrix} 1 & \tan\theta\sin\phi & \tan\theta\cos\phi \\ 0 & \cos\phi & -\sin\phi \\ 0 & \frac{\sin\phi}{\cos\theta} & \frac{\cos\phi}{\cos\theta} \end{pmatrix} \begin{pmatrix} \omega_x \\ \omega_y \\ \omega_z \end{pmatrix}. \tag{7.21}$$

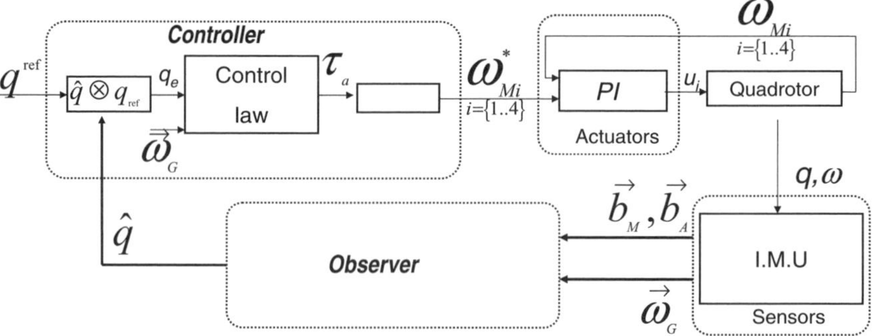

Figure 7.3. *Quadrotor control principle*

In the hover condition ($\phi \approx \theta \approx \psi \approx 0$), equation (7.20) is approximated to

$$(\dot{\phi}, \dot{\theta}, \dot{\psi})^T = (\omega_x, \omega_y, \omega_z)^T. \tag{7.22}$$

The dynamical model in terms of roll, pitch, and yaw angles is derived from (7.13), (7.20), and (7.22):

$$\ddot{\phi} = \dot{\theta}\dot{\psi}\left(\frac{I_{f_y} - I_{f_z}}{I_{f_x}}\right) + \frac{1}{I_{f_x}}\tau_a^\phi$$

$$\ddot{\theta} = \dot{\phi}\dot{\psi}\left(\frac{I_{f_z} - I_{f_x}}{I_{f_y}}\right) + \frac{1}{I_{f_y}}\tau_a^\theta \tag{7.23}$$

$$\ddot{\psi} = \dot{\phi}\dot{\theta}\left(\frac{I_{f_x} - I_{f_y}}{I_{f_z}}\right) + \frac{1}{I_{f_z}}\tau_a^\psi$$

The state variables are defined by

$$x^T = (x_1 \; x_2 \; x_3 \; x_4 \; x_5 \; x_6)^T = (\phi \; \dot{\phi} \; \theta \; \dot{\theta} \; \psi \; \dot{\psi})^T. \tag{7.24}$$

The linearization of system (7.23) around the hover condition yields

$$\dot{x} = Ax + Bu, \tag{7.25}$$

where

$$A = \begin{pmatrix} 0 & 1 & 0 & 0 & 0 & 0 \\ 0 & 0 & 0 & 0 & 0 & 0 \\ 0 & 0 & 0 & 1 & 0 & 0 \\ 0 & 0 & 0 & 0 & 0 & 0 \\ 0 & 0 & 0 & 0 & 0 & 1 \\ 0 & 0 & 0 & 0 & 0 & 0 \end{pmatrix} \quad B = \begin{pmatrix} 0 & 0 & 0 \\ \frac{1}{I_{f_x}} & 0 & 0 \\ 0 & 0 & 0 \\ 0 & \frac{1}{I_{f_y}} & 0 \\ 0 & 0 & 0 \\ 0 & 0 & \frac{1}{I_{f_z}} \end{pmatrix}. \tag{7.26}$$

Note that the linearized model (7.26) consists of three decoupled double integrators. The discretization of (7.25), with sampling period h, leads to

$$x_{k+1} = \Phi x_k + \Gamma u_k, \tag{7.27}$$

where

$$\begin{pmatrix} \Phi & \Gamma \\ 0 & I \end{pmatrix} = \exp\left\{\begin{pmatrix} A & B \\ 0 & 0 \end{pmatrix} h\right\}. \tag{7.28}$$

The problem to solve here is the stabilization of the quadrotor in a desired constant orientation. As a consequence, the angular velocity vector is brought to zero and must remain null once the desired attitude is reached. The three angle references are set to zero (hover condition), thus

$$\lim_{t \to \infty} x(t) = 0. \tag{7.29}$$

The discrete LQ controller minimizes

$$J = \sum_{k=0}^{N-1} [x_k^T Q x_k + u_k^T R u_k] + x_N^T Q_0 x_N. \tag{7.30}$$

The solution to (7.30) is a control that is linear in the state

$$u_k = -L x_k. \tag{7.31}$$

Matrices Q, R, and Q_0 are symmetric definite positive. They express the weight that is put either on the state error or on the input energy. They are chosen in order to obtain a suitable transient response, with feasible control signals.

For a sampling period $h = 10$ ms, the optimal gain matrix is given by

$$L = \begin{pmatrix} 0.0347 & 0.0280 & 0 & 0 & 0 & 0 \\ 0 & 0 & 0.0347 & 0.0280 & 0 & 0 \\ 0 & 0 & 0 & 0 & 0.0347 & 0.0280 \end{pmatrix}. \tag{7.32}$$

7.2.4. *The attitude observer*

7.2.4.1. *Nonlinear observer*

The measurements provided by the IMU feed a nonlinear observer, with its output $\hat{q}$ used by the controller. The observer proposed in [GUE 08] is depicted in Figure 7.4.

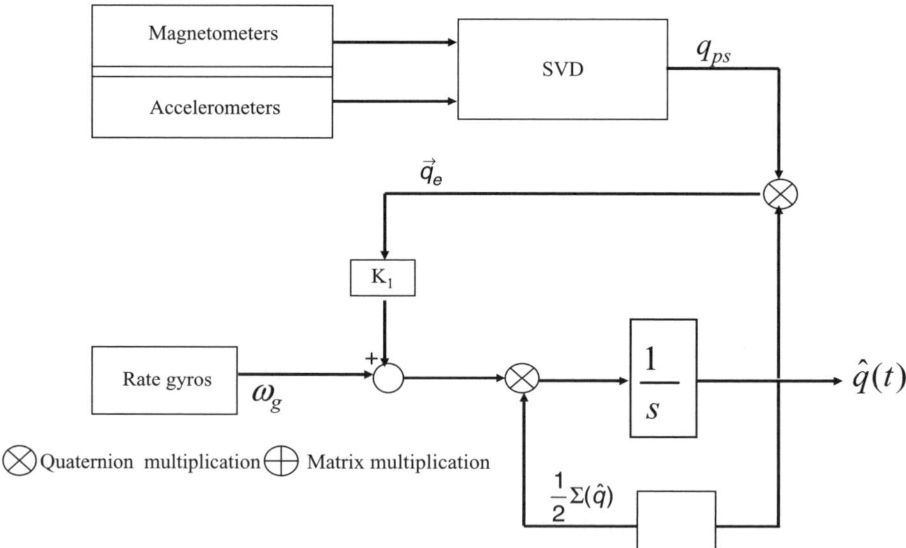

Figure 7.4. *Nonlinear observer scheme*

The observer principle is as follows. A "pseudomeasurement" quaternion q_{ps} is computed from b_{mag} and b_{acc}, based on the nonlinear static measurement equations (7.18) and (7.19)

$$q_{ps} = \arg\min \left[\frac{1}{2} \parallel [b_{\mathrm{acc}}^T \ b_{\mathrm{mag}}^T]^T - h(q) \parallel_2^2 \right] \quad \text{subject to } \parallel q_{ps} \parallel_2 = 1, \qquad (7.33)$$

where $h(q)$ is derived from (7.18) and (7.19). A Sequential Quadratic Programming (SQP) algorithm is used at this step (e.g. "fmincon" in the Matlab environment). The discrepancy between $\widehat{q}$ and q_{ps} is computed as (7.10)

$$q_e = \widehat{q} \otimes q_{ps}^{-1} = [q_{e_0} \ \vec{q}_e^{\,T}]^T, \qquad (7.34)$$

where $\widehat{q}$ is then obtained with the kinematic equation (7.11) using ω_g (7.17) and a correction term

$$\dot{\widehat{q}} = \frac{1}{2} \Xi(\widehat{q}) \left[\vec{\omega}_g + K_1 \vec{q}_e \right]. \qquad (7.35)$$

Thus, as usual, the observer has a prediction part based on the model, and a correction part based on a subset of measurements, which are those that are related to the state q by a static equation.

7.2.4.2. *Extended Kalman filter*

The Kalman filter (KF) is a well-known observer formulation that allows for the taking into account of noises that perturb the system's state and its measurements. The KF basic equations are similar to those of a traditional observer (see section 7.2.4 in chapter 6), with a prediction part based on the model and a correction part based on the measurements. The main difference is that the correction gain is not computed in order to provide "good" eigenvalues to the observer (and thus good dynamic properties): this is a function of the noise correlation matrices, in order to minimize their effect. The EKF is an extension to nonlinear systems and it makes use of the model linearization.

Consider the discrete time version of the quaternion attitude kinematic equation (7.11)

$$q_{k+1} = F_k q_k, \qquad (7.36)$$

where

$$F_k = \exp(\frac{1}{2}\Omega(\omega_k)h). \qquad (7.37)$$

$F_k \in \Re^{4 \times 4}$ is obtained under the assumption that ω_k remains constant over the sampling period h. Actually, ω_k is not directly known: it is measured (7.17), with the measurement being disturbed by noise η_{g_k}. Thus, matrix F_k becomes

$$F_k = F_k^o + \Delta F_k, \qquad (7.38)$$

with $F_k^o = \exp(\frac{1}{2}\Omega(\omega_{g_k})h)$. The error matrix ΔF_k can be expressed as a matrix power series depending on η_{g_k}. However, when the noise η_{g_k} and the sampling time h are small enough, ΔF_k can be approximated by

$$\Delta F_k \simeq \frac{1}{2}\Omega(\eta_{g_k})h, \tag{7.39}$$

where matrix $\Omega(\eta_{g_k})$ is defined as in (7.12). Therefore, (7.36) is rewritten as

$$\begin{aligned}
q_{k+1} &= F_k^o q_k + \frac{h}{2}\Omega(\eta_{g_k})q_k = F_k^o q_k + \frac{h}{2}\Sigma(q_k)\eta_{g_k} \\
&= F_k^o q_k + \eta_{q_k}, \tag{7.40}
\end{aligned}$$

with $\eta_{q_k} = G_k \eta_{g_k}$ and $G_k = \frac{h}{2}\Sigma(q_k)$. The detailed computation of the covariance matrix Q_k associated with noise η_{q_k} can be found in [LES 09].

The measurement equation $y_k \in \Re^6$ is built from the accelerometer (7.18) and magnetometer (7.19) measurements:

$$\begin{aligned}
y_k &= \begin{pmatrix} b_{\mathrm{acc}_k} \\ b_{\mathrm{mag}_k} \end{pmatrix} = \begin{pmatrix} C(q_k) & 0 \\ 0 & C(q_k) \end{pmatrix} \begin{pmatrix} g \\ h_m \end{pmatrix} + \eta_k \\
&= h(q_k) + \eta_k, \tag{7.41}
\end{aligned}$$

with $\eta_k = \left(\eta_{\mathrm{acc}_k}^T \; \eta_{\mathrm{mag}_k}^T \right)^T$. The noise covariance matrix R_k is directly computed with the measurement noise covariance matrices R_{acc} and R_{mag}:

$$R_k = E[\eta_k \eta_k^T] = \begin{pmatrix} R_{\mathrm{acc}_k} & 0 \\ 0 & R_{\mathrm{mag}_k} \end{pmatrix}. \tag{7.42}$$

Thus, the state and measurement equations are defined by (7.40) and (7.41), respectively. The linearization of the measurement equation (7.41) gives rise to matrix $C_k \in \Re^{6 \times 4}$ with

$$C_k = \left[\frac{\partial h(q_k)}{\partial q_k} \right]_{q_k = \widehat{q}_{k/k}}. \tag{7.43}$$

Its calculation is obvious and is not reported here.

The various steps of the EKF are now summarized.

– Initialization

$$\widehat{q}_{k=0} = E[q_{k=0}] \quad P_0 = E[q_{k=0}q_{k=0}^T]. \tag{7.44}$$

Usually, $q_{k=0}$ can take any value (with $\|q_{k=0}\|_2 = 1$) and P_0 is initialized to a diagonal matrix σI_4 where ($\sigma \gg 1$, expressing the very high imprecision on q_0.

– Prediction

$$\widehat{q}_{k+1/k} = F_k^o \widehat{q}_k \quad P_{k+1/k} = F_k^o P_{k/k} F_k^{oT} + Q_k. \tag{7.45}$$

– Computation of the EKF gain

$$K_k = P_{k/k-1} C_k^T \left[C_k P_{k/k-1} C_k^T + R_k \right]^{-1}, \tag{7.46}$$

R_k is given by (7.42) and C_k by (7.43).

– Correction

$$\begin{aligned}
\widehat{q}_{k+1/k+1} &= \widehat{q}_{k+1/k} + K_{k+1} \left(y_{k+1} - g(\widehat{q}_{k+1/k}) \right) \\
P_{k+1/k+1} &= P_{k+1/k} - P_{k+1/k} C_{k+1}^T \\
&\quad \times \left[C_{k+1} P_{k+1/k} C_{k+1}^T + R_{k+1} \right]^{-1} C_{k+1} P_{k+1/k}.
\end{aligned} \tag{7.47}$$

Note that the quaternion normalization $\widehat{q}_{k+1}/\|\widehat{q}_{k+1}\|_2$ must be performed at each step to ensure the estimation of a unitary quaternion.

7.2.4.3. *Simulation results*

Firstly, the simulation of the quadrotor with its control and its observer has been implemented with Matlab/Simulink. The quadrotor model is continuous while the control and observer are discretized with the sampling period $h = 10$ ms, which is very small compared to the system time response (order of magnitude of several seconds). Figure 7.5 corresponds to a scenario where the nonlinear control and nonlinear observer are used. The attitude is first stabilized to the equilibrium value $[\phi, \theta, \psi] = [0, 0, 0]°$ from an initial value $[30, -10, -25]°$. Then, a disturbance on one torque is added at time $t = 4$ s.

7.2.5. *The quadrotor diagnosis*

7.2.5.1. *Sensor diagnosis*

A bank of generalized observers has been designed for the quadrotor sensor diagnosis, based on the observer presented in the previous section ([TAN 07a; BER 08]). The background for generalized observer schemes is detailed in Chapter 6. Each observer uses a subset of the nine IMU sensors. Its estimated quaternion $\widehat{q}_i$ ($i = 1, \ldots, 9$) is compared to the quaternion computed with the theoretical model q_{model} fed by the four motor velocity references ω_{Mi}^*. The resulting error quaternion q_i^e (7.10) is thus insensitive to faults in the sensor that is discarded for the attitude estimation. The signature table obtained in this way is strongly isolable. Actually, q_i^e is not directly considered as residual. Thanks to (7.1), the residual $r_i = \beta_i$ is computed. If $r_i \approx 0$,

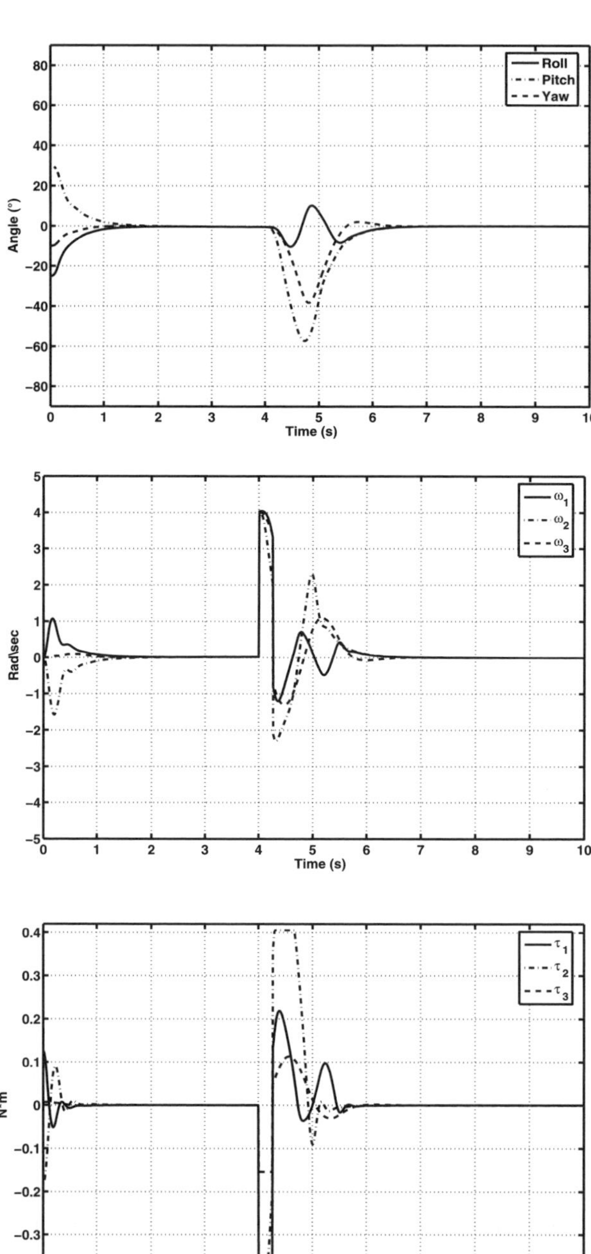

Figure 7.5. *Evolution of the angles, angular velocities and torques when the quadrotor is subject to a reference change at time* $t = 0$ *s and to a disturbance at time* $t = 4$ *s*

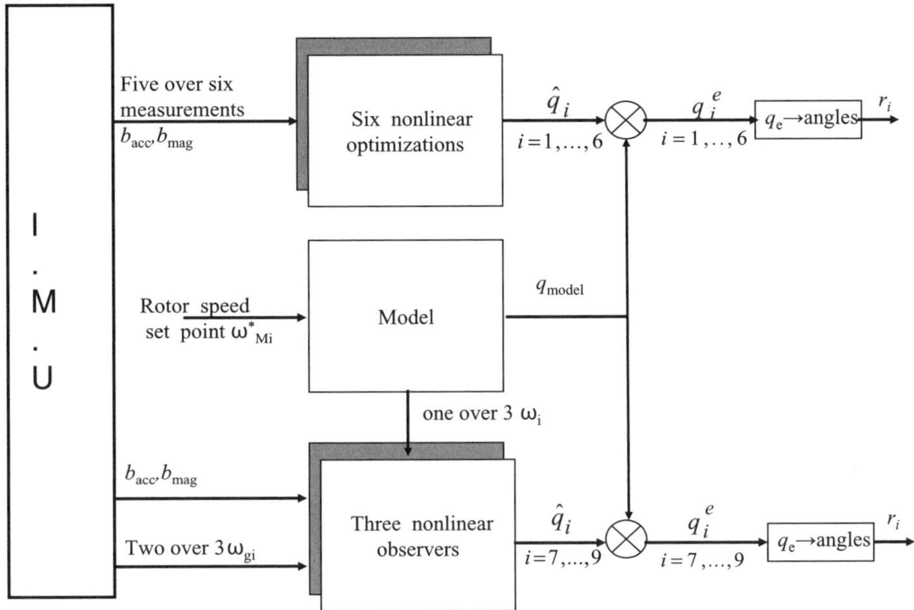

Figure 7.6. *Generalized observer scheme for the diagnosis of the IMU*

then $\hat{q}_i \approx q_{model}$. This transformation from q_i^e to r_i allows the detection thresholds to be easily fixed. From a practical viewpoint, the residual thresholds are set to $2°$ this value being deducted from the noise level.

For the design of this generalized observer bank, two different techniques have been implemented, depending on the subset of sensors considered. The first technique computes $\hat{q}_i$ using two rate gyros and all the magnetometer and accelerometer measurements. The second technique is only based on the static nonlinear measurement equations. It makes use of five out of six measurements acquired with the accelerometer and magnetometer tri-axes. The principle of the observer bank is depicted in Figure 7.6 and each technique is briefly presented below.

– Attitude estimation using the accelerometer and magnetometer tri-axes and two rate gyro measurements. As can be seen at the bottom of Figure 7.6, the measurements provided by the IMU feed three nonlinear observers based on the scheme presented in section 7.2.4, Figure 7.4. The discarded rate gyro measurement is replaced with the corresponding value that is computed with the mechanical model (7.13). Therefore, the attitude estimated with each observer in this bank is sensitive to faults in all the accelerometer and magnetometer measurements and in two rate gyros out of three. From these observers, three residual vectors, denoted r_i ($i = 7, \ldots, 9$), are obtained.

	f_{a1}	f_{a2}	f_{a3}	f_{m1}	f_{m2}	f_{m3}	f_{g1}	f_{g2}	f_{g3}
r_1	0	1	1	1	1	1	0	0	0
r_2	1	0	1	1	1	1	0	0	0
r_3	1	1	0	1	1	1	0	0	0
r_4	1	1	1	0	1	1	0	0	0
r_5	1	1	1	1	0	1	0	0	0
r_6	1	1	1	1	1	0	0	0	0
r_7	1	1	1	1	1	1	0	1	1
r_8	1	1	1	1	1	1	1	0	1
r_9	1	1	1	1	1	1	1	1	0

Table 7.3. *Sensor signature table*

– *Attitude estimation with measurements from the accelerometer and magnetometer tri-axes but without rate gyro measurements.* To estimate the residuals r_i, $(i = 1, \ldots, 6)$ as depicted at the top of Figure 7.6, only five out of six accelerometer and magnetometer measurements are considered. Therefore, five out of six equations in (7.18) and (7.19) are considered. The quaternion estimation $\widehat{q}_i$ based on these equations is obtained by solving a nonlinear optimization problem similar to the one in (7.33), but with five equations instead of six. Six estimators are designed in this way, and the quaternion errors q_i^e, $(i = 1, \ldots, 6)$ are sensitive to faults in all of the accelerometers and magnetometers except the discarded one.

The resulting signature table is given in Table 7.3, while Figures 7.7 and 7.8 display some diagnostic results. For instance, f_{a1} stands for faults in accelerometer $a1$ while r_1 is computed without measurement b_{acc1}. Thus r_1 is insensitive to faults in $a1$.

7.2.5.2. *Actuator diagnosis*

The four actuators are independent. The model of each electrical motor has been used for its diagnosis: the measured velocity is compared to its reference. The signature table is found in Table 7.4. A fault in the rotational velocity sensor or in the corresponding motor cannot be isolated. This limitation could be removed if extra sensors are added to measure the current or the voltage. However, this solution is incompatible with the weight constraint.

7.3. Simulation of the network

7.3.1. *Architecture of the networked control system*

The network is operated in a closed loop (Figure 7.9):

1) between the sensors and the controller: the sensors are associated with the numerical information provided by AD converters to generate the sensor flow (this is called sensor task);

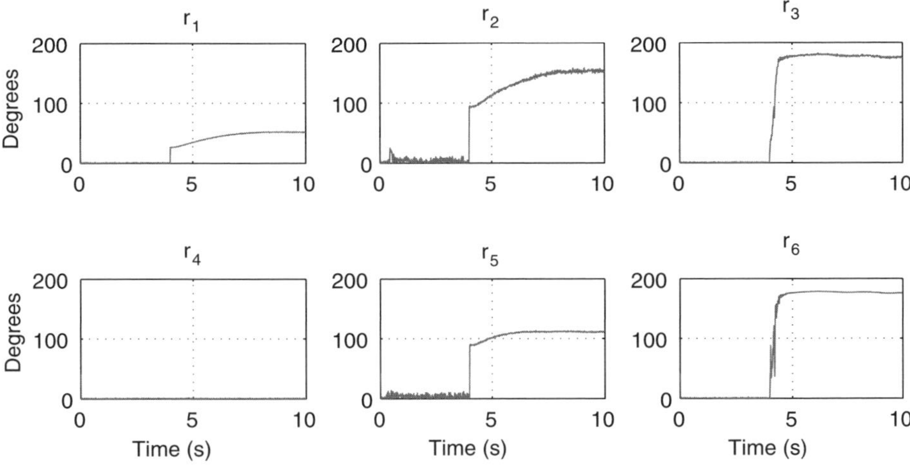

Figure 7.7. *Residuals generated by the six estimators when a breakdown f_{m1} in sensor $m1$ (along axis x_b) occurs at time $t = 4$ s*

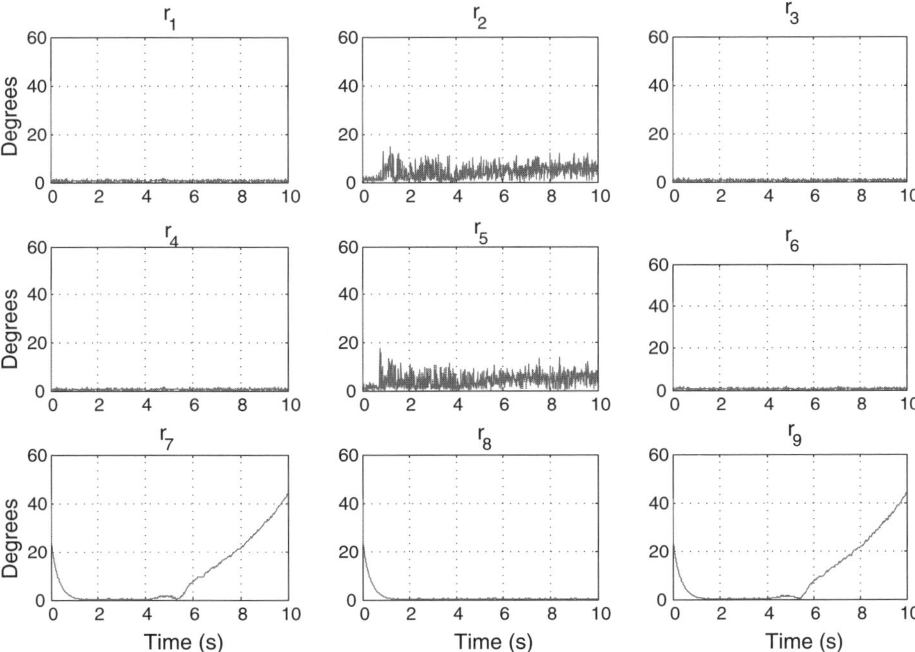

Figure 7.8. *Residuals generated by the nine observers when a breakdown f_{g2} in the rate gyro $g2$ (along axis y_b) occurs at time $t = 4$ s*

	f_{motor1}	f_{motor2}	f_{motor3}	f_{motor4}
r_{mot1}	1	0	0	0
r_{mot2}	0	1	0	0
r_{mot3}	0	0	1	0
r_{mot4}	0	0	0	1

Table 7.4. *Actuator signature table*

2) between the controller and the actuators (the four motors) driven through DA conversion (zero-order hold): the controller task generates the controller flow.

The model of the network has been integrated with the model of the quadrotor. This can be considered as a hybrid problematic since the model of the quadrotor is in continuous time; the control, observer and diagnostic algorithms are in discrete time, whereas the model of the network is event-driven.

7.3.2. *Network design*

A controller area network (CAN) has been chosen because it is a common choice for control applications. A CAN uses carrier sense multiple access with arbitration on message priority (CSMA-AMP) arbitration. If the network is busy, the sender waits until the network is free. If a collision occurs (two transmissions are being started within 1 ms), the message with the highest priority will continue its transmission.

The network in Figure 7.9 is characterized by the traffic of 17 periodic data flows. The order of priority of the flows is: the four flows from the main control unit ω^*_{Mi} ($i = 1, \ldots, 4$, set points for each local motor control), the nine flows from the IMU ω_{gi},

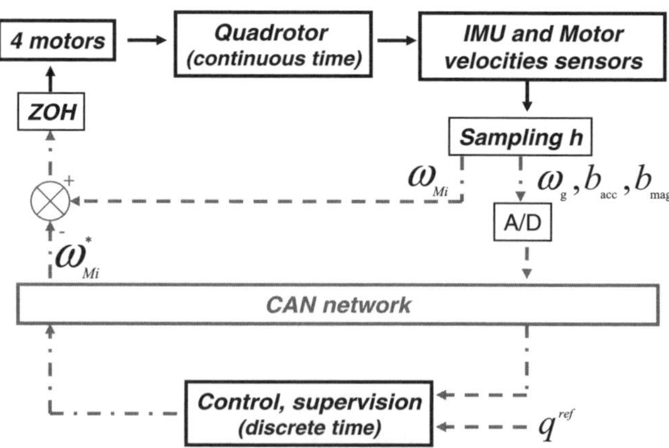

Figure 7.9. *Closed loop of the networked control quadrotor*

$b_{\mathrm{acc}i}$, $b_{\mathrm{mag}i}$, $(i = 1, \ldots, 3)$ and the four flows from the motor velocity measurements ω_{Mi} $(i = 1, \ldots, 4)$. The sensor task is time triggered. The sampling time is $h = 10$ ms. Note that the IMU data acquisition is synchronized. The data rate is 1 Mbits s^{-1}. The data length is 64 bits for all the periodic flows. Each flow is transmitted in 64 μs. Thus all the periodic flows are transmitted in 1 ms (17×64 μs), corresponding to 10% of the sampling period, which is negligible.

The observer and control algorithms have been discretized. The controller task and the diagnostic task are event triggered: they must wait for all the sensor values before any computation is performed [BER 07]. ω_{Mi}^{*} is sent through the network to the local motor control as soon as the controller has finished its computation.

7.3.3. *Tool implemented in the network simulation*

To go a step ahead in the design of this benchmark, the simulation of the network is introduced in the Matlab simulator, using the TrueTime toolbox [OHL 07]. TrueTime is a free, open, Matlab/Simulink compliant toolbox used to perform the simulation of distributed real-time control systems.

With TrueTime, the network is simulated at a high level of abstraction, in contrast to other tools such as SimEvent, a Matlab toolbox, or Opnet [Opn], which is not Matlab compliant but allows users to simulate the network in a more detailed way. The TrueTime library provides specific blocks for network interface modeling. This library is developed in C++ and compiled with an external C++ compiler. TrueTime allows users to simulate several types of networks (Ethernet, CAN, Round Robin, TDMA, FDMA, Switched Ethernet, and WLAN or ZigBee Wireless networks) and new models can be added.

The control performance can be severely affected by the properties of the network, and particularly by its load. For example, a data packet dropout can result from limited bandwidth and large amount of data packets transmitted over one line. To simulate theses properties, it is convenient to simulate a network that is not dedicated solely to the quadrotor but that is shared between different applications. To study the influence of the flows from other applications on the quadrotor stabilization, another task is introduced in the simulation. It is a periodic task with sampling period T_e. This task, called an external task, generates an external flow and represents overall the use of the network by other applications. Changing the value of T_e allows for the simulation of different network loads: the smaller the T_e, the more loaded the network is.

7.4. Hardware in the loop architecture

The utilization of Matlab/Simulink together with TrueTime remains pure simulation. To provide more realistic results on the influence of the network on the system,

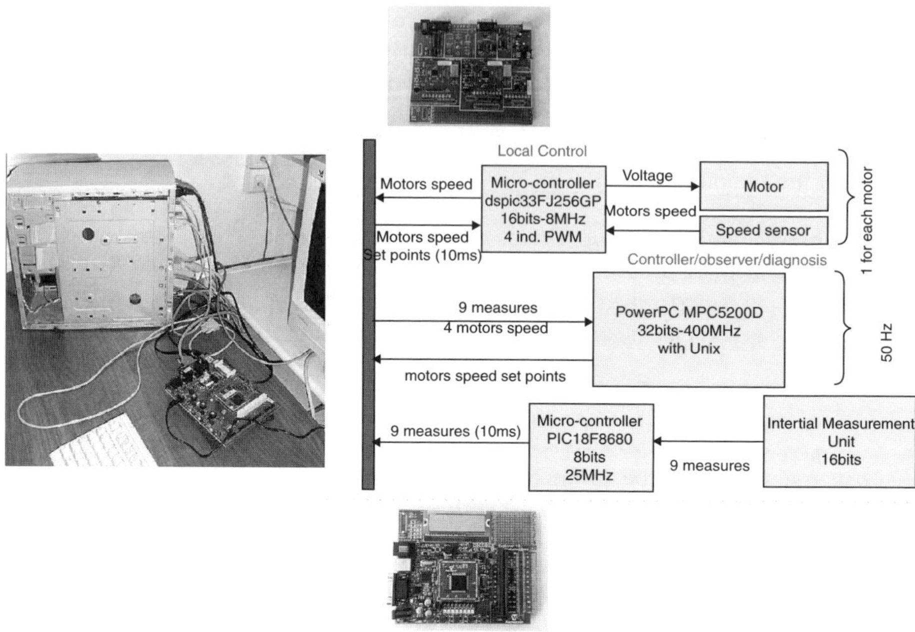

Figure 7.10. *Hardware-in-the-loop setup*

a hardware-in-the-loop experiment has been set up (Figure 7.10). Now, the mechanical behavior of the quadrotor is still simulated. The quadrotor model is handled by a numerical integrator running on an external PC running on Linux. This integration must be fast enough for the simulated model to run faster than the real-time model, and for the induced disturbances to be negligible w.r.t. computing and networking delays. The control, observation, and diagnostic algorithms are implemented in the real embedded hardware, a Phycore MPC5200B-tiny[1] embedded board running on Linux on a Freescale MPC603e CPU. Both computers communicate via a real CAN.

In the particular case of the quadrotor, hardware-in-the-loop experiments provide a safe environment for the validation of all algorithms and software, prior to any experiments with a real – and very fragile – quadrotor.

7.4.1. *The ORCCAD approach*

ORCCAD is a model-based software environment dedicated to the design, the verification and the implementation of real-time control systems. In addition to control

---

1. http://www.phytec.com/

law design, the specification and validation of complex missions involving the logical and temporal cooperation of various controllers along the life of a control application [BOR98; TÖR 06] can be achieved.

The ORCCAD methodology is bottom-up, starting from the design of control laws by control engineers, to the design of more complex missions.

The first step in designing a control application is to identify all the necessary elementary tasks involved. Then, for each of the tasks, various issues are considered, both with an automatic control viewpoint (such as regulation problem definition, control law design, choice of relevant events, specification of recovery behaviors, etc.) or with an implementation viewpoint (such as the decomposition of the control law into real-time tasks, and selection of timing parameters). Finally, all the real-time tasks are mapped on a target architecture. During this design, the control engineer may take advantage of many degrees of freedom to meet the end-user requirements, and ORCCAD aims at allowing the designer to safely exploit these degrees of freedom through guided design.

ORCCAD proposes a controller architecture which is naturally open, since the access to every level by different users is allowed: the application layer is accessed by the end-user (mission specialist), the control layer is used by the control expert, and the system layer is accessed by the system engineer. ORCCAD provides formalized control structures, which are coordinated using the synchronous paradigm, specifically using the Esterel language: while the control laws are often periodic (or more generally cyclic) and programmed using real-time tasks under control of a real-time scheduler, the discrete-event controller manages the set of control laws and handles exceptions and mode switching. Both activities run under the control of a real-time operating system (RTOS).

The main entities used in the ORCCAD framework are as follows:

– Modules which represent functions (e.g. controllers, filters, etc.), encoded as pieces of C code;

– module tasks (MT), the real-time tasks which implement modules (several modules can be gathered in a single MT);

– robot tasks (RT), the control tasks representing basic control actions encapsulated by a discrete-event controller;

– robot procedures (RP), a hierarchical composition of already existing RTs and RPs, to incrementally build more complex structures, from elementary executable actions to the full control application.

The RTs characterize continuous-time closed-loop control laws, along with their temporal features and the management of associated events. From the application perspective, the RT set of signals and associated behavior represent the external view

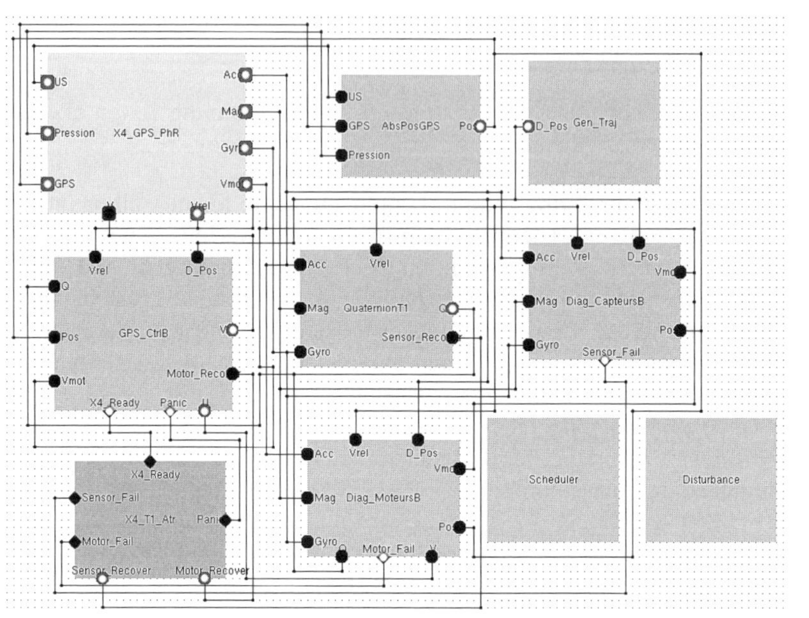

Figure 7.11. *Control and diagnosis block diagram*

of the RTs, hiding all specification and implementation details of the control laws. The RPs, which are more complex actions, can then be composed from RTs and other RPs in a hierarchical fashion leading to structures of increasing complexity. At the top level, RPs are used to describe and implement a full mission specification. At mid level, they can be used to fulfil a single basic goal through several potential solutions, e.g. a nominal controller supplemented by the recovery substitutions associated with fault detection and diagnosis.

Once a control application has been entirely designed, and for some parts formally verified, a run-time code can be automatically generated for various real-time operating systems, such as Linux in the present case[2].

7.4.2. *Quadrotor simulation setup*

Figure 7.11 describes the control and diagnostic setup used for testing purposes. In this block-diagram, the gray boxes represent the user-provided modules (i.e. functions) interconnected by their input/output ports (respectively black and white dots).

---

2. It is assumed that the underlying RTOS supports preemption and fixed priorities, so that the run-time can be easily ported on systems such as Posix.

The main parts of the functions network are described as follows:

– The attitude control path starts from the drivers of the quadrotor sensors (accelerometers Acc, rate gyros Gyr and magnetometers Mag), which are the outputs of the interface module *X4-GPS-PhR*. The box *X4_GPS_PhR* (where PhR stands for Physical Resource) is the interface between the controller and the device to control: sending or reading data on its ports actually calls the drivers, i.e. the functions used to interface the real-time controller with the real hardware, or with the real-time simulator. The raw measurements are used by the Quaternion module *QuaternionT1* to estimate the drone attitude, which is forwarded to the GPS_CtrlB module to perform the attitude control. The computed motor velocities that are desired are then sent to the quadrotor via the *V* port.

– Provision is given for future enhancements of the sensor set, since a GPS-like position sensor and ultrasonic sensors are expected to be integrated in the future. Therefore, a trajectory generator module *GEN_Traj* and a position estimator module *AbsPosGPS* are integrated in the control architecture to evaluate position control.

– The *Diag_CapteursB* module runs the diagnostic algorithm that isolates sensor failures. A failure is signaled by the Sensor_Fail weak exception to the X4_T1_Atr module and it is forwarded to the Quaternion module, so that the quaternion estimation algorithm can be adapted according to the reported failure.

– Similarly, the Diag_MoteursB module forwards motor failures to be handled by the X4_T1_Atr module.

– The scheduler module implements a feedback scheduler: it monitors the controller's real-time activity and may react by setting on-the-fly the task-scheduling parameters, e.g. their firing intervals. For example, it can been used to implement a (m,k)-firm dropping policy [JIA 07], and to dynamically adapt the priorities of messages on the CAN bus [JUA 08].

– A disturbance task allows users to generate an extra load either on the CPU or on the CAN bus, or to generate corruptions on the CAN bus.

Both computers communicate via a CAN bus: the driver ports located in the *X4_GPS_PhR* interface send and receive data using the Socket-CAN protocol[3]. From the real-time point of view, each module is implemented by a real-time task having its own programmable timer. Therefore, all the modules can be run asynchronously at their own (and possibly varying) sampling frequency. The task priorities are set according to their relative importance. Data integrity between asynchronous modules is provided by asynchronous lock-free buffers [SIM 97] which are automatically inserted at the code generation time.

---

3. http://developer.berlios.de/projects/socketcan/

7.5. **Experiments and results**

7.5.1. *Basic attitude control*

In this section, the basic attitude control and IMU diagnosis are presented when the network is dedicated only to the quadrotor application. To study the influence of the network, this case is compared to a case in which no network is implemented. All the scenarios presented start with an initial attitude $[-25, \ -35, \ -10]°$, and the objective is to stabilize the quadrotor in the hover conditions $[\phi, \ \theta, \ \psi] = [0, \ 0, \ 0]°$.

It can be observed in Figures 7.12 and 7.13 that the behavior of the quadrotor is slightly modified by the CAN, but the observer and the control converge at almost the same time as in the case without a network. The small differences are probably due to the small transmission delays. In [DIO 08], other networks are considered (Ethernet and switched Ethernet) and their performances compared to those of the CAN.

Figure 7.14 shows the behavior of the diagnostic algorithm with and without the CAN, when a breakdown f_{m1} in sensor $m1$ is considered at time $t = 4$ s (the same fault as in the experiment illustrated in Figure 7.7).

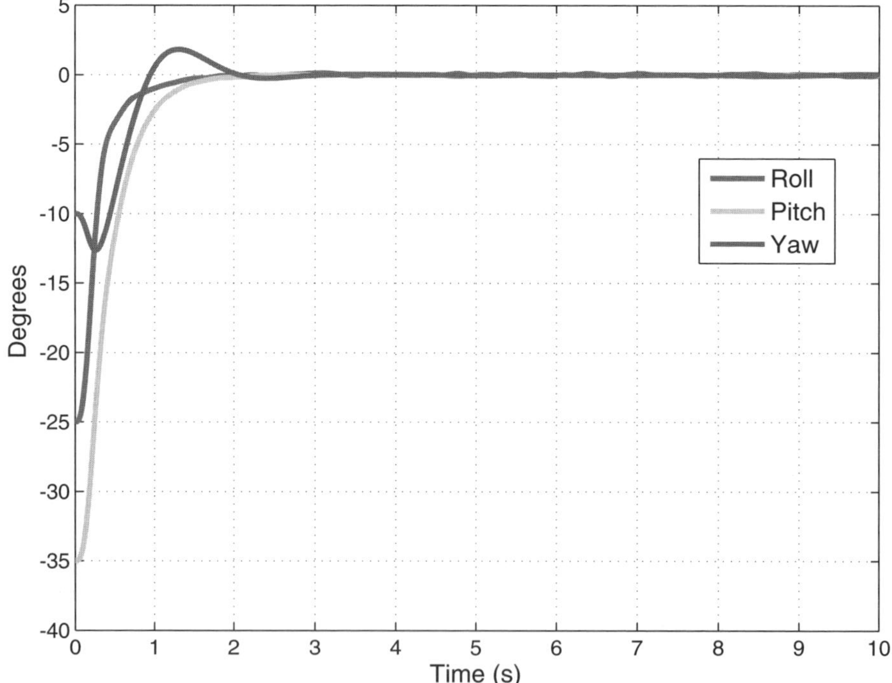

Figure 7.12. *Attitude with the dedicated CAN*

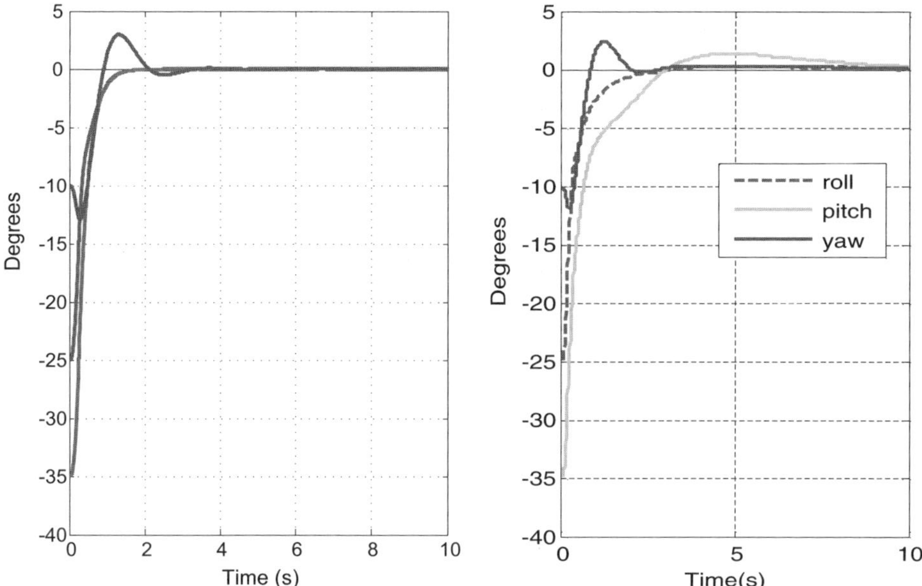

Figure 7.13. *Attitude estimated by the nonlinear observer*
(left: without network, right: with the CAN)

7.5.2. *Packet loss*

Packet loss is one well-known network drawback. Thanks to the TrueTime capabilities, packet losses can be easily introduced in the simulation. Let us now take as an example a packet loss of 10%. False alarms are observed if the diagnostic task is triggered every time new data is received [BER 07]. A number of solutions have been presented in this book to take into account this drawback, and some of them will now be presented below.

7.5.2.1. *Pragmatic solution*

The first solution consists of running the observer, controller and diagnostic tasks only once all the measurements have been received. A lost packet can be detected with the indicator r_{network}, the design of which is given in Chapter 6. This indicator is equal to 1 when the data is not received on time by the control and diagnostic modules. When such a loss is detected, the control and observer tasks are not run and the last values received by the actuators are maintained during the new sampling period. Figure 7.15 shows the influence of a 10% loss in packets. The quadrotor is still stabilized in the hover condition. However, during the transient time, the attitude is disturbed, compared to the case where no losses have occurred. Concerning the

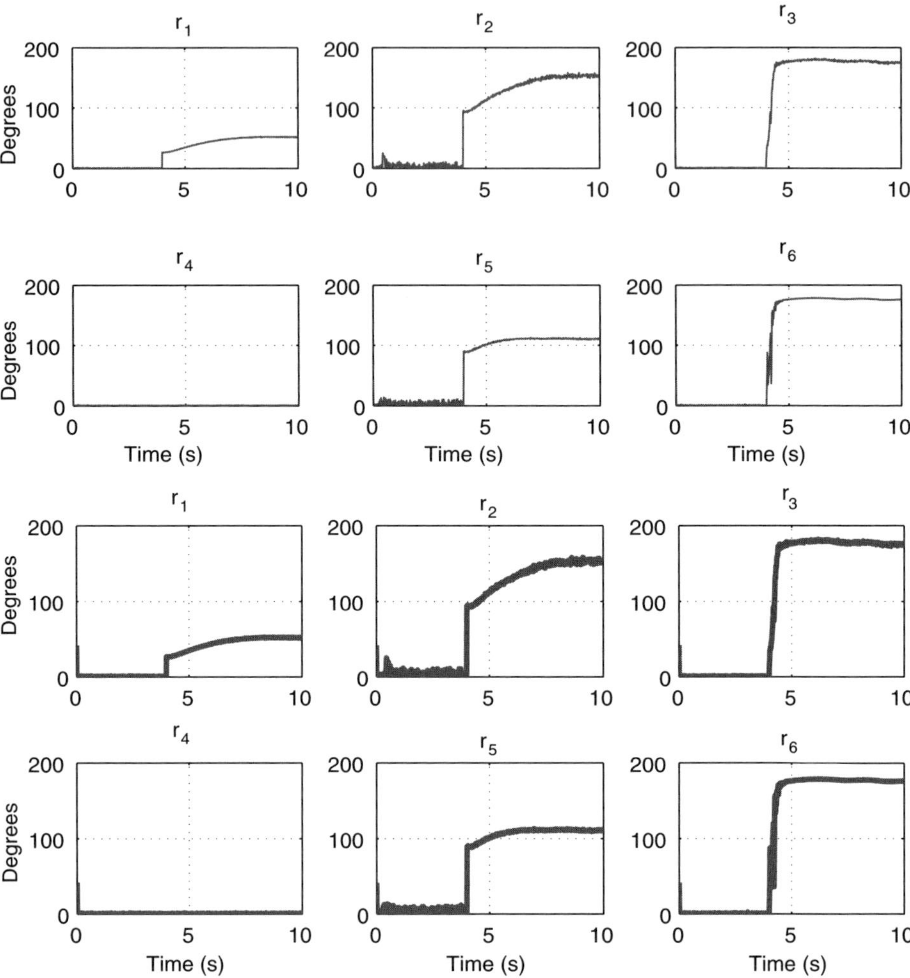

Figure 7.14. *Residuals generated by the six estimators when a breakdown f_{m1} in sensor $m1$ occurs at time $t = 4$ s (top: without network, bottom: with the CAN)*

diagnostic task, Figure 7.16 shows the simulation result. Every time the indicator r_{network} is equal to "1" (data loss), the diagnosis is not computed.

7.5.2.2. (m, k)-firm solutions

Now the policy introduced in Chapter 5 will be used to stabilize the quadrotor in the hover conditions. For the sake of simplicity, the LQ controller (7.2.3) is used and the angles are supposed to be provided directly by measurements rather than by an

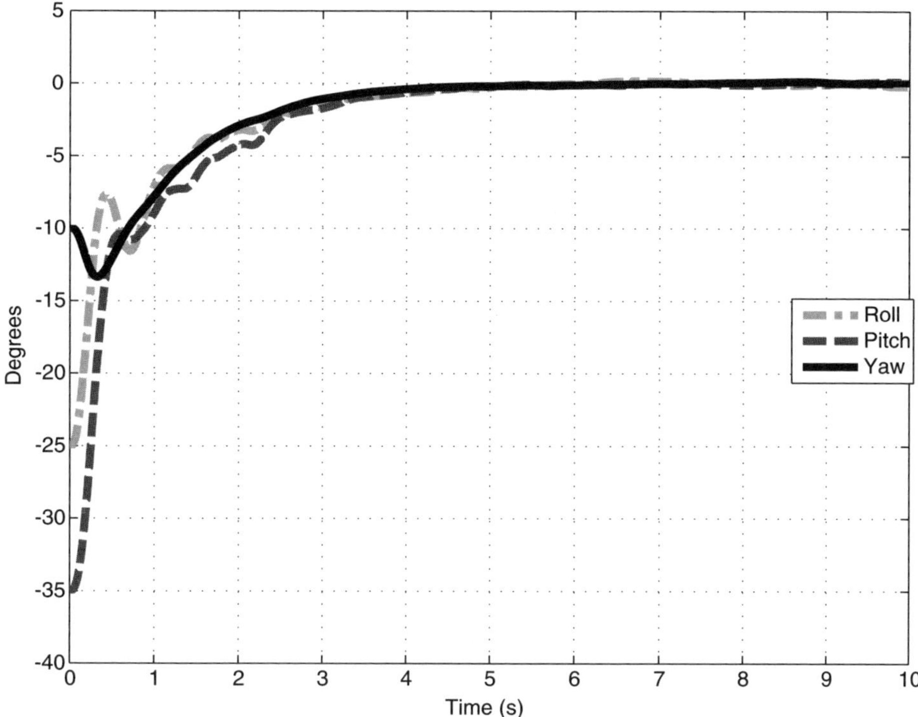

Figure 7.15. *Attitude in the presence of a 10% loss in packets*

observer. If this kind of policy is implemented, it is important to determine whether the system remains stable when packets are transmitted at a lower rate (i.e. with a larger sampling period). A lower bound on the packet transmission rate that ensures the stability of the system must be computed.

The stability of the linearized closed-loop system is analyzed thanks to the computation of its eigenvalues, based on a control delay of l sampling periods. This computation is simple because the system is composed of three independent double integrators, and therefore characterized by three pairs of identical eigenvalues. It can be observed that the system can tolerate the loss of the control update up to $l_{\mathrm{max}} = 55$ [GUE 09]. This value corresponds to the maximum delay the system can tolerate without losing stability: its eigenvalues remain inside the unit circle (Figure 7.17). The high l_{max} value is easily understandable because with a sampling period of $h = 10\ ms$ the system is oversampled.

The analysis introduced previously provides the maximum number of packet losses the NCS can tolerate. Nevertheless, it is also interesting to observe the behavior of the closed-loop system cost function J (7.30) as a function of the control update intervals

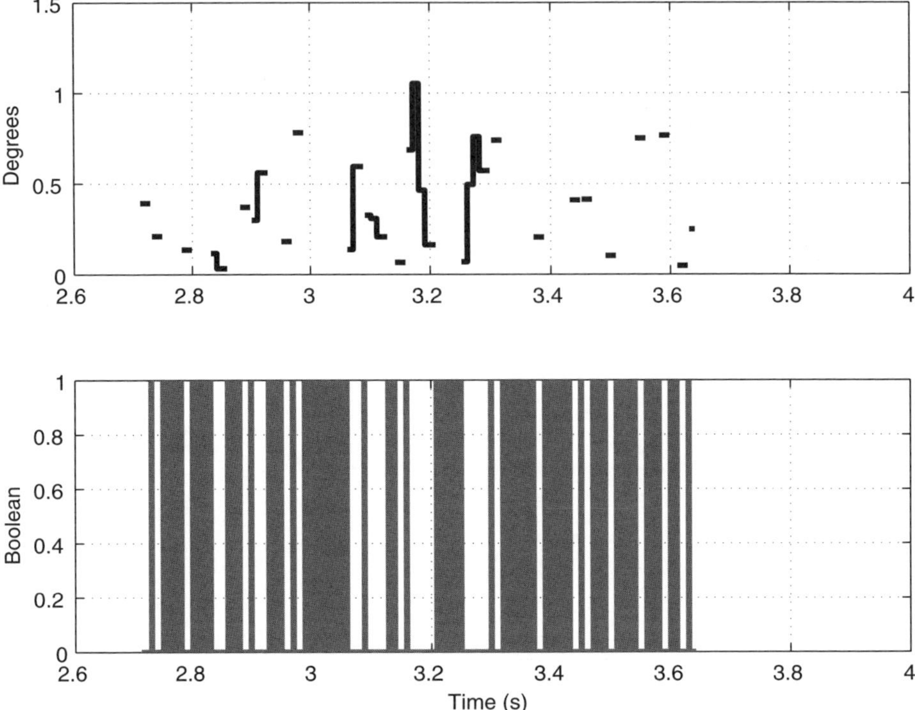

Figure 7.16. *Diagnosis without sensor fault and with a 10% loss in packets (top: residual r_1, bottom: indicator r_{network})*

(Figure 7.18). J increases slowly up to $l = 45$, the QoC being slightly modified. Then J increases dramatically, and the QoC is no longer acceptable. From the evolution of the cost function, it can be seen that for $l = 19$, the criterion J is ten times the optimal value obtained for $l = 1$. This degradation is reasonable for the quadrotor application. Therefore, during the overload period, $m = 3$ is chosen for the attitude control of the quadrotor and $(k - m) = 52$ packets can be dropped systematically every $k = 55$ samples. The resulting (m, k)-pattern, for $m = 3$, $k = 55$ is

$$\underbrace{0\,0\,0\ldots01}_{f_1=18}\underbrace{0\,0\,0\ldots01}_{f_2=18}\underbrace{0\,0\,0\ldots01}_{f_3=19} \tag{7.48}$$

The optimal control law can be redesigned to take into account the change in the sampling period and the linear time varying characteristic of the system

$$u_i = -L_i x(kh), \quad (i = 1, \ldots, 3) \tag{7.49}$$

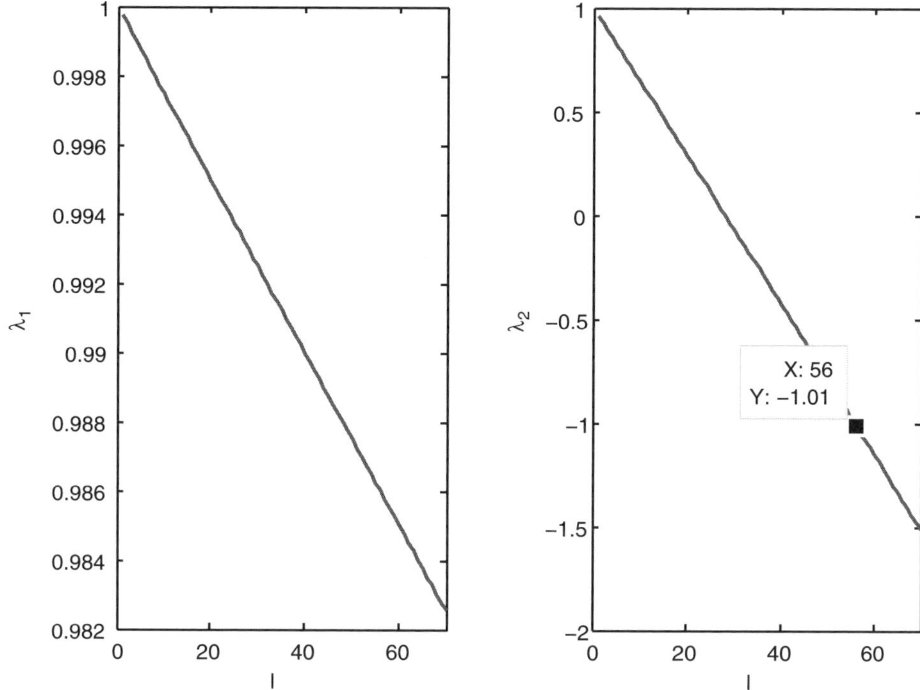

Figure 7.17. *System eigenvalues with l ∈ [0,70]*

with

$$L_{1,2} = \begin{pmatrix} 0.0253 & 0.0229 & 0 & 0 & 0 & 0 \\ 0 & 0 & 0.0253 & 0.0229 & 0 & 0 \\ 0 & 0 & 0 & 0 & 0.0253 & 0.0229 \end{pmatrix} \qquad (7.50)$$

$$L_3 = \begin{pmatrix} 0.0258 & 0.0231 & 0 & 0 & 0 & 0 \\ 0 & 0 & 0.0258 & 0.0231 & 0 & 0 \\ 0 & 0 & 0 & 0 & 0.0258 & 0.0231 \end{pmatrix} \qquad (7.51)$$

Figure 7.19 shows the behavior of the system with this control law.

7.5.2.3. *Dynamic priorities*

The use of dynamic priorities presented in Chapter 3 will now be applied to the quadrotor example. To know if the control application is urgent, information about the system's performance is needed. The error signal (difference between the reference and the output) is generally used to provide this information. In fact, if a system is

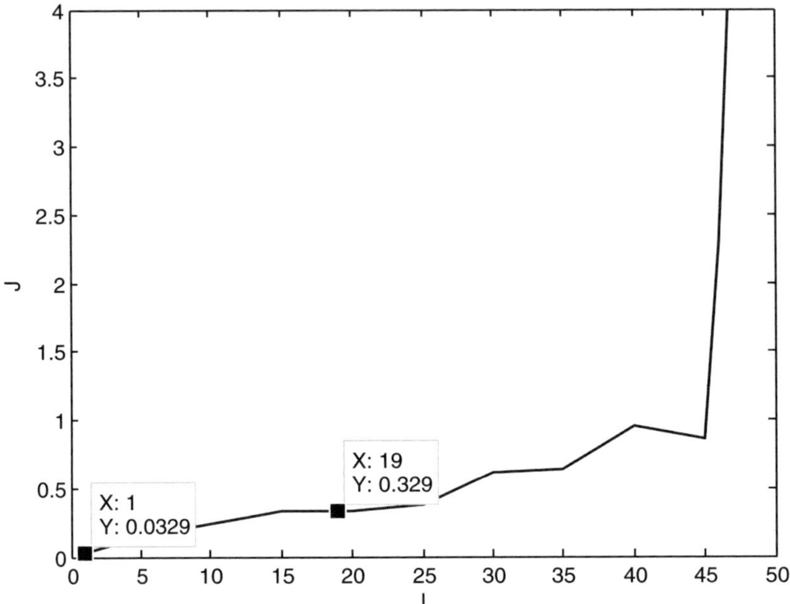

Figure 7.18. *Cost function J as a function of the interval control updates l*

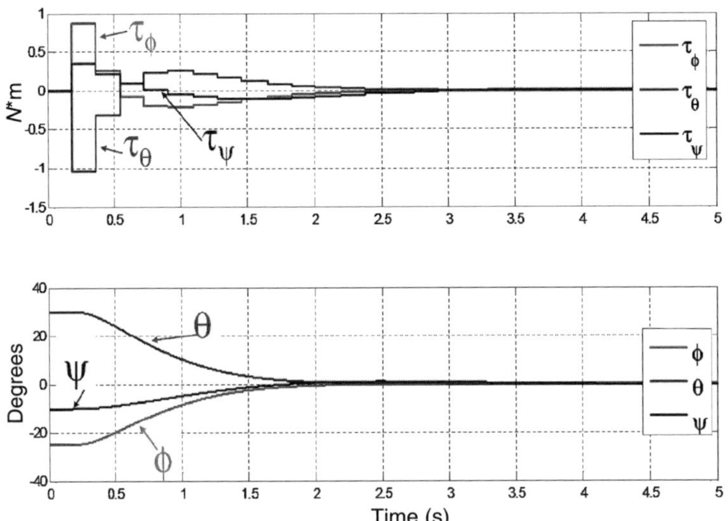

Figure 7.19. (m, k)-*firm policy*
(top: Control torque, bottom: attitude)

in transient behavior (reference change or disturbance effect), then the error signal will be high, whereas if the system is stabilized, the error signal will be close to zero. The quadrotor is a Multi-Input Multi-Output (MIMO) system that must be stabilized in relation to roll, pitch and yaw angles [BER 09], thus the strategy must take into account the three angles. Four priorities are implemented with the function *Prio*:

$$
\text{Prio} = \begin{cases}
P_{\min}, \text{if} & \begin{cases} e_\phi < & e_{\text{threshold}} & \text{and} \\ e_\theta < & e_{\text{threshold}} & \text{and} \\ e_\psi < & e_{\text{threshold}}. \end{cases} \\[2ex]
P_2, \text{if} & \begin{cases} e_\phi \text{ and } e_\theta < & e_{\text{threshold}} & \text{or} \\ e_\phi \text{ and } e_\psi < & e_{\text{threshold}} & \text{or} \\ e_\theta \text{ and } e_\psi < & e_{\text{threshold}}. \end{cases} \\[2ex]
P_1, \text{if} & \begin{cases} e_\phi \text{ and } e_\theta > & e_{\text{threshold}} & \text{or} \\ e_\phi \text{ and } e_\psi > & e_{\text{threshold}} & \text{or} \\ e_\theta \text{ and } e_\psi > & e_{\text{threshold}}. \end{cases} \\[2ex]
P_{\max}, \text{if} & \begin{cases} e_\phi > & e_{\text{threshold}} & \text{and} \\ e_\theta > & e_{\text{threshold}} & \text{and} \\ e_\psi > & e_{\text{threshold}}. \end{cases}
\end{cases}
\tag{7.52}
$$

To evaluate $e_{\text{threshold}}$, the "QoC" is considered. As the quadrotor is a critical application, an error higher than $e_{\max} = 1.10^{-3}$ degrees for each angle is considered unacceptable. The value $e_{\text{threshold}}$ is chosen as equal to $\frac{2}{3}e_{\max}$.

Figure 7.20 (top panel) shows the value of the three errors (i.e. signals used to switch the priority). Between $t = 0$ s and $t = 1.7$ s, the priority of the quadrotor control is $P_{\max}$ because all errors are over the threshold (Figure 7.20). Between $t = 1.7$ s and $t = 2.1$ s, the priority of the quadrotor control is P_1 because error e_ψ is inferior to the threshold. Between $t = 2.2$ s and $t = 3.3$ s, the priority of the quadrotor control is P_2 because two out of three errors are below the threshold (e_ϕ and e_θ). Between $t = 3.3$ s and $t = 5$ s, the priority of the quadrotor control is $P_{\min}$ because all the errors are below the threshold. Between $t = 5$ s and $t = 5.8$ s, the maximal priority is given to the quadrotor control ($P_{\max}$, because a disturbance appeared at time $t = 5$ s (Figure 7.20, bottom panel) and all the errors are higher than the threshold. After $t = 7$ s, the priority is again minimal because all the errors are below the threshold. The quadrotor is well controlled during the entire experiment.

7.5.2.4. *Extended Kalman filter*

The experiment reported in this section illustrates the design of the EKF presented in chapter 6 allowing users to cope with data loss. The filter used is presented in section 7.2.4. It takes into account a data loss of 20% for all the measurements from the accelerometers and magnetometers. The losses are assumed to be independent for

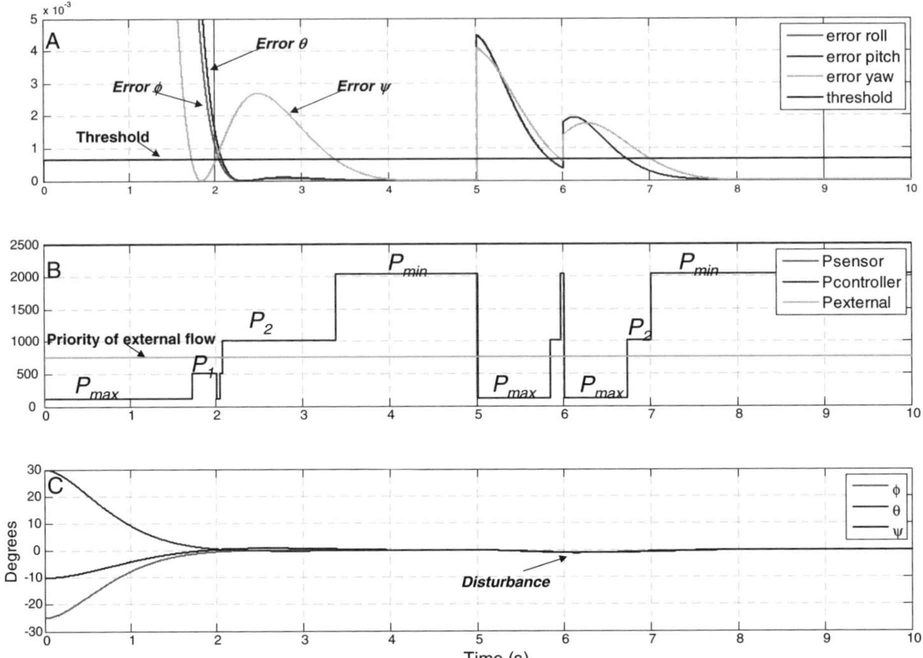

Figure 7.20. *Hybrid priority policy*
(top: errors; middle: priorities, bottom: attitude)

each of the sensor axes. The initial attitude is $[-25, \ -35, \ -10]°$ and the reference is first $[10, \ 4, \ 15]°$ and then changed to $[0, \ 0, \ 0]°$, at time $t = 3$ s. The attitude estimation filter is initialized at $[-9, \ 5, \ 57]°$. The real and the estimated attitudes are shown in Figure 7.21. Figure 7.22 gives the attitude estimation error and the data loss indicator for the accelerometer a_1.

7.5.3. *Hardware-in-the loop experiment*

7.5.3.1. *Basic scenario*

In this scenario, the fault-free case will be considered and the quadrotor stabilization will be shown. The quadrotor simulation starts with an initial attitude equal to $[120°; -10°; 50°]$ and the reference attitude equal to $[0°; 0°; 0°]$.

The hardware-in-the-loop result is shown in Figure 7.23 (left). The time response of the system can be compared to the one obtained with Matlab/Simulink and TrueTime (Figure 7.23, right). The time response is equal to 2.5 seconds for the roll and pitch angles, and 2.8 seconds for the yaw angle.

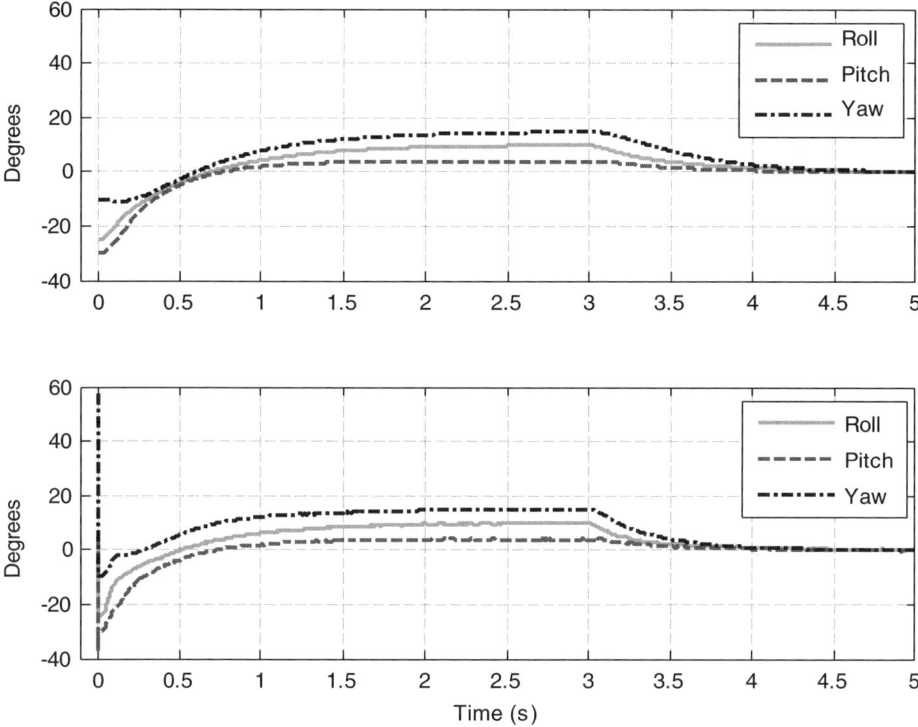

Figure 7.21. *EKF policy with a 20% loss in packets*
(top: attitude, bottom: estimated attitude)

7.5.3.2. *Packet loss*

In this scenario, the same initial and reference positions as in the previous subsection will be used. The objective here is to study the influence of packet losses on the system's behavior. A 10% loss in packets from a_1 will be observed. The fault indicator r_{network} [BER 07] mentioned previously (section sec:pragmatic) is used to make the difference between a sensor fault and a packet loss (Figure 7.24, bottom). When the data is lost at $t = (kh)$, the quaternion $\hat{q}(kh)$ is not computed and the control algorithm holds the value ω_{Mi}^{ref} computed at time $t = (k-1)h$. The results are shown in Figure 7.24. Small differences can be noted with respect to Figure 7.23 but it can be seen that the control law is robust to 10% in packet losses on this sensor. Several other simulations have been made with other packet loss scenarios, and the results are quite similar.

7.5.3.3. *Sensor failure*

In this scenario (Figure 7.25), a bias failure in the rate gyro ω_{g1} will be considered. The fault is simulated at time $t = 5$ s. Before the fault appearance, all the residuals

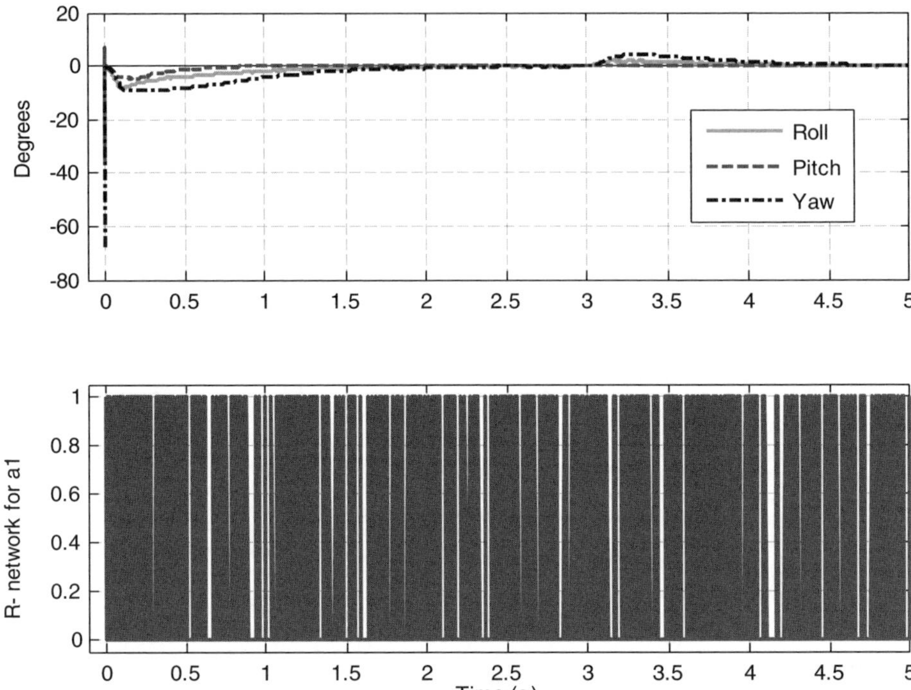

Figure 7.22. *EKF policy with a 20% loss in packets (top: Euler angles error; bottom: indicator of data loss)*

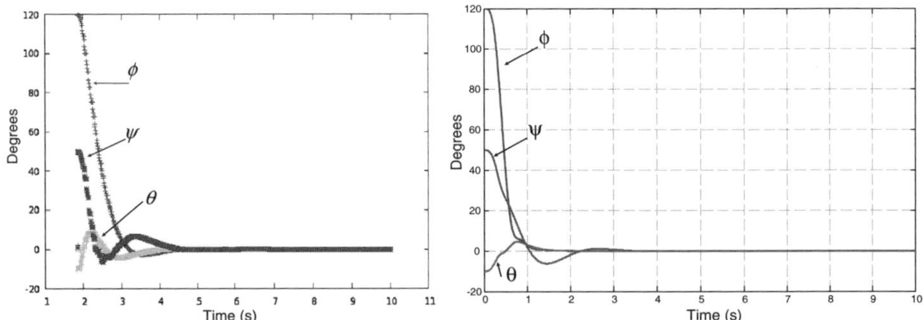

Figure 7.23. *Hardware in the loop experimentation (left: quadrotor attitude; right: Matlab/Simulink TrueTime simulation)*

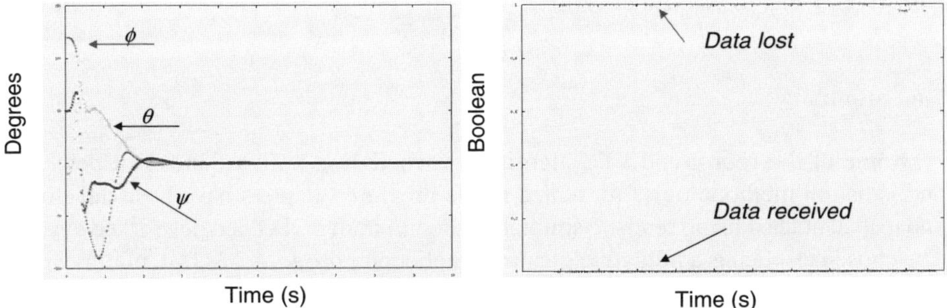

Figure 7.24. *Indicator r_{network} and attitude of the quadrotor with a 10% loss in packets. (left: attitude; right: indicator of packet loss)*

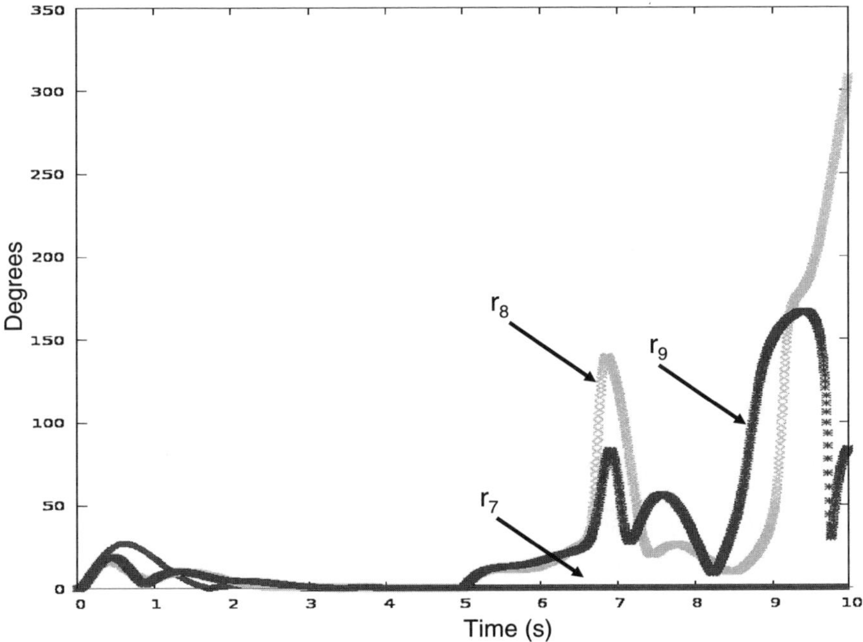

Figure 7.25. *Residual $r_i\,(i = 7, \ldots, 9)$, when a breakdown f_{g1} occurs at time $t = 5$ s*

are close to zero. After $t = 5$ s, residuals $r_i (i = 8, 9)$ are sensitive to the fault and residual r_7, computed with the observer that discards this sensor value, still stands at zero.

7.6. Summary

Some of the robust and FTC algorithms, scheduling policies and fault detection and isolation methodologies presented in the previous chapters have been developed and tested, at least using realistic simulations, on a small embedded networked system. The chosen test bed is a miniature quadrotor helicopter drone. It has fast, nonlinear dynamics, is open-loop unstable, and the particular disposition of its actuators essentially makes its attitude stabilization a difficult control problem. Beyond the usual modeling, control and FDI design problems, the hardware used to implement the control and diagnostic algorithms is made of a small network of micro-controllers, distributed over a CAN bus. Therefore, this test-bed gathers embedded control design that is subjected to network-induced disturbances, with dependability concerns in mind. A successful integration of control laws and FDI algorithms on such a real-time distributed platform requires a careful and intelligent incorporation of the control methodologies into the hardware and software engineering components. The challenge has been handled by following a progressive approach, to gradually integrate control, computing and networking features and constraints using the appropriate tools.

As usual, the control design process starts with the modeling of the physical devices. Note that dependability and safety concerns can be considered from this very early design stage: for example, the choice of quaternions to model the quadrotor's attitude allows for the bypassing of a number of singularity problems, which later will avoid unpleasant run-time issues. The complexity and the feasibility of the control laws and estimation algorithms on an embedded low-power platform also need to be taken into account in the early stages.

As far as the control algorithms are concerned, discretization, scheduling, and networking must be studied altogether, because traditional simulation tools are no longer appropriate. To this end, the TrueTime toolbox no only handles models of real-time scheduling policies and of some standard networks such as CAN and switched Ethernet, but it also allows us to simulate the execution of control laws on the modeled real-time platform. Note that this toolbox is open, so that the authors of this book could have enhanced some of its features along the lines of the SafeNecs project.

The next step before experimentation uses a "hardware-in-the-loop" real-time simulation concept; the real-time software runs on the real target, and the network is no longer simulated, whereas the physical controlled process still is. The development of the run-time software was made easier using ORCCAD, a model-based development environment dedicated to control design and code generation. Therefore, the coupling interaction between the control algorithms and the real-time execution can be examined and fine-tuned at no risk for the real plant, and even before the real plant is completed.

This integration approach allowed the authors to progressively develop, implement and evaluate most of the control, FDI and FTC methods studied in the previous chapters: these methods are applied on a the challenging quadrotor test-bed. During all the stages of the assessment process, feedback has been provided to enhance the methodologies, to evaluate their practical feasibility and to check or calibrate the models, e.g. as in Figure 7.23 where the simulated CAN bus in TrueTime closely matches the experimental data.

Therefore, control, diagnosis, computing and networking co-designed algorithms, such as selective data dropping, QoC aware prioritized messages, and EKF sustaining data loss, have been successfully evaluated and show an improved fault tolerance w.r.t. networked and scheduling-induced disturbances. As usual, the results obtained from a particular case study, even if within the framework of a significantly difficult case, cannot be generalized easily. However, this book presents a set of building blocks, methods and tools which are expected to provide effective control, computing and networking co-design and integration for a number of relevant categories of control and diagnostic applications.

7.7. Bibliography

[BER 07] BERBRA C., GENTIL S., LESECQ S., AND THIRIET J.-M., Co-design for a safe networked control DC motor, *3rd IFAC Workshop on networked control systems tolerant to faults Necst*, Nancy, France, June 2007.

[BER 08] BERBRA C., LESECQ S., GENTIL S., AND THIRIET J.-M., Co-design of a safe networked control quadrotor, *IFAC World Congress*, Seoul, Korea, July 2008.

[BER 09] BERBRA C., GENTIL S., AND LESECQ S., Hybrid priority scheme for networked control quadrotor, *17th Mediterranean Conference on Control & Automation*, Thessaloniki, Greece, June 2009.

[BOR 98] BORRELLY J.-J., COSTE-MANIÈRE E., ESPIAU B., KAPELLOS K., PISSARD-GIBOLLET R., SIMON D., AND TURRO N., The ORCCAD architecture, *International Journal of Robotics Research*, vol. 17, num. 4, p. 338–359, April 1998.

[CHO 92] CHOU J., Quaternion Kinematic and Dynamic Differential Equations, *IEEE Transactions on Robotics and Automation*, vol. 8, p. 53–64, 1992.

[DIO 08] DIOURI I., BERBRA C., GEORGES J.-P., GENTIL S., AND RONDEAU E., Evaluation of a switched Ethernet network for the control of a quadrotor, *16th Mediterranean Conference on Control and Automation, MED'08*, Ajaccio, France, July 2008.

[GUE 08] GUERRERO-CASTELLANOS J.-F., Estimation de l'attitude et commande bornée en attitude d'un corps rigide: Application à un mini hélicoptère à quatre rotors, PhD thesis, Joseph Fourier-Grenoble I University, France, 2008.

[GUE 09] GUERRERO-CASTELLANOS J.-F., BERBRA C., GENTIL S., AND LESECQ S., Quadrotor attitude control through a network with (m-k)-firm policy, *European Control Conference ECC09*, Budapest, Hungary, August 2009.

[JIA 07] JIA N., SONG Y.-Q., AND SIMONOT-LION F., Graceful degradation of the quality of control through data drop policy, *European Control Conference, ECC'07*, Kos, Greece, July 2007.

[JUA 08] JUANOLE G., MOUNEY G., AND CALMETTES C., On different priority schemes for the message scheduling in Networked Control Systems, *16th Mediterranean Conference on Control and Automation*, Ajaccio, France, July 2008.

[LES 09] LESECQ S., GENTIL S., AND DARAOUI N., Quadrotor attitude estimation with data losses, *European Control Conference ECC09*, Budapest, Hungary, August 2009.

[OHL 07] OHLIN M., HENRIKSSON D., AND CERVIN A., TrueTime 1.5 – *Reference Manual*, January 2007.

[Opn] OPNET TECHNOLOGIES INC, OPNET, http://www.opnet.com/

[SIM 97] SIMPSON H.-R., Multireader and multiwriter asynchronous communication mechanisms, *IEE Proceedings-Computer and Digital Techniques*, vol. 144, num. 4, p. 241–244, 1997.

[TAN 07a] TANWANI A., GALDUN J., THIRIET J.-M., LESECQ S., AND GENTIL S., Experimental networked embedded minidrone. Part I: consideration of faults., *European Control Conference ECC'07*, Kos, Greece, July 2007.

[TAN 07b] TANWANI A., GENTIL S., LESECQ S., AND THIRIET J., Experimental networked embedded minidrone. Part II: distributed FDI., *European Control Conference ECC'07*, Kos, Greece, July 2007.

[TöR 06] TÖRNGREN M., HENRIKSSON D., REDELL O., KIRSCH C., EL-KHOURY J., SIMON D., SOREL Y., ZDENEK H., AND ÅRZÉN K.-E., Co-design of control systems and their real-time implementation – A tool survey, Report no. TRITA - MMK 2006:11, Royal Institute of Technology, KTH, Stockolm, Sweden, 2006.

Glossary and Acronyms

CAN Controller Area Network (ISO 11898), a serial bus using prioritized messages, grounded in automotive applications.

COTS Commercial Off-The-Shelf.

CSMA/CD Carrier Sense Multiple Access with Collision Detection, a medium access protocol for Ethernet (IEEE 802.3).

DBP Distance Based Priority, a dynamic priority assignment scheme where the priority of a task is the function of the distance to a failure state defined by the (m, k)-firm model.

DTTS Discrete Time Switched System.

EDF Earliest Deadline First, a scheduling policy which dynamically assign the highest priority to the task with the earliest deadline.

FDI Fault Detection and Isolation.

FIF Fault Isolation Filter.

FIFO First In, First Out.

FTC Fault Tolerant Control.

LFT Linear Fractional Transform.

LMI Linear Matrix Inequality.

LQ Linear Quadratic, a state feedback controller for linear systems minimizing a quadratic criterion involving the system's state and control vectors.

LQG Linear Quadratic Gaussian, a LQ controller associated with a Kalman estimator when the state and measurements of the system are disturbed by additive white Gaussian noise.

LPV Linear Parameter Varying, methods to control linear, systems taking into account the variations of some parameters of the plant.

LTI related to Linear Time Invariant systems.

MIMO a dynamic system with Multiple Inputs, Multiple Outputs.

(_m,k_)-firm a scheduling policy enforcing the completion of at least **m** tasks execution (or message transmission) every **k** scheduling (or transmission) slots.

MPC Model Predictive Control, a feedback controller which predict and apply at each sample, a model-based control signal to a linear or non-linear process.

NCS Networked Control System, a control system where the sensors, controllers and actuators are distributed over a network.

NP-hard Non-deterministic Polynomial-time hard, a class of problems that are harder than those that can be solved by a non-deterministic Turing machine in polynomial time.

OS Operating System, the set of software utilities to interface a computer's hardware and user space applications.

PN Petri Net, a place/transition graph describing a discrete events system.

PID Proportional-Integral-Derivative, a very popular controller for SISO systems.

QoC Quality of Control, a measure of performance for a controller.

QoS Quality of Service, a measure of performance for a networked system.

RTOS Real Time Operating System, an OS with some predictability for tasks' execution time and memory space, thanks to a real-time scheduler.

RM Rate Monotonic, a scheduling policy which statically assign the highest priority to the the periodic task with shortest execution time.

SafeNecs Safe Networked Control Systems, is a joint academic research project funded by the "Agence Nationale de la Recherche" under grant ANR-05-SSIA-0015-03, from 2006 to 2009; it is devoted to research on fault tolerant control and diagnosis for networked control systems.

SISO a dynamic system with Single Input, Single Output.

TCHPN Timed Coloured Hierarchy Petri Net.

TDS Time-delay systems, systems whose state is a function taken over an interval including past values.

WCET Worst Case Execution Time.

WRR Weighted Round Robin, a scheduling policy where CPU (resp. bandwidth) is reserved for tasks (resp. data flows) according to their respective weight.

ZOH Zero-Order Hold, a mathematical model of a digital-to-analog converter holding a constant signal value for all the sample interval.

List of Authors

Christophe AUBRUN
CRAN
Nancy University
France

Mongi BEN GAID
IFP
Paris
France

Cédric BERBRA
National Polytechnic Institute and GIPSA-lab
Grenoble University
France

Belynda BRAHIMI
Altran
Paris
France

Flavia FELICIONI
National University of Rosario
Argentina

Sylviane GENTIL
National Polytechnic Institute and GIPSA-lab
Grenoble University
France

Jean-Philippe GEORGES
CRAN
Nancy University
France

Ning JIA
SEBIA
Paris
France

Guy JUANOLE
LAAS (CNRS)
Paul Sabatier University
Toulouse
France

Suzanne LESECQ
CEA-LETI MINATEC
Grenoble
France

Gérard MOUNEY
LAAS (CNRS)
Paul Sabatier University
Toulouse
France

Xuan Hung NGUYEN
LAAS-CNRS
Paul Sabatier University
Toulouse
France

David ROBERT
National Polytechnic Institute and GIPSA-lab
Grenoble University
France

Eric RONDEAU
CRAN
Nancy University
France

Dominique SAUTER
CRAN
Nancy University
France

Olivier SENAME
National Polytechnic Institute and GIPSA-lab
Grenoble University
France

Alexandre SEURET
CNRS, GIPSA-lab
Grenoble University
France

Daniel SIMON
INRIA
Grenoble Rhône-Alpes
France

Françoise SIMONOT-LION
LORIA
Nancy University
France

Ye-Qiong SONG
LORIA
Nancy University
France

Index